# FAST TRACK TO A 5

## Preparing for the AP*
# Chemistry Examination

To Accompany
**Chemistry**
**6th and 7th Editions**
by Steven S. Zumdahl and Susan A. Zumdahl

**Tina Ohn-Sabatello**
Maine East High School, Park Ridge, Illinois

**Gordan Morlan**
Grosse Pointe North High School and Wayne State University, Michigan

with assistance from
**Sheldon Knoespel**
Michigan State University and Mott Community College, Michigan

**McDougal Littell**  A Houghton Mifflin Company
Evanston, Illinois   Boston   New York

*AP and Advanced Placement Program are registered trademarks of the College Board,
which was not involved in the production of and does not endorse this product.

ISBN: 0-618-22171-9

123456789-CRS-09 08 07 06 05

# CONTENTS

# ABOUT THE AUTHORS

TINA OHN-SABATELLO has been teaching honors and AP Chemistry for twelve years at Maine East High School in Park Ridge, Illinois. Each year, her students achieve mostly 4's and 5's on the AP Chemistry exam. She also has taught for five summer terms at Triton Community College. Tina has B.S. and M.S. degrees from the University of Illinois at Chicago. She was one of five members of the SAT II Chemistry Test Development Committee. During her five-year term working for ETS, she wrote and revised chemistry questions for the subject matter test with one other high school teacher, three college professors, and two ETS consultants.

GORDON MORLAN taught chemistry for forty years; the last twenty years included advanced placement chemistry. He served as Science Department Chairman at Grosse Pointe North High School in Michigan for twenty-nine years. Gordon was also a College Board consultant and was a reader for the AP Chemistry Examination twelve times. He is an active competitive sailor and also enjoys reading, travel, and jazz.

# PREFACE

To learn chemistry well, you need to have a good text and a teacher to help you. No doubt, you already have a good text. I can't be there to help you review for the AP exam, but I can bring you the teaching techniques and ideas that have arisen from my classroom discussions. The chapter summaries, questions, and practice tests in this book will help lead you to success in the AP Chemistry exam.

I am very grateful to Houghton Miffflin for giving me the opportunity to reach students beyond my classroom. I couldn't have completed this task without the help and advice of reviewers Annis Hapkiewicz and Sheldon Knoespel. Most of all, I would like to thank my coauthor, Gordon Morlan, who worked diligently on the questions for this guide. I am very thankful for the guidance of Margaret Lannaman, my Houghton Mifflin liaison.

Lastly, I dedicate this book to my patient and loving family, Rick, Rina, and Antonio, who allowed me the time to write this book, and to my students for inspiring me to be creative in my teaching and to look forward to each year of teaching chemistry as if it were the first.

Tina Ohn-Sabatello
September 2005

---

Writing a book of this sort is never the result of only one or two people. The frank (sometimes brutally frank!) and carefully considered reviews of Annis Hapkiewicz are especially appreciated. So also is the continual upbeat attitude and encouragement of editor Margaret Lannamann. Credit for the entire final sample test belongs to Sheldon Knoespel.

My wife, Judy, has brought her understanding of how students learn to this project, as well as an understanding of the magnitude of this undertaking. Her encouragement and support were essential.

Gordon Morlan
September 2005

# Part I

## Strategies for the AP Test

# PREPARING FOR THE AP* CHEMISTRY EXAMINATION

Advanced Placement can be exhilarating. Whether you are taking an AP course at your school or you are working on AP independently, the stage is set for a great intellectual experience.

But sometime after New Year's Day, when the examination begins to loom on a very real horizon, Advanced Placement can seem downright intimidating—in fact, offered the opportunity to take the examination for a lark, even adults long out of high school refuse. If you dread taking the test, you are in good company.

The best way to deal with an AP examination is to master it, not let it master you. If you can think of these examinations as a way to show off how much chemistry you know, you have a leg up. Attitude *does* help. If you are not one of those students, there is still a lot you can do to sideline your anxiety. This book is designed to put you on a fast track. Focused review and practice time will help you master the examination so that you can walk in with confidence and score a 5.

## WHAT'S IN THIS BOOK

This book is keyed to *Chemistry* by Steven and Susan Zumdahl, 6th and 7th editions, but because it follows the College Board Topic Outline, it is compatible with all textbooks. It is divided into three sections. Part I offers suggestions for getting yourself ready, from signing up to take the test and sharpening your pencils to organizing a free-response essay. At the end of Part I, you will find a Diagnostic Test. This test has all of the elements of the AP Chemistry Examination, but the seventy-five multiple choice questions are organized according to the College Board Topic Outline. When you go through the answers at the end of the Diagnostic Test, you will see how the examination is weighted for each content area.

Part II is made up of sixteen chapters—again following the College Board Topic Outline. These chapters are not a substitute for your textbook and class discussion; they simply review the AP chemistry course. At the end of each chapter, you will find ten multiple-choice questions and two free-response questions based on the material in that chapter. Again, you will find page references at the end of each answer directing you to the discussion on that particular point in *Chemistry*.

Part III has two complete AP Chemistry examinations. At the end of each test, you will find the answers, explanations, and references to *Chemistry* for the multiple choice and the free-response questions.

---

## SETTING UP A REVIEW SCHEDULE

If you have been doing your homework steadily and keeping up with the course work, you are in good shape. Organize your notes, homework, and handouts from class by topic. Reference these materials as well as your textbook and this study guide when you have difficulty in a specific section. But even if you've done all that—or if it's too late to do all that—there are some more ways to get it all together.

To begin, read Part I of this book. You will be much more comfortable going into the test if you understand how the test questions are designed and how best to approach them. Then take the Diagnostic Test and see where you are right now.

Take out a calendar and set up a schedule for yourself. If you begin studying early, you can chip away at the review chapters in Part II. You'll be surprised—and pleased—by how much material you can cover in half an hour a day of study for a month or so before the test. Look carefully at the sections of the Diagnostic Test; if you missed a number of questions in one particular area, allow more time for the chapters that cover that area of the course. The practice tests in Part III will give you more experience with different kinds of multiple-choice questions and the wide range of free-response questions.

If time is short, skip reading the review chapters (although you might read through the chapter subheadings) and work on the multiple-choice and free-response questions at the end of each review. This will give you a good idea of your understanding of that particular topic. Then take the tests in Part III.

If time is *really* short, go straight from Part I to Part III. Taking practice tests over and over again is the fastest, most practical way to prepare. You cannot study chemistry by reading it like a novel. You must actively do problems to gain understanding and excel in your performance. Athletes don't perform well just by reading books about their sport or by watching others. They must get up and practice. So, you too, just like athletes, must practice, practice, practice if you want to do your best!

## BEFORE THE EXAMINATION

By February, long before the exam, you need to make sure that you are registered to take the test. Many schools take care of the paperwork and handle the fees for their AP students, but check with your teacher or the AP coordinator to make sure that you are on the list. This is especially important if you have a documented disability and need test accommodations. If you are studying AP independently, call AP Services at the College Board for the name of the local AP coordinator, who will help you through the registration process.

The evening before the exam is not a great time for partying. Nor is it a great time for cramming. If you like, look over class notes or drift through your textbook, concentrating on the broad outlines, not the small details, of the course. You might also want to skim through this book and read the AP tips.

The evening before the exam *is* a great time to get your things together for the next day. Sharpen a fistful of no. 2 pencils with good erasers; bring a scientific calculator with fresh batteries. You may bring a programmable calculator; the memory will not be erased or cleared by the test administrator. It must not have a typewriter-style keyboard. Wind your watch and turn off the alarm if it has one; get a piece of fruit or a power bar and a bottle of water for the break. Make sure you have your Social Security number and whatever photo identification and admission ticket are required. Then relax. And get a good night's sleep.

On the day of the examination, it is wise not to skip breakfast—studies show that students who eat a hot breakfast before testing get higher grades. Be careful not to drink a lot of liquids, necessitating a trip to the bathroom during the test. Breakfast will give you the energy you need to power you through the test—and more. You will spend some time waiting while everyone is seated in the right room for the right test. That's before the test has even begun. With a short break between Section I and Section II, the AP Chemistry exam lasts for over two and a half hours. So be prepared for a long morning. You do not want to be distracted by a growling stomach or hunger pangs.

Be sure to wear comfortable clothes, taking along a sweater in case the heating or air-conditioning is erratic. Be sure, too, to wear clothes you like—everyone performs better when they think they look better—and by all means wear your lucky socks.

You have been on the fast track. Now go get a 5!

# TAKING THE AP CHEMISTRY EXAMINATION

The AP Chemistry examination consists of two sections: Section I has 75 multiple-choice questions; Section II has six free-response questions. You will have 90 minutes for the multiple-choice portion. You will not be allowed to use a calculator for the multiple-choice questions. The questions are collected, and you will be given a short break. You then have 90 minutes for the free-response section which is broken into two portions.

You will be allowed to use a calculator in Part A, the first 40-minute segment. The first question in Part A is mandatory and has been a question about equilibrium for many years. Next, you will be allowed to choose to answer question 2 or question 3. Time will be called at 40 minutes, at which time calculators must be put away for the remaining 50 minutes.

In Part B, you will be asked to determine the products and write unbalanced net-ionic equations for your choice of five out of eight chemical reactions. You will then answer two mandatory essays, one of which is lab-based. Finally, you will choose to answer one of two essay questions.

The table on the next page summarizes each part of the AP Chemistry examination.

## STRATEGIES FOR THE MULTIPLE-CHOICE SECTION

Here are some rules of thumb to help you work your way through the multiple-choice questions:

■ **Guessing penalty** There are 5 possible answers for each question. Each correct answer is worth 1 point, and there is a 1/4-point guessing penalty for each incorrect answer. If you cannot narrow down the answers at all, it is against the odds to guess, so leave the answer sheet blank. However, if you can narrow down the answers even by eliminating one response, it is advantageous to guess. If you skip a question, be very careful to skip down that line on the answer sheet.

| AP Chemistry Exam – Distribution of Questions | | | |
|---|---|---|---|
| Weighting | | Possible Topics | Time Allowed |
| **Section I**    45% | | Multiple-choice | 90 minutes |
| **Section II**    55 % | | | |
| **Part A** | | | 40 minutes |
| Question 1 | | Equilibrium | |
| Question 2 or 3 | | Thermochemistry, Thermodynamics, Kinetics, Stoichiometry, Redox | |
| **Part B** | | | 50 minutes |
| Question 4—choose 5 of 8 | | Reactions | |
| Question 5—mandatory | | Essay—Lab-based question | |
| Question 6—mandatory | | Essay/Minor Calculations—topics same as those listed for Questions 7 & 8 | |
| Question 7 or 8 | | Essay—gases, states of matter, kinetics, redox, thermochemistry, electrochemistry, periodicity, bonding, forces, organic, colligative properties | |

- **Read the question carefully** Pressured for time, many students make the mistake of reading the questions too quickly or merely skimming them. By reading a question carefully, you may already have some idea about the correct answer. You can then look for it in the responses. Careful reading is especially important in EXCEPT questions.
- **Eliminate any answer you know is wrong** You can write on the multiple-choice questions in the test book. As you read through the responses, draw a line through any answer you know is wrong.
- **Read all of the possible answers, then choose the most accurate response** AP examinations are written to test your precise knowledge of a subject. Some of the responses may be partially correct but there will only be one response that is completely true.
- **Avoid absolute responses** These answers often include the words "always" or "never." For example, the statement "all chlorides are always soluble in water" is incorrect because compounds such as silver chloride and lead (II) chloride are insoluble in water.
- **Mark and skip tough questions** If you are hung up on a question, mark it in the margin of the question book. You can come back to it later if you have time. Make sure you skip that question on your answer sheet too.

## TYPES OF MULTIPLE-CHOICE QUESTIONS

There are various kinds of multiple-choice questions. Here are some suggestions for approaching each kind:

CLASSIC/BEST ANSWER QUESTIONS This is the most common type of multiple-choice question. It simply requires you to read the question and select the most correct answer. For example:

1.  Which of the following molecules is tetrahedral in shape?
    (A) $NH_3$
    (B) $H_2O$
    (C) $CH_4$
    (D) $SF_4$
    (E) $XeF_4$

ANSWER: C. Eliminate (A) and (B) since these do not have 4 atoms attached as in a tetrahedrally shaped molecule. On your booklet or scratch paper, draw the Lewis structures for the remaining molecules. A molecule with tetrahedral shape has 4 atoms attached and no lone pairs. (D) and (E) have lone pairs.

CLASSIFICATION SET In this type of question, there is a list of five lettered headings consisting of possible answers followed by a list of numbered phrases. You must select the heading that is most closely related to each numbered phrase. Headings may be used once, more than once, or not at all.

Questions 1–3 refer to 0.10 M solutions.
    (A) $NH_3$, ammonia
    (B) NaOH, sodium hydroxide
    (C) $CH_3COOH$, acetic acid
    (D) $NaNO_3$, sodium nitrate
    (E) $CH_3COOK$, potassium acetate

1.  Has a pH equal to 7.0.

2.  Has a pH less than 7.0.

3.  Can be mixed with 0.10 M sodium acetate to make a buffer.

For the classification set, answer each question one at a time. Some of the answers can be used more than once or not at all. Begin by knowing what is meant by each phrase in 1 through 3. For example, what is a buffer made of?

1.  Look for a neutral solution since the pH equals 7. Salts that have parent acids that are strong will be neutral since their conjugate base will be too weak to undergo hydrolysis. $NO_3^-$ is the conjugate base of the strong acid $HNO_3$. Answer is (D).

2.  Look for an acid or the conjugate acid of a weak base since the pH is less than 7. Answer is (C).

3.  A buffer is made of a weak acid and its conjugate base (or a weak base and its conjugate acid). Sodium acetate contains acetate, which is the conjugate base of the weak acid acetic acid. Answer (C).

LIST AND GROUP QUESTIONS In this type of question, there is a list of possible answers, and you must select the answer that contains the correct group of responses.

1.  The concentration of a solution can be determined by which of the following methods?
    I.   Titration with a solution of known concentration
    II.  Calorimetry
    III. Spectrometric/colorimetric

    (A) I only
    (B) II only
    (C) III only
    (D) I and II only
    (E) I and III only
    (F) I, II, and III

ANSWER E. To approach the question, draw a line through choice II, because calorimetry is used to determine the change in enthalpy of a reaction. Continue to cross out items that are wrong and the responses that contain them. Draw a line through III and answers (D) and (F) which both contain choice II. Now you've narrowed down the possible responses.

NONCALCULATOR COMPUTATIONS These questions require computations without the use of a calculator. Simple mathematics or the choice of the correct algebraic setup will be involved.

EXAMPLE: A weak acid, HA, has a $K_a$ value of $1.0 \times 10^{-6}$. Calculate the pH after the complete reaction of 40.0 mL of 0.0100 $M$ HA with 10.0 mL of 0.0200 $M$ NaOH.

(A) 2.0
(B) 6.0
(C) 7.0
(D) 8.0
(E) 14.0

To answer this question, you must consider at what point in the titration the question is referring to. Since the base, NaOH is twice as concentrated as the acid, HA, it will take half as much NaOH to reach the equivalence point, or 20.0 mL. We are not at the equivalence point. Half the amount of the NaOH, 10.0 mL, needed to reach the equivalence point has been added. Halfway to the equivalence point, the pH equals the $pK_a$ which equals $-\log (K_a)$ or 6.0. Answer is (B).

EXAMPLE: $2C(s) + O_2(g) \rightarrow 2CO(g)$

Into a 3.0-L container at 25°C are placed 1.2 g of carbon, graphite, and 3.2 g oxygen gas.

If the carbon and the oxygen react completely to form CO(g), what will be the final pressure in the container at 25°C?

(A) $\dfrac{0.05(0.082)(298)}{3.0}$ atm

(B) $\dfrac{0.10(0.082)(25)}{3.0}$ atm

(C) $\dfrac{0.15(0.082)(298)}{3.0}$ atm

(D) $\dfrac{0.20(0.082)(298)}{3.0}$ atm

(E) $\dfrac{0.10(0.082)(298)}{3.0}$ atm

To answer this question, 1.2 g of carbon is 0.10 mol of carbon (1 mole of carbon equals 12 g). Similarly, 3.2 g of $O_2$ is 0.10 mol of $O_2$. The limiting reactant is carbon producing 0.10 mol of CO. $O_2$ is present in excess. (0.10 mol of CO requires 0.05 mol of $O_2$ leaving 0.05 mol of $O_2$.) When the reaction is complete, there is 0.10 mol CO + 0.05 mol $O_2$ left which equals 0.15 mol total. Answer is (C).

## STRATEGIES FOR THE FREE-RESPONSE SECTION

Section II of the AP exam comes with a periodic table, a table of standard reduction potentials, $E^\circ_{red}$, and a table of equations and constants.

■ Scan all of the questions in the section you are working in and mark those that you know you can answer correctly. Do these problems first.

■ You will be provided with an answer booklet as well as an insert that contains the same questions as the answer booklet, but without the spaces. You can remove the insert for reference. All of your work and answers for each problem must be shown in the answer booklet. No credit will be given for work shown on the insert, but you may write on it.

■ Show all of your work. Partial credit will be awarded for problems if the correct work is shown but the answer is not present or is incorrect. In problems involving calculations, circle your final answer.

■ Cross out incorrect answers with an "X" rather than spending time erasing.

■ Be clear, neat, and organized in your work. If a grader cannot clearly understand your work, you may not receive full credit.

■ Some free-response questions have four parts: a, b, c, and d. Attempt to solve each part. Even if your answer to "a" is incorrect, you still may be awarded points for the remaining parts of the question if the work is correct for those parts.

■ Units are important in your answer. Keeping track of your units throughout calculations and performing unit cancellation where possible, will help guide you to your answer. Points will be deducted for missing or incorrect units in the answer.

■ You do not need to work the questions in order. Make sure you put the number of the question in the corner of each page of your essay booklet. In addition, questions are sometimes broken into parts, such as (a) and (b). When this is the case, label each part of your response.

## STRATEGIES FOR ANSWERING ESSAY QUESTIONS

Essay questions will ask you to explain, compare, and predict. Minor calculations, showing mathematical relationships, or drawing graphs or structures, may also be involved. Usually these are not traditional essay questions. Most free-response questions do not require an introduction or conclusion. Most do not even require a thesis. Many of these questions may be written in a bulleted or short-answer format. Although this may sound easier than writing a traditional essay, it is important that you know the material very well because these are targeted questions. Examination readers want specifics. They are looking for accurate information presented in clear, concise prose. You cannot mask vague information with elegant prose.

To be successful in writing essays for the AP Chemistry exam, be sure to get straight to the answer and use key terms in your explanations. Sometimes, if you ramble on and on, you might accidentally state an incorrect fact. Points will be deducted for incorrect or extraneous information.

Techniques to write essays include chart format, bullet format, and outline format. None of these styles requires that you write complete sentences. In each of these styles, restate the question in simple terms, using your own words. Restate each part (a, b, and so on) separately, not together. In your restatement and response, underline key words or concepts.

- **Fill in a chart to answer the question**. This style is helpful in answering questions about electronic and molecular structure. This method is also used in the example on identification of the set of four quantum numbers for each electron in an element in this chapter.
- **In bullet format, make a list**, using a bullet (■) for each new concept. Leave room between concepts because you may want to come back later to fill them in.
- **Outline format** is more traditional, using Roman numerals, letters, etc. This takes more time to organize ideas, but it does show progression of ideas in a logical sequence. As in the bulleted format, leave room between concepts because you may want to come back later to fill them in.
- Lastly, **the freestyle method** sometimes results in rambling and incomplete answers. Writing in paragraphs does not allow room for additional ideas to be added.

# Scoring for Free-Response Questions

## Section II—Part A

Question number one is usually a multistep equilibrium problem.

### Sample Problem #1

1. Answer the following questions related to the equilibrium of benzoic acid, $HC_7H_5O_2$ .
   (A) A sample of 0.100 M benzoic acid, $HC_7H_5O_2$, ionizes by 2.5%. Calculate the equilibrium constant, $K_a$, for benzoic acid.
   (B) A 30.0 mL sample of 0.100 M $HC_7H_5O_2$ (aq) is titrated with 5.0 mL of 0.20 M KOH(aq). Calculate the pH of the resulting solution after the base has been added.
   (C) A 30.0 mL sample of 0.100 M $HC_7H_5O_2$ (aq) is titrated with 7.5 mL of 0.20 M KOH(aq). Calculate the pH of the resulting solution after the base has been added.
   (D) A 30.0 mL sample of 0.100 M $HC_7H_5O_2$ (aq) is titrated with 15.0 mL of 0.20 M KOH(aq). Calculate the pH of the resulting solution after the base has been added.
   (E) A 30.0 mL sample of 0.100 M $HC_7H_5O_2$ (aq) is titrated with 20.0 mL of 0.20 M KOH(aq). Calculate the pH of the resulting solution after the base has been added.
   (F) What mass of sodium benzoate, $NaC_7H_5O_2$ must be added to 100.0 mL of 0.100 M $HC_7H_5O_2$ (aq) to make a buffer with a pH equal to 5.00? Assume no volume change (molar mass of $NaC_7H_5O_2$ = 144 g/mol).

### Scoring Guidelines – Sample Problem 1

(A) $\dfrac{x}{0.100} \times 100\% = 2.5\%$; $x = 0.0025$

1 point for correct significant figures
1 point for the correct value of $x$

$$HC_7H_5O_2 \rightleftharpoons H^+ + C_7H_5O_2^-$$

| | | | |
|---|---|---|---|
| I | 0.100 | 0 | 0 |
| C | $-x$ | $x$ | $x$ |
| E | $0.100 - x$ | $x$ | $x$ |

$$K_a = \frac{[H^+][C_7H_5O_2^-]}{[HC_7H_5O_2]} = \frac{x^2}{0.10-x} =$$

$$\frac{(0.0025)^2}{0.100-0.0025} = \frac{6.25\times10^{-6}}{0.098} = 6.4\times10^{-5}$$

1 point for correct value of $K_a$

(B) 
$$HC_7H_5O_2 \rightleftharpoons H^+ + C_7H_5O_2^-$$

| | $HC_7H_5O_2$ | $H^+$ | $C_7H_5O_2^-$ |
|---|---|---|---|
| initial mol | 0.0030 | 0 | 0 |
| Add OH⁻ | −0.0010 | 0 | +0.0010 |
| new initial mol | 0.0020 | 0 | 0.0010 |
| I | 0.0350 L | | 0.0350 L |
| | 0.057 M | 0 | 0.029 M |
| C | −$x$ | +$x$ | +$x$ |
| E | 0.057 − $x$ | $x$ | 0.029 + $x$ |
| | 0.057 − $x$ | | |

1 point for correct calculation of moles or molarity of $HC_7H_5O_2$ and $C_7H_5O_2^-$

$$6.4 \times 10^{-5} = \frac{[H^+](0.029 + x)}{0.057 + x} \approx \frac{[H^+]0.029}{0.057}$$

$$1.3 \times 10^{-4} M = [H^+] \quad pH = -\log[H^+] = 3.89$$

Alternately, you can use the Henderson Hasselbach equation. Volume cancels.

1 point for correct pH

$$pH = pK_a + \log \left[ \frac{0.0010 \, \text{mol}}{0.0020 \, \text{mol}} \right] = 4.19 + (-.30) = 3.89$$

(C)  0.00750 L × 0.200 mol/L = 
    0.0015 mol OH⁻ = moles $C_7H_5O_2^-$
Initial moles of acid = 0.0030 mol
Half of the acid has been neutralized; pH = $pK_a$ = 4.19

1 point for calculation of mol OH⁻ = moles $C_7H_5O_2^-$
1 point for correct pH

(D)  0.015 L x 0.20 mol / L = 0.0030 mol OH⁻
mol acid = mol OH⁻
equivalence point only $C_5H_7O_2^-$ present
0.0030 mol/0.045 L = 0.067 M

$$C_5H_7O_2^- + H_2O \rightleftharpoons OH^- + HC_5H_7O_2$$

| | $C_5H_7O_2^-$ | $OH^-$ | $HC_5H_7O_2$ |
|---|---|---|---|
| I | 0.67 M | 0 | 0 |
| C | −$x$ | +$x$ | +$x$ |
| E | 0.067 − $x$ | $x$ | $x$ |

$$K_b = \frac{K_w}{K_a} = \frac{1.0 \times 10^{-14}}{6.4 \times 10^{-5}} = 1.6 \times 10^{-10}$$

1 point for correct $K_b$ value

$$K_b = \frac{[OH^-][HC_7H_5O_2]}{[C_7H_5O_2^-]} = \frac{x^2}{0.067 - x} \sim \frac{x^2}{0.067}$$

$$x = [(1.6 \times 10^{-10})(0.67)]^{1/2} = 3.3 \times 10^{-6} M \ OH^-$$
$$pOH = 5.48 \ ; \ pH = 14.00 - 5.48 = 8.52$$

1 point for correct pH

(E)  0.0300 L × 0.10 mol/L = 0.0030 mol $HC_5H_7O_2$

0.0200 L × 0.20 mol/L = 0.0040 mol OH⁻

Excess OH⁻ = 0.0040 mol − 0.0030 mol/0.0500 L = 0.020 M OH⁻

pOH = 1.70 ; pH = 14 − pOH = 12.30

1 point for correct molarity of excess ion

1 point for correct pH; alternatively, can solve using $K_w$ = [H⁺][OH⁻], solve for H⁺ then, pH.

(F)   pH = p$K_a$ + log ([C$_7$H$_5$O$_2$$^-$]/[HC$_7$H$_5$O$_2$])

5.00 = 4.19 + log ([mol C$_7$H$_5$O$_2$$^-$]/[0.010 mol])

0.81 = log([mol C$_7$H$_5$O$_2$$^-$]/[0.010 mol])

antilog 0.81 = (mol C$_7$H$_5$O$_2$$^-$)/0.010 mol

0.065 = mol C$_7$H$_5$O$_2$$^-$

Mass of NaC$_7$H$_5$O$_2$ = 0.065 × 144 g/mol = 9.4 g

*1 point for correct values for p$K_a$ and mol of acid*

*1 point for correct answer*

## SAMPLE PROBLEM #2 OR 3

You will get to choose to do either problem #2 or #3.

$$C_2H_4(g) + 3 O_2(g) \rightarrow 2 CO_2(g) + 2H_2O(g)$$

Information about the substances involved in the reaction presented above is tabulated below.

| Substance | $S^0$(J/mol•K) | $\Delta G^\circ_f$(kJ/mol) |
|---|---|---|
| C$_2$H$_4$(g) | ? | 68 |
| O$_2$(g) | 205 | 0 |
| CO$_2$(g) | 213.6 | −394 |
| H$_2$O(g) | 189 | −229 |

| Bond | Bond Energy (kJ/mol) |
|---|---|
| O – H | 467 |
| O – O | 146 |
| O = O | 495 |
| C – H | 413 |
| C – C | 347 |
| C = C | 614 |
| C ≡ C | 839 |
| C – O | 358 |
| C = O | 799 |

(a) Calculate the value for the standard free energy change, $\Delta G^\circ$, at 25°C for the reaction. What does the sign of $\Delta G^\circ$ indicate about the reaction?

(b) Calculate the value for the standard free energy change, $\Delta H^\circ$, at 25°C for the reaction. What does the sign of $\Delta H^\circ$ indicate about the reaction?

(c) Calculate the value for the standard entropy change, $\Delta S^\circ$, at 25°C for the reaction. What does the sign of $\Delta S^\circ$ indicate about the reaction?

(d) Calculate the value for the absolute entropy of C$_2$H$_4$(g) at 25°C.

SCORING GUIDELINES—SAMPLE PROBLEM #2 OR 3

(A) $\Delta G° = 2(-394) + 2(-237) - 68 = -1330$ kJ/mol.

1 point for correct value of $\Delta G°$

(B)

1 point for meaning of sign

1 point for correct Lewis Structures of Reactants and products

| Bonds Broken | Bonds Formed |
|---|---|
| 4 C–H    4(413) | 4 C=O    4 (799) |
| 1 C=C    1 (614) | 4 O-H    4 (467) |
| 3 O=O   3(495) | −5064 kJ |
| +3751 kJ | |

$\Delta H° = 3751$ kJ + $-5064$ kJ = $-1313$ kJ

1 point for correct value of $\Delta H°$

The reaction is exothermic since $\Delta H°$ is negative.

(C) $\Delta G° = \Delta H° - T\Delta S°$ ;   $\Delta S° = (\Delta H° - \Delta G°) / T$

1 point for meaning of sign

$\Delta S° = -1313 - (-1330$ kJ$) / 298$ K $= 0.057$ kJ/K

1 point for correct value of $\Delta S°$

A positive sign of change in entropy indicates that entropy is increasing in the reaction.

1 point for meaning of sign

Here, it barely changes.

(D) $\Delta S° = 2S°_{CO_2} + 2 S°_{H_2O} - (S°_{C_2H_4} + 3S°_{O_2})$

1 point for value of absolute $S°$

$S°_{C_2H_4} = 2 S°_{CO_2} + 2 S°_{H_2O} - 3 S°_{O_2} - \Delta S°$

$= 2 (213.6) + 2 (189) - 3(205) - 57 = 133$ J/K

SAMPLE PROBLEM #4 REACTIONS—WEIGHTED 15%

You will be given 8 chemical reactions of which you can choose 5 to answer. Solutions are aqueous unless otherwise indicated. In all cases, a reaction occurs. If a substance is extensively ionized, such as a strong acid or a completely soluble compound, write it as separate ions. Be sure to cancel out spectator ions or molecules that are unchanged by the reaction. You need not balance the equation. Circle your final answer!

Each reaction is worth 3 points, 1 point for the correct reactants and 2 points for correct product(s). To earn the second of two points in the products, all of the products must be correct.

(a) Equal volumes of equimolar solutions of sodium hydroxide and acetic acid are mixed.

$CH_3COOH + OH^- \rightarrow HOH + CH_3COO^-$

1 point for correct reactants
2 point for correct products
Only 1 product point is awarded if water is written as $H^+ + OH^-$ or if $CH_3COONa$ is written as one of the products

SAMPLE PROBLEM #5—WEIGHTED 15%

This mandatory essay is usually a lab-based question. There are twenty-two recommended experiments required by the College Board. The question usually asks you to describe a laboratory procedure for one of these experiments or analyze sources of errors. Your response is graded for its accuracy and relevance. You should be clear and well organized in answering.

A student is instructed to identify an unknown solid by conducting four qualitative laboratory tests. The solid substance could either be potassium carbonate, $K_2CO_3$, or benzoic acid, $C_6H_5COOH$. Describe each of the four possible lab tests which can be performed. For each test, identify the observations for both potassium carbonate, $K_2CO_3$, and benzoic acid, $C_6H_5COOH$.

| | |
|---|---|
| 1) Perform a flame test. The potassium ion in $K_2CO_3$ will burn a characteristic lavender color. The flame resulting from burning $C_6H_5COOH$ will not be lavender. | 1 point for identification of test<br>1 point for listing observations for both possible substances |
| 2) Dissolve solid in water and perform a conductivity test. Potassium carbonate is a strong conductor and the bulb on the tester will be brightly lit. If the unknown is $C_6H_5COOH$, the bulb will barely be lit since benzoic acid is a weak acid, so it is also a weak conductor of electricity. | 1 point for identification of test<br>1 point for listing observations for both possible substances |
| 3) Dissolve solid in water and measure the pH. A basic pH (>7) reveals that the unknown is $K_2CO_3$. An acidic pH identifies the unknown as $C_6H_5COOH$. | 1 point for identification of test<br>1 point for listing observations for both possible substances |
| 4) Add an acid like HCl, which contains a source of hydrogen ion. If the unknown is $K_2CO_3$, the mixture will bubble (release gas). If the unknown is $C_6H_5COOH$, there will be no bubbles. | 1 point for identification of test<br>1 point for listing observations for both possible substances |

SAMPLE PROBLEM #6—MANDATORY WEIGHTED 15%

This problem usually combines chemical calculations with essay writing. Questions of the type mentioned in Sample Problem #7 have also been found here.

$$A(g) + 2 B(g) \rightarrow C(g) + D(g)$$

Consider the equation above,
  (a) Using the table below, Calculate the value of $\Delta H°$ for the reaction at 25°C. Show all of your work.

| | $\Delta H°_f$ (kJ/mol) |
|---|---|
| A | −1275 |
| B | −166 |
| C | −393 |
| D | −286 |

(b) Sketch the potential energy diagram for the reaction above. Label the activation energy for the forward reaction.

(c) Explain how the magnitude of the standard entropy of the reaction, $\Delta S^\circ$, will change for the reaction at 25°C.

(d) Predict the sign of the standard free energy change, $\Delta G^\circ$, using your answers to parts (a) and (c).

(e) Write the rate law for the reaction using the information in the table below. Explain how you obtained your result.

| Trial # | Initial [A] (mol/L) | Initial [B] (mol/L) | Initial Rate of Formation of C (mol L⁻¹s⁻¹) |
|---|---|---|---|
| 1 | 0.10 | 0.10 | 0.20 |
| 2 | 0.20 | 0.10 | 0.40 |
| 3 | 0.40 | 0.20 | 6.4 |

(a)  $\Delta H^\circ = [-393 + -286] - [-1275 + 2(-166)] = -679 - (-1607) = 928 \text{ kJ}$

                                                                                      1 point for correct value of $\Delta H^\circ$

(b)

1 point for showing that energy of products > energy of reactants

1 point for labeling the activation energy

(c) The standard entropy change will decrease because the moles of gas decrease from reactants to products (3 mol -> 2 mol).

1 point must say why entropy decreases

(d) $\Delta G^\circ = \Delta H^\circ - T\,\Delta S^\circ$

    = positive – (negative) = positive $\Delta G^\circ$

therefore the reaction is nonspontaneous at all temperatures.

1 point for determination of order in A: must show work

(e) $\dfrac{\text{Rate 2}}{\text{Rate 1}} = \dfrac{k[A]^x[B]^y}{k[A]^x[B]^y}$

$$= \frac{k[0.20]^x \,\cancel{[0.10]^y}}{k[0.10]^x \,\cancel{[0.10]^y}} = \frac{0.40}{0.20}$$

$$= 2^x = 2; \; x = 1$$

$$\frac{\text{Rate 3}}{\text{Rate 2}} = \frac{k[A]^x[B]^y}{k[A]^x[B]^y}$$

$$\frac{k[0.40]^x[0.20]^y}{k[0.20]^x[0.10]^y} = \frac{6.4}{0.40}$$

$$= 2^x \times 2^y = 16$$

$$= 2^1 \times 2^y = 16; \; 2^y = 8; \; y = 3$$

**1 point for determination of order in B: must show work**

## SAMPLE PROBLEM #7 OR #8

You choose to answer either question 7 or 8. Be sure to read each question before deciding which one to answer. This essay usually asks you to explain chemical phenomena.

Use appropriate chemical principles to account for the following observation. Your answer must include references to both substances.

(a)  At 25°C and 1.0 atm, ethane ($C_2H_6$) is a gas and ethanol ($CH_3CH_2OH$) is a liquid.

There are usually 3 more questions, parts (b), (c), and (d), which may or may not be related to (a) or each other, but all are scored in a similar way to (a) below.

(a)  Ethane, $C_2H_6$, has dispersion forces. Ethanol, $CH_3CH_2OH$, has hydrogen bonding. Hydrogen bonding is stronger than dispersion forces.

**1 point for indicating the correct intermolecular force for each molecule**

**1 point for indicating that hydrogen bonding is stronger than dispersion forces**

# A Diagnostic Test

The purpose of this test is to give you an indication of how well you will perform on the AP Chemistry exam. These questions are representative of the AP Chemistry examination, but bear in mind it is impossible to predict exactly how well you will do on the actual exam. Calculators may not be used for answering questions in the first section of this test. The first section is 45% of your total test grade. Time yourself to finish this part in 90 minutes. There are two types of multiple-choice questions used in this examination. The first type consists of five lettered headings followed by a listing of number phrases. For each phrase you are to select the one heading that is most closely related to it. Headings may be used once, more than once, or not at all. The majority of multiple-choice questions consist of a question or incomplete statements followed by five possible answers. Select the one that is best in each case.

## AP CHEMISTRY EXAMINATION
### Section I: Multiple-Choice Questions
### Time: 90 minutes
### Number of Questions: 75

No calculators are to be used in this section; no tables (except Periodic Table) permitted.

Questions 1–3 refer to the following elements:
- (A) Chromium
- (B) Sodium
- (C) Copper
- (D) Phosphorus
- (E) Beryllium

1. Forms colored ions having the formulas $X_2O_7^{2-}$ and $XO_4^{2-}$

2. Forms an oxide that yields aqueous solutions that are acidic

3. Forms hydrogen gas when it reacts with cold water

Questions 4–7 refer to the following substances:
- (A) Hydrofluoric acid, HF
- (B) Hydrobromic acid, HBr
- (C) Sodium hydroxide, NaOH
- (D) Nitrous acid, $HNO_2$
- (E) Carbon dioxide, $CO_2$

4. Will dissolve or etch glass

5. Is a strong acid

6. Will neutralize acids

7. Contributes to the greenhouse effect

Noble gases

Alkali metals · Alkaline earth metals · Halogens · Transition metals · metals ← → nonmetals

| 1 1A | 2 2A | 3 | 4 | 5 | 6 | 7 | 8 | 9 | 10 | 11 | 12 | 13 3A | 14 4A | 15 5A | 16 6A | 17 7A | 18 8A |
|---|---|---|---|---|---|---|---|---|---|---|---|---|---|---|---|---|---|
| 1 **H** 1.008 | | | | | | | | | | | | | | | | | 2 **He** 4.003 |
| 3 **Li** 6.941 | 4 **Be** 9.012 | | | | | | | | | | | 5 **B** 10.81 | 6 **C** 12.01 | 7 **N** 14.01 | 8 **O** 16.00 | 9 **F** 19.00 | 10 **Ne** 20.18 |
| 11 **Na** 22.99 | 12 **Mg** 24.31 | | | | | | | | | | | 13 **Al** 26.98 | 14 **Si** 28.09 | 15 **P** 30.97 | 16 **S** 32.07 | 17 **Cl** 35.45 | 18 **Ar** 39.95 |
| 19 **K** 39.10 | 20 **Ca** 40.08 | 21 **Sc** 44.96 | 22 **Ti** 47.88 | 23 **V** 50.94 | 24 **Cr** 52.00 | 25 **Mn** 54.94 | 26 **Fe** 55.85 | 27 **Co** 58.93 | 28 **Ni** 58.69 | 29 **Cu** 63.55 | 30 **Zn** 65.38 | 31 **Ga** 69.72 | 32 **Ge** 72.59 | 33 **As** 74.92 | 34 **Se** 78.96 | 35 **Br** 79.90 | 36 **Kr** 83.80 |
| 37 **Rb** 85.47 | 38 **Sr** 87.62 | 39 **Y** 88.91 | 40 **Zr** 91.22 | 41 **Nb** 92.91 | 42 **Mo** 95.94 | 43 **Tc** (98) | 44 **Ru** 101.1 | 45 **Rh** 102.9 | 46 **Pd** 106.4 | 47 **Ag** 107.9 | 48 **Cd** 112.4 | 49 **In** 114.8 | 50 **Sn** 118.7 | 51 **Sb** 121.8 | 52 **Te** 127.6 | 53 **I** 126.9 | 54 **Xe** 131.3 |
| 55 **Cs** 132.9 | 56 **Ba** 137.3 | 57 **La*** 138.9 | 72 **Hf** 178.5 | 73 **Ta** 180.9 | 74 **W** 183.9 | 75 **Re** 186.2 | 76 **Os** 190.2 | 77 **Ir** 192.2 | 78 **Pt** 195.1 | 79 **Au** 197.0 | 80 **Hg** 200.6 | 81 **Tl** 204.4 | 82 **Pb** 207.2 | 83 **Bi** 209.0 | 84 **Po** (209) | 85 **At** (210) | 86 **Rn** (222) |
| 87 **Fr** (223) | 88 **Ra** 226 | **Ac†** (227) | 104 **Rf** | 105 **Db** | 106 **Sg** | 107 **Bh** | 108 **Hs** | 109 **Mt** | 110 **Ds** | 111 **Uuu** | 112 **Uub** | | 114 **Uuq** | | | | |

*Lanthanides

| 58 **Ce** 140.1 | 59 **Pr** 140.9 | 60 **Nd** 144.2 | 61 **Pm** (145) | 62 **Sm** 150.4 | 63 **Eu** 152.0 | 64 **Gd** 157.3 | 65 **Tb** 158.9 | 66 **Dy** 162.5 | 67 **Ho** 164.9 | 68 **Er** 167.3 | 69 **Tm** 168.9 | 70 **Yb** 173.0 | 71 **Lu** 175.0 |
|---|---|---|---|---|---|---|---|---|---|---|---|---|---|

†Actinides

| 90 **Th** 232.0 | 91 **Pa** (231) | 92 **U** 238.0 | 93 **Np** (237) | 94 **Pu** (244) | 95 **Am** (243) | 96 **Cm** (247) | 97 **Bk** (247) | 98 **Cf** (251) | 99 **Es** (252) | 100 **Fm** (257) | 101 **Md** (258) | 102 **No** (259) | 103 **Lr** (260) |
|---|---|---|---|---|---|---|---|---|---|---|---|---|---|

Group numbers 1–18 represent the system recommended by the International Union of Pure and Applied Chemistry.

Most of the multiple-choice questions are questions or incomplete statements followed by five suggested answers or completions. You should select the best one in each case.

For questions 8–9, consider the electrochemical cell:

$$Ag \mid Ag^+(1\ M) \parallel Cu^{2-}(1M) \mid Cu$$

The cell reaction is
$$2Ag^+(aq) + Cu\ (s) \rightarrow Cu^{2+}(aq) + 2Ag(s).$$

The measured voltage is +0.46 volts.

8. Increasing the concentration of silver ions will
   (A) cause a decrease in blue color in the cell.
   (B) increase the cell voltage above +0.46 volts.
   (C) decrease the concentration of copper (II) ions.
   (D) cause no change in the cell voltage.
   (E) cause a change in the direction of electron flow through the external circuit.

9. The reaction at the anode is
   (A) $Ag^+(aq) + e^- \rightarrow Ag(s)$.
   (B) $Ag\ (s) \rightarrow Ag^+(aq)$.
   (C) $Cu(s) \rightarrow Cu^{2+}(aq) + 2e^-$.
   (D) $Cu^{2+}(aq) + 2e^- \rightarrow Cu(s)$.
   (E) $Cu(s) + 2e^- \rightarrow Cu^{2+}(aq)$.

10. A spark coil is used to begin the reaction between hydrogen and oxygen in a balloon. The spark
    (A) allows the gases to obtain the activation energy.
    (B) causes the hydrogen to vaporize, then react.
    (C) causes both the hydrogen and the oxygen to vaporize, then react.
    (D) provides the enthalpy for this reaction.
    (E) increases the entropy of both oxygen and hydrogen.

11. The amount of silver which will be formed when 0.00200 mol of $Ag_2S$ reacts completely with excess zinc is
    (A) 0.00100 mol.
    (B) 0.00200 mol.
    (C) 0.0040 mol.
    (D) 0.00248 grams.
    (E) 0.00456 grams.

12. The flame test color for a solution of sodium nitrate is
    (A) pale yellow.
    (B) blue.
    (C) violet.
    (D) purple.
    (E) crimson.

13. In the expression $PV = nRT$,
    (A) $R$ is a temperature dependent constant.
    (B) $R$ is a pressure dependent constant.
    (C) $R$ is valid except for gases at very high temperature and/or very low pressure.
    (D) $T$ must be expressed in Celsius degrees.
    (E) the ratio, $PV/nRT$, is a very large value @ STP.

14. The aqueous solution with the lowest freezing temperature is
    (A) 0.100 m $AlCl_3$.
    (B) 0.200 m $AlCl_3$.
    (C) 0.300 m NaCl.
    (D) 0.300 m KF.
    (E) 0.350 m KF.

15. Calcium oxide, CaO, has a lower melting temperature than magnesium oxide, MgO, due to the
    (A) higher charge density of $Mg^{+2}$ than of Ca.
    (B) higher charge density of $Ca^{+2}$ than of $O^{-2}$.
    (C) greater atomic volume of $Mg^{+2}$ than of $O^{-2}$.
    (D) greater atomic volume of $Mg^{+2}$ than of $Ca^{+2}$.
    (E) greater positive charge on Ca than on Mg.

16. The general classification for the compound $C_3H_7OH$ is
    (A) an alcohol.
    (B) an aldehyde.
    (C) a carboxylic acid.
    (D) a ketone.
    (E) an ester.

17. A solution of 50.0 mL of 0.0010 M $Ba(NO_3)_2$ is slowly titrated with 50.0 mL of 0.0030 M $H_2SO_4$. The conductivity of this solution will
    (A) decrease to near zero, then increase.
    (B) decrease to near zero and remain very low.
    (C) increase as the acid is added.
    (D) increase as the acid is added then become constant at a high value.
    (E) increase as the acid is added and then slowly become very low.

18. Which of the following pairs illustrates the Law of Multiple Proportions?
    (A) $SO_2$, $SO_3$
    (B) $CO_2$, $CCl_4$
    (C) NaCl, NaBr
    (D) $NH_4Cl$, $NH_4Br$
    (E) $SO_2$, $CO_2$

19. The scientist who is remembered for the alpha particle scattering properties of gold foil, concluding that the nucleus is small, dense, and positively charged, is
    (A) Ernest Rutherford.
    (B) J. J. Thomson.
    (C) John Dalton.
    (D) Robert Boyle.
    (E) Antoine Lavoisier.

20. Isotopic forms of the same element
    (A) differ in the number of neutrons in the nucleus.
    (B) are formed by gaining electrons.
    (C) always have a positive charge.
    (D) are found only in metals.
    (E) have the same number of neutrons.

21. The number of moles of oxygen atoms in one mole of iron (II) phosphate is
    (A) 1.
    (B) 2.
    (C) 3.
    (D) 4.
    (E) 8.

22. Magnesium fluoride, a salt of low solubility in water, has a $K_{sp}$ of 6.4 x $10^{-9}$. The concentration of $Mg^{2+}$ ions in this solution would be
    (A) $\sqrt{6.4 \times 10^{-9}}$ M .
    (B) $\sqrt{(6.4 \times 10^{-9}/2)}$ M .
    (C) $\sqrt[3]{(6.4 \times 10^{-9}/3)}$ M .
    (D) $\sqrt[3]{(6.4 \times 10^{-9}/4)}$ M .
    (E) $\sqrt[3]{6.4 \times 10^{-9}}$ M .

Questions 23 and 24 refer to this pH curve for the titration of 50.0 mL of 0.100 M acid with 0.100 M base.

Vol added (mL)

23. This is the curve
    (A) for a strong acid/strong base titration.
    (B) for a strong acid/weak base titration.
    (C) for a weak acid/strong base titration.
    (D) for either a strong acid/strong base or a weak acid/strong base titration.
    (E) for an acid/base titration but it is impossible to describe the strength of the acid and of the base used without more information.

24. Buffering occurs
    (A) between points F and G.
    (B) between points G and H.
    (C) at point H between points I and J.
    (D) between both F and G and between I and J.
    (E) none of these

25. The systematic (Stock Nomenclature) name for $Ca(OCl)_2$ is
    (A) calcium (II) hypochlorite.
    (B) calcium hypochlorite.
    (C) calcium dioxochloride.
    (D) calcium monoxodichloride.
    (E) calcium dichloride.

26. Bond angle data shows the following angles for three compounds:

    $CH_4$ 109.5°,
    $NH_3$ 107°,
    $H_2O$ 104.5°.

    This trend is basically because
    (A) lone pairs of electrons require more room than bonding pairs.
    (B) hydrogen atoms repel each other more in water than in $CH_4$ or $NH_3$.
    (C) oxygen has a higher electronegativity than does N, and C has even less.
    (D) of the attempt of all central atoms to achieve the tetrahedral shape.
    (E) only the positions of nuclei determine molecular shapes and bond angles.

27. The acid $HClO_4$ is given the name
    (A) perchloric acid.
    (B) chloric acid.
    (C) chlorous acid.
    (D) hypochlorous acid.
    (E) hypochlorotetroxide.

28. When 100.0 mL of 2.0 M $NH_3$ and 100.0 mL of 1.0 M $AgNO_3$ are mixed, but before any reaction occurs, the major species in solution are
    (A) $Ag^+$, $NO_3^-$, $NH_3$, and $H_2O$.
    (B) $Ag^+$, $NO_3^-$, and $Ag(NH_3)^+$.
    (C) $Ag(NH_3)^+$ and $Ag(NH_3)_2^+$.
    (D) $NH_3$ and $Ag(NH_3)_2^+$.
    (E) $NH_3$, $Ag(NH_3)_2^+$ and $H_2O$.

29. $H_3PO_4$ is a triprotic acid with equilibrium constants of

    $K_{a_1} = 7 \times 10^{-3}$,
    $K_{a_2} = 6 \times 10^{-8}$,
    $K_{a_3} = 5 \times 10^{-13}$.

    This information leads to the conclusion that
    I   $H_3PO_4$ is a weak acid.
    II  $H_3PO_4$ dissociates in three steps to form $PO_4^{3-}$.
    III Only the first dissociation will make an important contribution to the [$H^+$] when dissolved in water.

    Of these three statements
    (A) only I is valid.
    (B) only II is valid.
    (C) only III is valid.
    (D) both I and II are valid but not III.
    (E) all three statements (II, II, and III) are valid.

30. Burets can usually be read to the nearest ±0.01 mL. If the liquid volume in a buret is recorded as 22.00 mL, the number of significant figures in this value is
    (A) 1.
    (B) 2.
    (C) 3.
    (D) 4.
    (E) 5.

31. The mass of a sample of $KClO_3$ is determined with five weighings:

    1.391 g; 1.392 g; 1.299 g; 1.390 g; 1.388 g.

    The number of significant figures in the average of these five readings is
    (A) 1.
    (B) 2.
    (C) 3.
    (D) 4.
    (E) 5.

32. A weighing tray has a mass of 0.911 g. To this tray is added 3.2 g of NaOH. The total mass of tray plus NaOH is
    (A) 4.1 g.

(B) 4.11 g.
(C) 4.111 g.
(D) 4.100 g.
(E) 4.10 g.

33. The density of a new plastic is determined experimentally by first measuring the volume using water displacement; the mass of the dry sample is also found. The Data:

    Initial volume of water—13.0 mL

    Final volume, water and sample—27.1 mL

    Mass of sample —36.123 g

    The density of this sample of plastic is best expressed as
    (A) 2.5619 g/mL.
    (B) 2.562 g/mL.
    (C) 2.56 g/mL.
    (D) 2.6 g/mL.
    (E) 3. g/mL.

34. The density of diethyl ether is 0.714 $g/cm^3$. The volume occupied by 10.00 g of this liquid is
    (A) 1.4 $cm^3$.
    (B) 14. $cm^3$.
    (C) 14.0 $cm^3$.
    (D) 14.01 $cm^3$.
    (E) 71.4 $cm^3$.

35. When equal volumes of 0.150 M NaOH and 0.150 M $HC_2H_3O_2$ are mixed, the resulting solution has a pH about 9. This is due to
    (A) incomplete reaction of acetic acid in water.
    (B) an unequal number of moles of $OH^-$ (from the NaOH) and $H^+$ (from the $HC_2H_3O_2$).
    (C) the $C_2H_3O_2^-$ reacting with water to provide more $OH^-$.
    (D) the cation of the acid, which remains in solution at the equivalent point, and is a base.
    (E) both the $C_2H_3O_2^-$ reacting with water and the acidic nature of the anion of the acid.

36. An isotope has the atomic number of 8 and a mass number of 17. This element
    (A) is an isotopic form of oxygen.
    (B) has 8 neutrons.
    (C) has 9 protons if it is in ionic form.
    (D) is an isotopic form of fluorine.
    (E) has 17 electrons.

37. The rate of reaction in a collection of gas particles is always much lower than the calculated collision frequency would indicate. This means that only in a small percentage of the collisions does a reaction result. This is because

    I   few particles have the required activation energy.
    II  few particles have formed the activated complex.
    III few particles have the correct orientation when they collide. Of these three statements, those that are valid are

    (A) only I.
    (B) only II.
    (C) only III.
    (D) both I and II only.
    (E) all three (I, II, and III).

38. The pH of a 0.00100 M HBr solution is
    (A) 0.00100M.
    (B) 3.0.
    (C) 7.0.
    (D) 11.0.
    (E) impossible to determine without more data.

39. Aqueous solutions of lead (II) nitrate and potassium chromate are allowed to react. The results show
    (A) both lead (II) chromate and potassium nitrate precipitate.
    (B) no solid forms from this reaction.
    (C) lead (II) chromate will precipitate from solution.
    (D) potassium nitrate precipitates.
    (E) potassium chromate is insoluble so there is no reaction.

40. A rigid cylinder contains $CO_2$ gas; some of the carbon dioxide is allowed to escape with temperature adjusted to a constant value. Which of the following applies to the $CO_2$?
    (A) The pressure of the gas increases.
    (B) The volume of the gas decreases.
    (C) The total number of gas molecules within the cylinder remains unchanged.
    (D) The average molecular speed decreases.
    (E) The distance between $CO_2$ molecules is increased.

41. Gases are less soluble in water at
    (A) high temperature, low pressure.
    (B) low temperature, high pressure.
    (C) high temperature, high pressure.
    (D) low temperature, low pressure.
    (E) pressure has no effect on the solubility of gases in water.

42. $2NH_3 \rightarrow N_2 + 3H_2$

    The above reaction occurs in a closed system of constant volume and temperature. What is the resultant pressure of the hydrogen if the partial pressure of the ammonia decreases by 0.40 atm.?
    (A) increases by 0.20 atm
    (B) increases by 0.40 atm
    (C) increased by 0.60 atm
    (D) decreased by 0.60 atm
    (E) impossible to calculate without more data

43. Air is pumped into a rigid steel cylinder at constant temperature. The increase in pressure is due to
    (A) increased molecular collisions.
    (B) the greater kinetic energy of the gas particles.
    (C) increase in the size of the individual molecules.
    (D) the greater force of attraction between gas molecules at high pressure.
    (E) the molecular contraction at higher pressure.

44. Real gases depart from ideal at conditions of
    (A) high pressure, low temperature.
    (B) low pressure, low temperature.
    (C) high pressure, high temperature.
    (D) low pressure, low temperature.
    (E) constant pressure and temperature.

45. Gas pressure is due to gas particles
    (A) slowing down at high temperature.
    (B) colliding with container walls.
    (C) neither attracting or repelling each other.
    (D) attracting each other.
    (E) moving in random patterns.

46. $2KClO_3(s) \rightarrow 2KCl(s) + 3O_2(g)$

    According to the above equation, 0.40 mol of solid $KClO_3$ completely decomposes, forming KCl and $O_2$. The dry gas is collected at STP. The volume of this oxygen gas would be most nearly
    (A) 1.4 L.
    (B) 14 L.
    (C) 140 L.
    (D) 1400 L.
    (E) 14,000 L.

47. A hollow steel cylinder of volume 24 L contains 1.0 mole of Ne and 2.0 mol of Ar. The partial pressure of the Ne is
    (A) 1/2 the total pressure.
    (B) 1/3 the total pressure.
    (C) 3 times that of the Ar.
    (D) 2 times that of the Ar.
    (E) equal to that of the Ar.

48. As the atomic number changes in Period Two of the Periodic Table from alkaline metals to halogens, the atomic radii
    (A) do not change.
    (B) increase.
    (C) decrease.
    (D) decrease, then slightly increase.
    (E) cannot be predicted because there is no trend.

49. Assume that $2A + B \rightarrow C$ is the rate determining step. 3.0 moles of A and 2.0 moles of B are placed in a 1.0-L flask; after five minutes the concentration of C reaches 1.0 M. After five minutes the rate will have
    (A) increased by a factor of 9.
    (B) increased by a factor of 18.
    (C) decreased by a factor of 9.
    (D) decreased by a factor of 10.
    (E) decreased by a factor of 18.

50. For a given reaction at a temperature of 27°C, the rate law is Rate = $k$ [X] [Y]. If the concentration of X and of Y are both 0.40 M, the rate is $4.0 \times 10^{-6}$ mol/L min. Determine the value of $k$ (the rate constant) at this temperature.
    (A) $2.5 \times 10^{-5}$ L/mol·min
    (B) $2.5 \times 10^{-3}$ L/mol·min
    (C) $2.5 \times 10^{-5}$ mol/L·min
    (D) $1.0 \times 10^{-6}$ L/mol·min
    (E) $1.0 \times 10^{-5}$ L/mol·min

51. The major reason an increase in temperature causes an increase in reaction rate is that
    (A) the activation energy changes with temperature.
    (B) the fraction of high energy molecules increases.
    (C) the reactant molecule pressure increases.
    (D) molecules collide with greater frequency.
    (E) catalysts become more effective.

52. Catalysts effectively increase reaction rate by
    (A) increasing the $K_{eq.}$
    (B) increasing the concentration of the reactant.
    (C) decreasing the concentration of the products.
    (D) lowering the activation energy requirements.
    (E) decreasing the reaction temperature.

53. For all zero-order reactions,
    (A) the reaction rate is independent of time.
    (B) the rate constant equals zero.
    (C) the concentration of reactants does not change over time.
    (D) activation energy is very low.
    (E) the concentration of reactants does not change and the rate is independent of time.

54. It is found that in a certain first order reaction the half-life is 1.4 minutes. The rate constant, $k$, for that same temperature is about
    (A) 0.35 min$^{-1}$.
    (B) 0.50 min$^{-1}$.
    (C) 0.71 min.
    (D) 2.0 min.
    (E) 2.0 min$^{-1}$.

55. To determine the order with respect to Br$^-$ in the reaction $BrO_3^-$ (aq) + 5Br$^-$(aq) + 6H$^+$(aq)
    $\rightarrow$ 3H$_2$O(l) + 3Br$_2$(g) solutions should be prepared which differ in
    (A) [$BrO_3^-$] and [Br$^-$].
    (B) [Br$^-$].
    (C) [H$^+$].
    (D) [$BrO_3^-$] and [Br$^-$] and [H$^+$].
    (E) [H$_2$O] and [Br$_2$].

56. The first-order rate constant for nuclear unstable $^{60}$Co is 0.13 yr$^{-1}$, and for $^{90}$Sr it is 0.24 yr$^{-1}$.
    (A) The half-life of Sr is longer than that of Co.
    (B) The half-life of Sr is shorter than that of Co.
    (C) The half-lives of Sr and Co are equal.
    (D) The half-lives of Sr and Co cannot be compared from these data.
    (E) There is no simple relationship between half-life and stability.

57. The activation energy for a reaction is +30 kJ/mol; therefore the activation energy for the reverse reaction must be
    (A) –30 kJ / mol.
    (B) greater than 30 kJ / mol.
    (C) less than 10 kJ / mol.
    (D) + 30 kJ / mol.
    (E) a value that cannot be calculated without more data.

58. The rate law expression is rate = $k[X]^2$ [Y] for a certain reaction. Both $X$ and $Y$ are reactant gases. If the volume of the system is reduced to 1/3 of the original volume, the relative new reaction rate will be
    (A) 3 times greater.
    (B) 9 times greater.
    (C) 27 times greater.
    (D) 27 times less.
    (E) 9 times less.

59. For the reaction $X + 2B \rightarrow Z$, the rate law is rate = $k$ [$X$] [$B$]$^2$. If the concentration of $B$ increases by a factor of 3 while [$X$] is held constant, the rate of the reaction will
    (A) increase by 3 times.
    (B) increase by 6 times.
    (C) increase by 9 times.
    (D) increase by 27 times.
    (E) decrease.

60. Spontaneous reactions at all temperatures are favored by values for enthalpy and entropy of

    |     | $\Delta H$ | $\Delta S$ |
    | --- | --- | --- |
    | (A) | + | + |
    | (B) | – | – |
    | (C) | 0 | 0 |
    | (D) | + | – |
    | (E) | – | + |

61. A certain liquid boils at 27° C at 1.0 atm. pressure, with a heat of vaporization of 60.0 kJ / mol. Determine the entropy change for the boiling of one mole of this liquid.
    (A) +200. J/mol deg.K
    (B) –200. J/mol deg.K
    (C) +2.00 J/mol deg.K
    (D) –2.00 J/mol deg.K
    (E) 0.050 J/mol deg.K

62. The $K_a$ for a weak acid is $5.0 \times 10^{-10}$ at 25°C. Determine the value of $K_b$ for this same weak acid.
    (A) $0.50 \times 10^{-5}$
    (B) $1.5 \times 10^{-5}$
    (C) $2.0 \times 10^{-5}$
    (D) $5.0 \times 10^{-5}$
    (E) $5.0 \times 10^{-10}$

63. Determine the pH of a 0.10 M solution of a base, B, with a $K_b$ of $1.0 \times 10^{-5}$.
    (A) 1.0
    (B) 3.0
    (C) 6.0
    (D) 8.0
    (E) 11.0

64. Which of the following is most likely to be a brittle compound with low conductivity as a solid, and have a high melting temperature?
    (A) RbF
    (B) $CCl_4$
    (C) $CS_2$
    (D) ICl
    (E) $SF_6$

65. Of the following, the species that has the correctly predicted largest radius is
    (A) Ar.
    (B) $Cl^-$.
    (C) $Br^-$.
    (D) $K^+$.
    (E) $Sr^{2+}$.

66. The species with the most polar bond is
    (A) Cl–Cl.
    (B) Si–Si.
    (C) Cr–Br.
    (D) P–Cl.
    (E) Ca–Ca.

67. The molecule whose Lewis structure requires resonance structures to best explain its bonding is
    (A) $BeCl_2$.
    (B) $CO_2$.
    (C) $PCl_5$.
    (D) $OF_5$.
    (E) $SO_2$.

68. The geometry of a $PH_3$ molecule is described by the VSEPR model as
    (A) linear.
    (B) trigonal planar.
    (C) tetrahedral.
    (D) bent or angular.
    (E) trigonal pyramidal.

69. The hybridization of an As atom in the $AsF_5$ molecule is
    (A) sp.
    (B) $sp^2$.
    (C) $sp^3$.
    (D) $dsp^3$.
    (E) $d^2sp^3$.

70. The following are in order of increasing boiling temperatures:
    (A) $RbCl > CH_3Cl > CH_3OH > CH_4$.
    (B) $CH_4 > CH_3Cl > CH_3OH > RbCl$.
    (C) $CH_3Cl > CH_3OH > RbCl > CH_4$.
    (D) $CH_4 > CH_3OH > CH_3Cl > RbCl$.
    (E) $RbCl > CH_3OH > CH_3Cl > CH_4$.

71. The following liquids are in order of increasing vapor pressure:
    (A) $CH_3CH_2OH < CH_3OCH_3 < CH_3CH_2CH_3$.
    (B) $CH_3CH_2OH < CH_3CH_2CH_3 < CH_3OCH_3$.
    (C) $CH_3CH_2CH_3 < CH_3OCH_3 < CH_3CH_2OH$.
    (D) $CH_3OCH_3 < CH_3CH_2OH < CH_3CH_2CH_3$.
    (E) $CH_3CH_2CH_3 < CH_3CH_2OH < CH_3OCH_3$.

72. The equilibrium system not affected by the pressure change which results from a volume change at constant temperature is
    (A) $2O_3(g) \rightleftharpoons 3O_2(g)$.
    (B) $PCl_5(g) \rightleftharpoons PCl_3(g) + Cl_2(g)$.
    (C) $2NaCl(s) \rightleftharpoons 2Na(s) + Cl_2(g)$.
    (D) $H_2(g) + I_2(g) \rightleftharpoons 2HI(g)$.
    (E) $2NO_2(g) \rightleftharpoons N_2O_4(g)$.

73. Given: $2X + Y \rightleftharpoons 3C + D$ (all are gases)

    Equal numbers of moles of X and Y are added to an empty tank; when equilibrium is achieved,
    (A) $[X] < [Y]$.
    (B) $[D] = [Y]$.
    (C) $[X] = [Y]$.
    (D) $[Y] < [X]$.
    (E) $[X] + [Y] < [C] + [D]$.

74. The $K_c$ for $A + B \rightleftharpoons C$ is 4.0. The $K_c$ for $2C \rightleftharpoons 2A + 2B$ is
    (A) $1 / 16$.
    (B) $1 / 4$.
    (C) 8.
    (D) 16.
    (E) 32.

75. The solubility of $Zn(OH)_2$ is $2.0 \times 10^{-6}$M at a certain temperature. Determine the value of the $K_{sp}$ at this same temperature.
    (A) $2.0 \times 10^{-6}$
    (B) $4.0 \times 10^{-6}$
    (C) $8.0 \times 10^{-6}$
    (D) $16 \times 10^{-18}$
    (E) $32 \times 10^{-18}$

## A5.5 Standard Reduction Potentials at 25°C (298 K) for Many Common Half-Reactions

| Half-Reaction | e° (V) | Half-Reaction | e° (V) |
|---|---|---|---|
| $F_2 + 2e^- \rightarrow 2F^-$ | 2.87 | $Cu^+ + e^- \rightarrow Cu$ | 0.52 |
| $Ag^{2+} + e^- \rightarrow Ag^+$ | 1.99 | $O_2 + 2H_2O + 4e^- \rightarrow 4OH^-$ | 0.40 |
| $Co^{3+} + e^- \rightarrow Co^{2+}$ | 1.82 | $Cu^{2+} + 2e^- \rightarrow Cu$ | 0.34 |
| $H_2O_2 + 2H^+ + 2e^- \rightarrow 2H_2O$ | 1.78 | $Hg_2Cl_2 + 2e^- \rightarrow 2Hg + 2Cl^-$ | 0.34 |
| $Ce^{4+} + e^- \rightarrow Ce^{3+}$ | 1.70 | $AgCl + e^- \rightarrow Ag + Cl^-$ | 0.22 |
| $PbO_2 + 4H^+ + SO_4^{2-} + 2e^-$ | | $SO_4^{2-} + 4H^+ + 2e^- \rightarrow H_2SO_3 + H_2O$ | 0.20 |
| $\rightarrow PbSO_4 + 2H_2O$ | 1.69 | $Cu^{2+} + e^- \rightarrow Cu^+$ | 0.16 |
| $MnO_4^- + 4H^+ + 3e^- \rightarrow MnO_2 + 2H_2O$ | 1.68 | $2H^+ + 2e^- \rightarrow H_2$ | 0.00 |
| $2e^- + 2H^+ + IO_4^- \rightarrow IO_3^- + H_2O$ | 1.60 | $Fe^{3+} + 3e^- \rightarrow Fe$ | −0.036 |
| $MnO_4^- + 8H^+ + 5e^- \rightarrow Mn^{2+} + 4H_2O$ | 1.51 | $Pb^{2+} + 2e^- \rightarrow Pb$ | −0.13 |
| $Au^{3+} + 3e^- \rightarrow Au$ | 1.50 | $Sn^{2+} + 2e^- \rightarrow Sn$ | −0.14 |
| $PbO_2 + 4H^+ + 2e^- \rightarrow Pb^{2+} + 2H_2O$ | 1.46 | $Ni^{2+} + 2e^- \rightarrow Ni$ | −0.23 |
| $Cl_2 + 2e^- \rightarrow 2Cl^-$ | 1.36 | $PbSO_4 + 2e^- \rightarrow Pb + SO_4^{2-}$ | −0.35 |
| $Cr_2O_7^{2-} + 14H^+ + 6e^- \rightarrow 2Cr^{3+} + 7H_2O$ | 1.33 | $Cd^{2+} + 2e^- \rightarrow Cd$ | −0.40 |
| $O_2 + 4H^+ + 4e^- \rightarrow 2H_2O$ | 1.23 | $Fe^{2+} + 2e^- \rightarrow Fe$ | −0.44 |
| $MnO_2 + 4H^+ + 2e^- \rightarrow Mn^{2+} + 2H_2O$ | 1.21 | $Cr^{3+} + e^- \rightarrow Cr^{2+}$ | −0.50 |
| $IO_3^- + 6H^+ + 5e^- \rightarrow \frac{1}{2}I_2 + 3H_2O$ | 1.20 | $Cr^{3+} + 3e^- \rightarrow Cr$ | −0.73 |
| $Br_2 + 2e^- \rightarrow 2Br^-$ | 1.09 | $Zn^{2+} + 2e^- \rightarrow Zn$ | −0.76 |
| $VO_2^+ + 2H^+ + e^- \rightarrow VO^{2+} + H_2O$ | 1.00 | $2H_2O + 2e^- \rightarrow H_2 + 2OH^-$ | −0.83 |
| $AuCl_4^- + 3e^- \rightarrow Au + 4Cl^-$ | 0.99 | $Mn^{2+} + 2e^- \rightarrow Mn$ | −1.18 |
| $NO_3^- + 4H^+ + 3e^- \rightarrow NO + 2H_2O$ | 0.96 | $Al^{3+} + 3e^- \rightarrow Al$ | −1.66 |
| $ClO_2 + e^- \rightarrow ClO_2^-$ | 0.954 | $H_2 + 2e^- \rightarrow 2H^-$ | −2.23 |
| $2Hg^{2+} + 2e^- \rightarrow Hg_2^{2+}$ | 0.91 | $Mg^{2+} + 2e^- \rightarrow Mg$ | −2.37 |
| $Ag^+ + e^- \rightarrow Ag$ | 0.80 | $La^{3+} + 3e^- \rightarrow La$ | −2.37 |
| $Hg_2^{2+} + 2e^- \rightarrow 2Hg$ | 0.80 | $Na^+ + e^- \rightarrow Na$ | −2.71 |
| $Fe^{3+} + e^- \rightarrow Fe^{2+}$ | 0.77 | $Ca^{2+} + 2e^- \rightarrow Ca$ | −2.76 |
| $O_2 + 2H^+ + 2e^- \rightarrow H_2O_2$ | 0.68 | $Ba^{2+} + 2e^- \rightarrow Ba$ | −2.90 |
| $MnO_4^- + e^- \rightarrow MnO_4^{2-}$ | 0.56 | $K^+ + e^- \rightarrow K$ | −2.92 |
| $I_2 + 2e^- \rightarrow 2I^-$ | 0.54 | $Li^+ + e^- \rightarrow Li$ | −3.05 |

# Advanced Placement Chemistry Equations and Constants

## ATOMIC STRUCTURE

$E = h\nu$

$\lambda = \frac{h}{mv}$

$c = \lambda\nu$

$p = mv$

$E_n = \frac{-2.178 \times 10^{-18}}{n^2}$ joule

## EQUILIBRIUM

$K_a = \frac{[H^+][A^-]}{[HA]}$

$E^\Omega =$

$K_w = [OH^-][H^+] = 1.0 \times 10^{-14}$ at 25°C

$\quad = K_a \times K_b$

$pH = -\log[H^+], pOH = -\log[OH^-]$

$14 = pH + pOH$

$pH = pK_a + \log\frac{[A^-]}{[HA]}$

$pOH = pK_b + \log\frac{[HB^+]}{[B]}$

$pK_a = -K_a, pK_b = -\log K_b$

$K_p = K_c(RT)^{\Delta n}$

where $\Delta n =$ moles product gas – moles reactant gas

## THERMOCHEMISTRY/KINETICS

$\Delta S° = \sum S°$ products $- \sum S°$ reactants

$\Delta H° = \sum \Delta H_f°$ products $- \sum \Delta H_f°$ reactants

$\Delta G° = \sum \Delta G_f°$ products $- \sum \Delta G_f°$ reactants

$\Delta G° = \Delta H° - T\Delta S°$

$\quad = -RT \ln K = -2.303 RT \log K$

$\quad = -n \Im E°$

$\Delta G = \Delta G° + RT \ln Q = \Delta G° + 2.303 RT \log Q$

$q = mc\Delta T$

$C_p = \frac{\Delta H}{\Delta T}$

$\ln[A]_t - \ln[A]_0 = -kt$

$\frac{1}{[A]_t} - \frac{1}{[A]_0} = kt$

$\ln k = \frac{-E_a}{R}\left(\frac{1}{T}\right) + \ln A$

$E$ = energy

$\lambda$ = wavelength

$v$ = velocity

$m$ = mass

$\nu$ = frequency

$p$ = momentum

$n$ = principal quantum number

Speed of light, $c = 3.0 \times 10^8$ ms$^{-1}$

Planck's constant, $h = 6.63 \times 10^{-34}$ Js

Boltzmann's constant, $k = 1.38 \times 10^{-23}$ J K$^{-1}$

Avogadro's number $= 6.022 \times 10^{23}$ mol$^{-1}$

Electron charge, $e = -1.602 \times 10^{-19}$ coulomb

1 electron volt per atom $= 96.5$ kJ mol $-1$

## EQUILIBRIUM CONSTANTS

$K_a$ (weak acid)

$K_b$ (weak base)

$K_w$ (water)

$K_p$ (gas pressure)

$K_c$ (molar concentrations)

$S°$ = standard entropy

$H°$ = standard enthalpy

$G°$ = standard free energy

$E°$ = standard reduction potential

$T$ = temperature

$n$ = moles

$m$ = mass

$q$ = heat

$c$ = specific heat capacity

$C_p$ = molar heat capacity at constant pressure

$E_a$ = activation energy

$k$ = rate constant

$A$ = frequency factor

Faraday's constant, $\Im = 96,500$ per mole of electrons

Gas constant, $R \quad = 8.31$ J mol$^{-1}$ K$^{-1}$

$\quad = 0.0821$ L atm mol$^{-1}$ K$^{-1}$

$\quad = 8.31$ volt coulomb mol$^{-1}$ K$^{-1}$

## GASES, LIQUIDS, AND SOLUTIONS

$PV = nRT$

$\left(P + \frac{n^2 a}{V^2}\right)(V - nb) = nRT$

$P_A = P_{total} \times X_A$, where $X_A = \frac{\text{moles A}}{\text{total moles}}$

$P_{total} = P_A + P_B + P_C + \dots$

$n = \frac{m}{M}$

$°K = °C + 273$

$\frac{P_1 V_1}{T_1} = \frac{P_2 V_2}{T_2}$

$D = \frac{m}{V}$

$$u_{rms} = \sqrt{\frac{3kT}{m}} = \sqrt{\frac{3RT}{M}}$$

$KE$ per molecule $= \frac{1}{2}mv^2$

$KE$ per mole $= \frac{3}{2}RT$

$$\frac{r_1}{r_2} = \sqrt{\frac{M_2}{M_1}}$$

molarity, $M$ = moles solute per liter solution

molality = moles solute per kilogram solvent

$\Delta T_f = iK_f \times$ molality

$\Delta T_b = iK_b \times$ molality

$\pi = MRT$

$A = abc$

## OXIDATION-REDUCTION; ELECTROCHEMISTRY

$Q = \dfrac{[C]^c[D]^d}{[A]^a[B]^b}$, where $a\,A + b\,B \rightarrow c\,C + d\,D$

$I = \dfrac{q}{t}$

$E_{cell} = E°_{cell} - \dfrac{RT}{n\Im}\ln Q = E°_{cell} - \dfrac{0.0592}{n}\log Q \,@\,25°C$

$\log K = \dfrac{nE°}{0.0592}$

$P$ = pressure
$V$ = volume
$T$ = temperature
$n$ = number of moles
$D$ = density
$m$ = mass
$v$ = velocity

$u_{rms}$ = root-mean-square speed

$KE$ = kinetic energy

$r$ = rate of effusion

$M$ = molar mass

$\pi$ = osmotic pressure

$i$ = van't Hoff factor

$K_f$ = molal freezing-point depression constant

$K_b$ = molal boiling-point elevation constant

$A$ = absorbance

$a$ = molar absorptivity

$b$ = path length

$c$ = concentration

$Q$ = reaction quotient

$I$ = current (amperes)

$q$ = charge (coulombs)

$t$ = time (seconds)

$E°$ = standard reduction potential

$K$ = equilibrium constant

Gas constant, $R = 8.31$ J mol$^{-1}$ K$^{-1}$

$= 0.0821$ L atm mol$^{-1}$ K$^{-1}$

$= 8.31$ volt coulomb mol$^{-1}$ K$^{-1}$

Boltzmann's constant, $k = 1.38 \times 10^{-23}$ J K$^{-1}$

$K_f$ for $H_2O = 1.86$ K kg mol$^{-1}$

$K_b$ for $H_2O = 0.512$ K kg mol$^{-1}$

$1$ atm $= 760$ mm Hg

$= 760$ torr

STP $= 0.000°C$ and $1.000$ atm

Faraday's constant, $\Im = 96,500$ coulombs per mole of electrons

# Section II: Free-Response Questions
## Time: 90 minutes
## Number of Questions: 6

Section II of the AP Chemistry Examination counts for 55% of the total test grade and involves several parts. In the actual exam, there will be a choice of questions, but here you should do all questions because the choices are omitted. Calculators may be used on the first two questions only. A Periodic Table and Equation Tables are provided elsewhere in this study guide and may be used.

## Part A: Time—40 minutes

1. Equilibrium (worth 20% of this part of the test)

$$2SO_3(g) \rightleftharpoons O_2(g) + 2SO_2(g)$$

   A 3.21 g sample of sulfur trioxide is placed in a 2.25-L cylinder and allowed to reach equilibrium at a constant temperature of 500.K, as shown in the above equation. Analysis shows $1.23 \times 10^{-2}$ mol of sulfur dioxide at equilibrium.

   (a) Write the equilibrium constant expression ($K_c$) for this system.
   (b) Calculate the concentration of all three gases at equilibrium.
   (c) Calculate the $K_c$ value for this system.
   (d) Calculate the value of $K_p$ at this same temperature.

2. The actual test will have a choice of two questions here (worth 20% of this part of the test).

   Perbromic acid, $HBrO_4$, can react with sulfuric acid, $H_2SO_4$, in the following manner: $HBrO_4 + H_2SO_4 \longrightarrow H_3SO_4^+ + BrO_4^-$.

   (a) Indicate the two conjugate acid base pairs in this reaction and label the acids in each pair.
   (b) In the $SO_4^{2-}$ ion, sulfur is surrounded by four oxygens in a tetrahedral structure. Indicate a Lewis structure for the $H_3SO_4^+$ ion.
   (c) Is a $H_4SO_4^{2+}$ ion stable (can it exist)? Support you answer with a Lewis structure and/or discussion.
   (d) Contrast the use of the term 'acid' as used in the Arrhenius model, in the Brønsted-Lowry model, and in the Lewis model of acids.

## Part B: Time—50 minutes

Calculators may not be used on this part of the test

3. Reactions (worth 15% of this part of the test)

   Write formulas of both the reactants and products for 5 of the following reactions. Unless otherwise indicated, all of the solutions are aqueous. Be sure to consider carefully if substances are ions or molecules. Do not show spectator ions. You are not required to balance the expressions. In all cases, a reaction does take place.

   1. Ammonia gas is added to boron trichloride gas.
   2. Iron (III) ions are reduced by iodide ions.
   3. Potassium oxide solid is added to water.
   4. Hydrogen peroxide solution is added to iron (II) sulfate solution.

5.  Ammonia water is added to iron (III) chloride.
6.  Solid potassium chlorate is heated.
7.  Concentrated ammonia solution is added to copper(II) nitrate solution.
8.  Magnesium oxide solid is added to carbon dioxide.

4.  Laboratory Question (worth 15% of this section of the test)

Analysis of an iron (II) solution by titration with a standardized potassium permanganate solution leads to the determination of the percent of iron in the original solution. When performing this laboratory work, the solid of known mass and containing iron (II) is dissolved in 25.0 mL of water.

(a) Explain how each of the following affects the reported percentage of iron in the unknown solid:
    i.   The student fills the buret with $KMnO_4$ after rinsing with only deionized water.
    ii.  An air bubble appears in the buret tip before titration begins.
    iii. More than 25.0 mL of water is used to dissolve the solid.
    iv.  Each time the volume of fluid is measured in the buret, measurement is made to the top of the meniscus, rather than the bottom of the meniscus.
(b) Assuming that the accepted value for the percentage of iron in the original sample was 7.77% and that your experiment gave a result of 6.896%, determine the percentage error in your work.
(c) Balance the reaction which occurs between iron (II) ions and the permanganate ion in acidic solution. Identify the reducing agent.

5.  General Discussion (worth 15% of this part of the test)

Explain the following observations:

(a) When a solid is heated at its melting temperature, the temperature does not increase.
(b) The melting temperature of KCl is less than that of NaF.
(c) Liquid bromine and ICl have almost the same molecular mass, but ICl boils at a temperature 40°C higher than $Br_2$.
(d) The boiling temperature of HCl is lower than that of HF.
(e) When alcohol (ethanol) is poured over your arm, your skin feels cold.

6.  General Discussion (worth 15% of this part of the test) The actual test will have a choice of two questions here.

For a given reaction, a proposed rate law is Rate = $k [X]^n [Y]^m$.

(a) Explain how the values for "$n$" and for "$m$" can be determined experimentally.
(b) If both "$n$" and "$m$" are 1, what is the overall order of this reaction?
(c) How may the value of the rate constant, $k$, be changed? Support your answer.
(d) Contrast the physical meaning of a first-order reaction with that of a zero-order reaction.
(e) Explain why knowing the rate law expression is essential when proposing a reaction mechanism.

*END OF EXAMINATION*

## ANSWERS TO DIAGNOSTIC TEST

### MULTIPLE-CHOICE ANSWERS

Using the table below, score your test.

Determine how many questions you answered correctly and how many you answered incorrectly. You will find explanations of the answers on the following pages.

| | | | | |
|---|---|---|---|---|
| 1.  A | 2.  D | 3.  B | 4.  A | 5.  B |
| 6.  C | 7.  E | 8.  B | 9.  C | 10. A |
| 11. E | 12. A | 13. C | 14. B | 15. A |
| 16. A | 17. A | 18. A | 19. A | 20. A |
| 21. E | 22. D | 23. C | 24. A | 25. B |
| 26. A | 27. A | 28. A | 29. E | 30. D |
| 31. D | 32. A | 33. C | 34. C | 35. C |
| 36. A | 37. E | 38. B | 39. C | 40. E |
| 41. A | 42. C | 43. A | 44. A | 45. B |
| 46. B | 47. B | 48. C | 49. E | 50. A |
| 51. B | 52. D | 53. A | 54. B | 55. B |
| 56. B | 57. E | 58. C | 59. C | 60. E |
| 61. A | 62. C | 63. E | 64. A | 65. C |
| 66. C | 67. E | 68. E | 69. D | 70. B |
| 71. A | 72. D | 73. A | 74. A | 75. E |

### CALCULATE YOUR SCORE:

Number answered correctly: _____

Adjust ¼ point for guessing penalty:
    Count the number of questions you
    answered incorrectly, multiply by .25, and subtract:   – _____

Determine your adjusted score: _____

### WHAT YOUR SCORE MEANS:

Each year, since the test is different, the scoring is a little different. But generally, if you scored 25 or more on the multiple-choice questions, you'll most likely get a 3 or better on the test. If you scored 35 or more, you'll probably score a 4 or better. And if you scored a 50 or more, you'll most likely get a 5. Keep in mind that the multiple-choice section is worth 45% of your final grade, and the free-response section is worth 55% of your final grade. To learn more about the scoring for the free-response questions, turn to the last page of this section.

## Answers and Explanations

### Multiple-Choice Answers

1. Answer: A

2. Answer: D

3. Answer: B

   In questions like 1, 2 and 3, it is helpful to consider what you do know about each of the listed responses. In this case you know that chromium is a transition element which suggests that the ion is capable of being multicharged. Transition metals are frequently colored in compounds. You might recall the polyatomic ion list which includes $CrO_4^{2-}$ (chromate ion) and $Cr_2O_7^{2-}$ (dichromate ion). Sodium, like all of the alkali metals, reacts with water to form hydrogen. You have most likely seen this reaction demonstrated in your classroom. Recall also that the oxides of metals form water solutions of hydroxide ions (bases), whereas the oxides of nonmetals form acidic solutions. Note also that an answer can be used more than once in this type of question (*Chemistry* 6th ed. pages 324–335, 619, 986–998 / 7th ed. pages 309–318, 588, 943–955).

4. Answer: A

5. Answer: B

6. Answer: C

7. Answer: E

   The term *descriptive chemistry* hides much information that you "just must know." Of course there is so much of it that it is essential to get some direction on what things are paramount. This includes knowing the six strong acids, the common solubility rules, the colors of several ions (usually encountered in your laboratory work), which ions tend to complex, and a few very specific properties. Knowing that hydrofluoric acid, although a weak acid, will dissolve (etch) glass and that carbon dioxide will not support combustion, are examples of properties that a well-educated chemist—that's you!—is expected to know (*Chemistry* 6th ed. pages 657–662, 148–156, 998–1003 / 7th ed. pages 626–631, 140–148, 955–959).

8. Answer: B   Increasing the $[Ag^+]$ will cause a shift toward products, causing an increase in the potential difference of this cell. This would, over time, cause an increase in the concentration of copper (II) ions, making the blue of the copper (II) ions darker. You could also consider this from the Nernst equation. In that case, as $[Ag^+]$ is increased, $[Cu^{2+}]/[Ag^+]^2 < 1$ such that $\log Q$ is a negative number. Subtracting a negative number from the standard potential then

gives a larger cell potential (*Chemistry* 6th ed. pages 841–846 / 7th ed. pages 803–808).

9. ANSWER: C By definition the anode is the site of oxidation. Oxidation is the loss of electrons. Copper must lose two electrons to become copper (II) ions. Half-reactions involving a decrease in positive charge, like responses (B) and (E), require a gain of electrons (*Chemistry* 6th ed. pages 828–837 / 7th ed. pages 791–800).

10. ANSWER: A It is necessary for the reactants to gain enough energy to form the activated complex; the difference in the energy of the reactants and the activated complex is called the activation energy [Section 12.7, especially Figure 12.11]. Of course, both hydrogen and oxygen have already vaporized; they exist as gases at room conditions. Enthalpy change represents the difference in energy between the reactants and the products (at constant pressure) [Section 6.2]. Entropy change represents the change in the disorder from reactants to products (*Chemistry* 6th ed. pages 582–588 esp. Figure 12.11, 248–256, 790 / 7th ed. pages 552–557 esp. Figure 12.11, 235–242, 755).

11. ANSWER: E This is a very straightforward use of the chemical formula and your understanding of moles. Since there are two moles of silver atoms per formula unit of silver sulfide, there are twice as many moles of silver atoms as moles of formula units of the silver sulfide. You can also quickly see that the grams formed would be about 0.4 g if you recall that silver has an atomic mass of about 108 g/mol (*Chemistry* 6th ed. pages 86–90 / 7th ed. pages 82–85).

12. ANSWER: A Some flame colors which you should know include lithium: bright red; sodium: pale-yellow; potassium: violet; rubidium: purple; cesium: blue; calcium: orange-red; strontium: crimson-red; and barium: yellowish-green (*Chemistry* 6th ed. pages 297–299 / 7th ed. pages 284–285).

13. ANSWER: C "*R*" is the general gas constant and applies as long as the gas behaves ideally. This means that the value of "*R*" and the expression $PV = nRT$ must be adjusted in the nonideal conditions of high temperature and/or low pressure [Section 5.8]. That is what the van der Waals equation is all about. The gas laws are based on Kelvin temperature change (not Celsius), since Kelvin is an index of the motion of the gas particles. The value of $PV/nRT$ for ideal gases is 1 (*Chemistry* 6th ed. pages 222–224 / 7th ed. pages 208–210).

14. ANSWER: B Freezing temperature depression is directly related to the number of particles (in this case, ions) in solutions. Since $AlCl_3$ provides four ions per formula unit, the higher concentration of ions for that salt would cause a greater freezing-temperature lowering (*Chemistry* 6th ed. pages 531–535 / 7th ed. pages 504–507).

15. **ANSWER: A**   Both CaO and MgO are ionically bonded. Both involve oxygen; therefore the difference in melting temperatures must be due to the difference in attraction developed by Mg compared to that of Ca. Both are 2+ ions, but magnesium ions are smaller, meaning that the charge is more concentrated in the $Mg^{2+}$ ion. The combination of charge and size factors, called the charge density, affects the force of attraction that the ion has; ionic charge / ionic radius = charge density (*Chemistry* 6th ed. pages 324–335 / 7th ed. pages 309–318).

16. **ANSWER: A**   The 'general formulas' for these common organic molecules are R–OH—alcohol; R–C=O–H—aldehydes; R–C=O–OH—carboxylic acids; R–C=O–R′—ketones; R–C=O–OH—esters (*Chemistry* 6th ed. pages 1052–1064 esp. Table 22.4 / 7th ed. pages 1005–1016 esp. Table 22.4).

17. **ANSWER: A**   The reaction forms $BaSO_4(s)$ which means that ions are removed from solution, thereby making the solution less conductive. However, continued addition of the acid causes the conductivity to increase again with the excess of ions (*Chemistry* 6th ed. pages 148–154 / 7th ed. pages 140–145).

18. **ANSWER: A**   The Law of Multiple Proportions is illustrated by two elements which form at least two compounds. It then compares the ratios of the masses of one element with a constant mass of the other element in the two (or more) compounds. These ratios always reduce to simple whole numbers (*Chemistry* 6th ed. pages 43–46 / 7th ed. pages 41–43).

19. **ANSWER: A**   In addition to remembering Rutherford for his work with the nucleus, recall Thomson for his work with the electron, Dalton for an early model of the atom, Boyle for the gas law dealing with volume and pressure, and Lavoisier for the conservation of mass in chemical reactions (*Chemistry* 6th ed. pages 42–53 / 7th ed. pages 39–49).

20. **ANSWER: A**   Isotopes are found for every element and differ from each other in mass for the same element due only to a different number of neutrons in the nucleus (*Chemistry* 6th ed. pages 53–56, 81–85 / 7th ed. pages 49–52, 77–81).

21. **ANSWER: E**   Begin by determining the formula for iron (II) phosphate: $Fe_3(PO_4)_2$. From this you can see that there are two phosphate ions with four oxygens in each, for a total of 8 moles of oxygen for each unit of $Fe_3(PO_4)_2$ (*Chemistry* 6th ed. pages 62–72 / 7th ed. pages 57–67).

22. **ANSWER: D**   When solving $K_{sp}$ problems, first write the equation with the solid on the left and the ions on the right: $MgF_2(s) \rightleftharpoons Mg^{2+}$ (aq) + 2F⁻(aq). If $s$ represents the mol/L of the solid that goes into solution, $[Mg^{2+}]$ also equals $s$, and $2s$ represents the $[F^-]$. Next write the $K_{sp}$ expression in terms of $s$: $K_{sp}=(s)(2s)^2 = 4s^3 =$

$6.4 \times 10^{-9}$. Solve for *s* (*Chemistry* 6th ed. pages 751–754 / 7th ed. pages 717–719).

23. ANSWER: C    A good clue as to the type of titration curve is the position of the equivalence point, which is the center of the vertical section of the graph showing the very rapid change of pH. In this case, that is in the basic region (above 9). At this point, the amount of added OH⁻ equals the original amount of acid. The pH exceeds 7 (neutral) due to the hydrolysis of the anion from the acid (*Chemistry* 6th ed. pages 729–730, 733–741 / 7th ed. pages 696–697, 700–707).

24. ANSWER: A    The leveling-off shown between F and G is caused by buffering. Optimal buffering occurs when [HA] = [A⁻], which would be at a volume of 25 mL of base in this case (*Chemistry* 6th ed. pages 726–727, 733–741 / 7th ed. pages 693–694, 700–707).

25. ANSWER: B    The systematic naming system, the Stock Nomenclature, uses Roman numerals only for transition elements whose ions display more than one charge (e.g., $Cu^{2+}$ is called copper (II) ion, and $Cu^{+}$ is known as copper (I) ion. Calcium is not a transition element; it is found only as a 2+ ion. OCl⁻ is the hypochlorite ion (*Chemistry* 6th ed. pages 62–72 esp. Table 2.5 / 7th ed. pages 57–67 esp. Table 2.5).

26. ANSWER: A    Lone pairs of electrons (unbonded pairs) are less associated with the attractive forces of positive nuclei, therefore will occupy more space than bonded pairs (*Chemistry* 6th ed. pages 392–393 / 7th ed. pages 369–371).

27. ANSWER: A    Acids which contain oxygen often give students some difficulty. These names are formed from the root name of the anion with a suffix of -ic or -ous. The -ous form always contains one less oxygen than the -ic form. For example, $H_2SO_4$ is sulfuric acid and $H_2SO_3$ is sulfurous acid (*Chemistry* 6th ed. Tables 2.7, 2.8, and Figure 2.25 on page 72 / 7th ed. Tables 2.7, 2.8, and Figure 2.25 on page 67).

28. ANSWER: A    Before reaction, no combination of $NH_3$ and $Ag^{+}$ has formed; the silver nitrate is found as ions, the ammonia is molecular, and water is also present (*Chemistry* 6th ed. pages 766–769 / 7th ed. pages 731–734).

29. ANSWER: E    The low value for $K_{a_1}$ indicates that $H_3PO_4$ is a weak acid. The three values suggest three steps in dissociation. The large differences in these $K_a$ values indicate that only step 1 will provide a significant [H⁺] (*Chemistry* 6th ed. pages 682–687, 715 / 7th ed. pages 650–655, 682–683).

30. ANSWER: D    Significant figures is a method of indicating to others how accurately you have measured a value; all measured values have some uncertainty in the measurement, (another method is to show the measured value followed by a ± value which indicates

how precisely you have measured or how much the measurement is limited by the equipment you have used to obtain it). All of the numerals are significant here. It is the zero which can be difficult. If this value were known only to the nearest 1 mL then it would be correctly written as 22. mL; since 22.00 mL were recorded there must have been a reason to record the values in the tenths and hundredths places. The reason is that the measurement was taken to the hundredths of a  mL (*Chemistry* 6th ed. pages 524–531 / 7th ed. pages 497–504).

31. **ANSWER: D**  The average is found by totaling the five readings. When adding (or subtracting), examine the position of the decimal point. In this case, all five values are known to the nearest thousandth, so the total can be known to the nearest thousandths. When this total is divided by five (an exact number in this case), a different rule applies. When multiplying or dividing you must count the number of significant figures; the least precise piece of data used to obtain a result determines the number of significant figures in that result. In this case, that would be four (*Chemistry* 6th ed. pages 11–18 / 7th ed. pages 10–16).

32. **ANSWER: A**  This answer is limited to the nearest tenths by the measured value 3.2 g. Note that this is an application of the rule for addition and subtraction (*Chemistry* 6th ed. pages 11–18, esp. "Rules for Significant Figures in Mathematical Operations" on page 16 / 7th ed. pages 10–16, esp. "Rules for Significant Figures in Mathematical Operations" on page 14).

33. **ANSWER: C**  Recall that $d = m/v$, 36.123g/(27.1 – 13.0) mL = 2.56 g/mL. Here the number of significant figures in the answer is limited by the divisor which has three significant figures (*Chemistry* 6th ed. pages 14–18, 25–26 / 7th ed. pages 13–16, 24).

34. **ANSWER: C**  This time $d = m/v$ is used in the form $v = m/d$. The answer is limited to three significant figures by the measured value of the density, 0.714 $g/cm^3$, which is known to only three significant figures. Note that in this as well as several of the other questions, it is not the mathematical processes that are being tested but your understanding of how to apply significant figures (look at all of the answers!). Significant figures are important on the AP Chemistry Examination too, especially in the second section (free-response problems) where one point is lost for each such error (*Chemistry* 6th ed. pages 25–26 / 7th ed. pages 24).

35. **ANSWER: C**  The anion of the acid is reacting with water (this is hydrolysis, of course): $A^- + HOH \rightarrow AH + OH^-$ and occurs when HA is a weak acid (*Chemistry* 6th ed. pages 733–741 / 7th ed. pages 700–707).

36. **ANSWER: A**  The atomic number indicates the number of protons for the atom (as well as for the ion!) and therefore identifies the element. Even though most oxygen has a mass of 16, this is still oxygen. It differs in the number of neutrons (17 – 8 = 9), so it is an

isotope of oxygen (*Chemistry* 6th ed. pages 53–56, 81–85 / 7th ed. pages 49–52, 77–81).

37. ANSWER: E   All three of these factors are essential, according to collision theory (*Chemistry* 6th ed. pages 582–586 / 7th ed. pages 552–555).

38. ANSWER: B   Since HBr is a strong acid (completely dissociated), the [H$^+$] would be 0.00100 M which gives a pH = –log 0.00100 = 3.0 (*Chemistry* 6th ed. pages 657–666 / 7th ed. pages 626–635).

39. ANSWER: C   The substances you might expect to form solids due to low solubility might be lead (II) chromate and potassium nitrate. There are two rules which suggest the latter is not going to precipitate: "most 1A compounds are soluble," and "most nitrate salts are soluble" (*Chemistry* 6th ed. pages 148–154 / 7th ed. pages 140–145).

40. ANSWER: E   If gas is allowed to escape, then there are fewer molecules left within the cylinder; these fewer molecules exert less pressure (gas pressure is due to the collision of gas particles with the sides of the container). These remaining gas molecules occupy the entire container, so there must be more room between the individual molecules (*Chemistry* 6th ed. pages 190–192, 212–219 / 7th ed. pages 179–181, 199–206).

41. ANSWER: A   Gas solubility in water depends on both temperature and pressure. Gases dissolve more completely at low temperatures and higher pressures (consider your experiences with carbonated drinks) (*Chemistry* 6th ed. pages 519–531 / 7th ed. pages 492–504).

42. ANSWER: C   While this may appear to be an equilibrium question, it is actually just asking you to show an understanding of the stoichiometry between ammonia and hydrogen gases as this reaction proceeds. If enough ammonia reacts to cause a decrease of 0.40 atm in pressure, then enough hydrogen is formed to cause an increase in pressure of 0.60 atm., (0.40 atm × 3 mol H$_2$ / 2 mol NH$_3$) (*Chemistry* 6th ed. pages 108–112 / 7th ed. pages 102–106).

43. ANSWER: A   Go back to first principles: What is pressure? The basic Kinetic Molecular Theory of Gases indicates that gas pressure is due to molecular collisions and more collisions mean greater pressure (*Chemistry* 6th ed. pages 212–219 / 7th ed. pages 199–206).

44. ANSWER: A   Real gases differ from ideal gases because of two factors: attractions between real gas molecules and the molecular volume of real gases. These factors become important only under conditions of high pressure (molecules closer together) and low temperature (lower average kinetic energy) (*Chemistry* 6th ed. pages 212–219, 222–224 / 7th ed. pages 199–206, 208–210).

45. ANSWER: B   Gas particles move faster at higher temperature. Real gas molecules do have attractions between each other. While they certainly do move in a random manner, it is the collisions that cause pressure (*Chemistry* 6th ed. pages 190–192, 212–219, 222–224 / 7th ed. pages 179–181, 199–206, 208–210).

46. ANSWER: B   This stoichiometry problem may cause you to reach for you calculator, but that tool is not needed; note the widely differing answers for the volume of the gas. 0.40 mol of $KClO_3$ will form 0.60 mol of oxygen gas. The molar volume of gases at STP is 22.4 L / mol, so the volume is around 14 L (0.60 mol × 22 L / mol) (*Chemistry* 6th ed. pages 108–112, 203–206 / 7th ed. pages 102–106, 190–194).

47. ANSWER: B   The part of the pressure (partial pressure) due to a gas in a mixture of gases is the same as the mole ratio of that gas to the total, which is the same as the ratio of the number of molecules of that gas is to the total number of molecules. In this case there are 3 moles of gas total of which Ne is 1 part, hence 1/3 of the total pressure is due to the Ne (*Chemistry* 6th ed. pages 206–211 / 7th ed. pages 194–199).

48. ANSWER: C   The atomic volume decreases due to an increase in nuclear charge; note that the 'added' electrons are going into the same gross energy level as you go from left to right across the same period (*Chemistry* 6th ed. pages 324–330 esp. fig. 7.35, 914–918 esp. fig. 19.2 / 7th ed. pages 309–314 esp. fig. 7.34, 875–880 esp. fig. 19.2).

49. ANSWER: E   Since this occurs as a one step reaction, we can write that the rate = $k[A]^2[B]$. Initially the rate will be proportional to $(3)^2(2) =18$. Later, [A] becomes $3 - 2 =1$, and [B] becomes $2 - 1 =1$, so the rate is then proportional to $(1)^2(1) =1$. That means a change in the rate by a factor of 18 to 1, or that the rate is now 1/18 as large as it was originally (*Chemistry* 6th ed. pages 563–582 / 7th ed. pages 534–552).

50. ANSWER: A   This time you need to use the form $k$ = rate/[X] [Y] to determine the value of the rate constant, $k$.

$k = 4.0 \times 10^{-6}$ mol/L · min / $(4.0 \times 10^{-1}$mol/L)$(4.0 \times 10^{-1}$ mol/L)

$= 1/4 \times 10^{-4}$ L / mol · min $= .25 \times 10^{-4}$ L / mol · min

$= 2.5 \times 10^{-5}$ L · mol$^{-1}$min$^{-1}$.

(Note that if set up in this way, the mathematics become rather easy to handle.) Do watch units for rate constants, which, unlike equilibrium constants, traditionally have units assigned (*Chemistry* 6th ed. pages 563–579 / 7th ed. pages 534–549).

51. ANSWER: B   There is almost always a question on kinetics like this on a test. Note that it is not just the greater number of molecular

collisions that cause the reaction to occur faster but an increase in the fraction of high energy molecules which can then obtain the activation energy requirement and reaction that cause this rate increase (*Chemistry* 6th ed. pages 582–595 esp. fig. 12.12 / 7th ed. pages 552–563 esp. fig. 12.12).

52. **ANSWER: D** Catalysts lower the activation energy barrier by forming a different activated complex (*Chemistry* 6th ed. pages 588–595, esp. fig. 12.15 and 12.16 / 7th ed. pages 557–563, esp. fig. 12.15 and 12.16).

53. **ANSWER: A** For a zero-order rate law, rate = $k[A]^0$ (i.e. $n$ = zero). Since any number taken to the zero power is equal to 1, rate = $k$ (the rate constant). This has the physical meaning that the rate does not speed up or slow down over time (like most reactions do); it either takes place at a constant rate or does not take place at all (*Chemistry* 6th ed. pages 568–578, Table 12.6 on page 578, 595–597 / 7th ed. pages 538–547, Table 12.6 on page 548, 564–565).

54. **ANSWER: B** Using the general half-life expression, $t_{1/2}$ = 0.693 / $k$, $k$ = 0.693 / 1.4 min = 0.50 min$^{-1}$. Note that this expression is valid only for first-order reactions, which would include nuclear decay processes (*Chemistry* 6th ed. pages 568–578 / 7th ed. pages 538–547).

55. **ANSWER: B** If you wish to determine how order is affected by a given concentration, then a series of reactions should be run with changes in only the concentration of that substance, with all other concentrations held constant (as well as all other conditions like temperature held constant) (*Chemistry* 6th ed. pages 568–578 esp. Sample Exercise 12.5 / 7th ed. pages 538–547 esp. Sample Exercise 12.5).

56. **ANSWER: B** This is another application of the half-life expression for first-order rate laws, $t_{1/2}$ = 0.693/$k$. Note that half-life (time) and the rate constant are inversely related; hence the element with the lower rate constant will have the longer half-life (*Chemistry* 6th ed. pages 568–578 / 7th ed. pages 538–547).

57. **ANSWER: E** Since the activation energy for the reverse reaction is the sum of the activation energy for the forward reaction plus the reverse of the enthalpy change for the forward reaction, and we do not know this last value, the value cannot be calculated (*Chemistry* 6th ed. pages 579–582, fig. 12.11 on page 584 / 7th ed. pages 549–552, fig. 12.11 on page 553).

58. **ANSWER: C** If the volume is reduced to 1/3 the original value, then the concentration of these gases will all be three times greater. Using the given rate law and increasing both concentrations by a factor of 3, gives a rate increase of $(3)^2(3)$ = 27 times (*Chemistry* 6th ed. pages 563–579 / 7th ed. pages 534–549).

59. **ANSWER: C** Since only the concentration of *B* is altered, you need to see how tripling [*B*] affects the rate: from $r = k[X][B]^2$ you can see that the rate increases by a factor of $(3)^2$, or 9 times (*Chemistry* 6th ed. pages 563–579 / 7th ed. pages 534–549).

60. **ANSWER: E** Use the equation $\Delta G = \Delta H + T\Delta S$, and remember that a negative value for the Gibb's Free Energy is required for a spontaneous reaction. Then determine the signs for $\Delta H$ and for $\Delta S$ which give a negative $\Delta G$ (*T* is always positive, of course) (*Chemistry* 6th ed. pages 784–795 / 7th ed. pages 749–759).

61. **ANSWER: A** For systems at the boiling temperature, the change in state does not offer any free energy. So *G* is zero. This allows you to use $\Delta S = \Delta H / T$ to find the entropy change:

$$\Delta S = +60,000 \text{ J / mol} / 300.\text{K} = +200. \text{ J/ mol} \cdot \text{deg. K.}$$

Note the change from kJ to J in this calculation (*Chemistry* 6th ed. pages 795–798 / 7th ed. pages 759–762).

62. **ANSWER: C** From $K_a \times K_b = K_w$; $K_b = 1.0 \times 10^{-14} / 5.0 \times 10^{-10} = 2.0 \times 10^{-5}$ (*Chemistry* 6th ed. pages 687–693 / 7th ed. pages 655–660).

63. **ANSWER: E** Using the general equation $B^- + HOH \rightarrow HB + OH^-$, $[HB][OH^-]/[B^-] = 1.0 \times 10^{-5}$, $[OH^-] = 1.0 \times 10^{-3}$ M, pOH = 3.0. pH = 14.0 − 3.0 = 11.0. $OH^-$ pOH = −log $1.0 \times 10^{-3}$ (*Chemistry* 6th ed. pages 681–682 / 7th ed. pages 649–650).

64. **ANSWER: A** Compounds with these properties are ionically bonded. This suggests elements from the far left side of the Periodic Table (1A or 2A) with elements from the far right side (6A or 7A) (*Chemistry* 6th ed. pages 366–367, 479–483 / 7th ed. pages 346–347, 456–459).

65. **ANSWER: C** Ions with negative charges have gained an electron and are therefore larger than their parent atoms. Between the chloride ions and the bromide ion you have outermost electrons in the third gross energy level versus the fourth gross energy level, hence we would expect $Br^-$ to be larger (*Chemistry* 6th ed. pages 357–362 esp. fig. 8.7 / 7th ed. pages 338–342 esp. fig. 8.8).

66. **ANSWER: C** While using actual electronegativities would be helpful for this question, a good generalization is that the further apart the elements are on the periodic table the greater the difference in control they have over the shared electron pair, and the more polar the bond. Of course, if an element from the 1A or 2A Group bonds with something from the 6A or 7A Group, the difference in control over the electron is so great that we call the bond ionic, but that possibility is not one of the answers in this question. Note also that this is a question dealing with the polarity of an individual bond and not with the polarity of an entire molecule (*Chemistry* 6th ed. pages 352–354 / 7th ed. pages 333–335).

67. **Answer: E** The need for resonance structures seems to be greatest for molecules in which the same two elements are bonded with different type bonds (e.g. one single, one double bond) in the same molecule. The usual Lewis structure for sulfur dioxide shows that the sulfur and oxygen are bonded with one single bond and, in the other sulfur–oxygen bond, doubly bonded. (Be sure you can draw this Lewis structure.) (*Chemistry* 6th ed. pages 383–388 / 7th ed. pages 363–367)

68. **Answer: E** The VSEPR Model is very helpful in describing almost all molecular shapes. Note that in $PH_3$ the shape may, at first, seem to be tetrahedral, but it does not have a hydrogen in one of the four apex positions, so it is a trigonal pyramid. Molecular shape is determined by the position of the nuclei and not by the position of just electrons, either pairs or single electrons (*Chemistry* 6th ed. pages 389–400 esp. fig. 8.15 / 7th ed. pages 367–379 esp. fig. 8.16).

69. **Answer: D** Hybridization is a way of describing the atomic orbitals used to share electrons and thereby form bonds. The careful student will become familiar with the five types of hybrid orbitals indicated by the five responses to this question (*Chemistry* 6th ed. pages 413–426 / 7th ed. pages 391–403).

70. **Answer: B** The temperature at which substances boil is a function of the forces between the molecules of that substance (intermolecular forces). Small, nonpolar molecules boil at the lowest temperatures since they have the weakest IMF to overcome. As polarity increases, so does the boiling temperature, from semi-polar to the extremely polar hydrogen 'bond.' Ionic substances have even higher boiling temperatures (*Chemistry* 6th ed. pages 450–453 / 7th ed. pages 426–429).

71. **Answer: A** The more easily a substance can obtain the vapor state, the more molecules will be available to exert vapor pressure. The lower the polarity of the molecule, the easier it will vaporize. The $CH_3CH_2OH$ is the most polar of this group, due to the hydrogen 'bonding' at the –OH site; it will have the lowest vapor pressure. Propane is the most symmetrical of these molecules, the least polar, and will have the highest vapor pressure (*Chemistry* 6th ed. pages 450–453, 483–492 / 7th ed. pages 426–429, 459–466).

72. **Answer: D** If you think about equilibrium shifting to attempt to achieve pressure equilibrium, then only when the same number of moles of gas are found on both sides of the reaction will the system not be affected. This is the case in this problem when you have two moles of gas on each side. This effect of a change in pressure on an equilibrium system is described by Le Châtelier's principle (*Chemistry* 6th ed. pages 636–642 esp. Sample Exercise 13.14 / 7th ed. pages 604–610 esp. Sample Exercise 13.14).

73. **Answer: A** Since the stoichiometry of this reaction indicates that twice as much X will react as will Y and that you have added equal numbers of moles of these materials initially, then there must be

less X remaining after any reaction takes place (*Chemistry* 6th ed. pages 610–613 / 7th ed. pages 579–582).

74. ANSWER: A  Note that the second equation is the reverse of the first, and has been doubled. Therefore the $K_c$ for the second equation will   be the reciprocal of the first squared: $(1 / 4)^2 = 16$ (*Chemistry* 6th ed. pages 613–617 especially Sample Exercise 13.2 and "We Can Summarize These Conclusions About the Equilibrium Expression" / 7th ed. pages 582–586 especially Sample Exercise 13.2 and "We Can Summarize These Conclusions About the Equilibrium Expression").

75. ANSWER: E   From the equation $Zn(OH)_2 \rightarrow Zn^{2+}(aq) + 2OH^-(aq)$ you can see that if $2.0 \times 10^{-6}$ mol/L of the zinc hydroxide dissolves, that will result in the same concentration of zinc ions ($2.0 \times 10^{-6}$ mol/L) and twice that much of hydroxide ions ($2.0 \times 10^{-6}$ mol/L $\times 2 = 4.0 \times 10^{-6}$ mol/L). Then: $K_{sp} = [Zn^{2+}] [OH^-]^2 = (2.0 \times 10^{-6}$ mol/L$)(4.0 \times 10^{-6}$ mol/L$)^2 = 32 \times 10^{-18}$ (usually units are not shown with this kind of value) (*Chemistry* 6th ed. pages 751–760 / 7th ed. pages 717–724).

# SECTION II FREE-RESPONSE ANSWERS

## Question 1: Answers
(a)  $K_C = [O_2] [SO_2]^2 / [SO_3]^2$
(b)  $[SO_2] = 1.23 \times 10^{-2}$ mol / 2.25 L $= 0.00547$ M
 $[O_2] = $ (one half as much as above) $= 0.00273$ M
 $[SO_3] = 0.0401$ mol $- 0.0123$ mol / 2.25 L $= 0.0124$ M
  note: 3.21 g / 80.1 g/mol $= 0.401$ mol of $SO_3$ initially
(c)  $K_C = (0.00273) (0.00547)^2 / (0.0124)^2 = 5.31 \times 10^{-4}$
(d)  $K_p = K_c(RT)^{\Delta n} = 5.31 \times 10^{-4} (0.0821)(500.)^1 = 2.18 \times 10^{-2}$
(*Chemistry* 6th ed. pages 618–619, 625–630 / 7th ed. pages 587–588, 593–598)

## Question 2: Answers
(a)  The pairs are $HBrO_4$ (acid) / $BrO_4^-$ (conjugate base) and $H_3SO_4^+$ (acid) / $H_2SO_4$ (base). Note that acid form always has one more $H^+$ than its conjugate base; this reaction takes place in pure acids; no water is involved.

(b)
$$\left[ \begin{array}{c} :\!O\!: \\ \| \\ H-\ddot{O}-S-\ddot{O}-H \\ | \\ O \\ | \\ H \end{array} \right]^+$$

(c)
$$\left[ \begin{array}{c} H \\ | \\ :\!O\!: \\ | \\ H-\ddot{O}-S-\ddot{O}-H \\ | \\ :\!O\!: \\ | \\ H \end{array} \right]^{2+}$$

(d) Arrhenius acids produce $H^+$ ions in water solutions. Brønsted-Lowry acids donate $H^+$ (protons) in any solvent; these acids need not be in an aqueous solution. Lewis acids are electron pair acceptors (in any solvent).

(*Chemistry* 6th ed. pages 655–656, 660, 696–699 / 7th ed. pages 624–626, 629, 663–665)

## Question 3: Answers

Note that physical states need not be included, and that the expressions need not be balanced. Do not include spectator ions.

1. $NH_3 + BCl_3 \rightarrow H_3N{:}BCl_3$ (this could be shown as $NH_3BCl_3$)
2. $Fe^{3+} + I^- \rightarrow Fe^{2+} + I_2$
3. $K_2O + HOH \rightarrow K^+ + OH^-$
4. $H_2O_2 + Fe^{2+} \rightarrow Fe^{3+} + H_2O$
5. $NH_3 + Fe^{3+} + H_2O \rightarrow Fe(OH)_3 + NH_4^+$
6. $KClO_3 \rightarrow KCl + O_2$
7. $NH_3 + Cu^{2+} \rightarrow Cu(NH_3)_4{}^{2+}$
8. $MgO + CO_2 \rightarrow MgCO_3$

(*Chemistry* 6th ed. pages 148–155 / 7th ed. pages 140–146)

## Question 4: Answers

a(i) The $KMnO_4$ is actually less concentrated than you believe it to be since it has been diluted with the water left in the buret. That means that more of the standardized solution will be needed to react with the iron (II) solution leading you to report that the iron concentration is higher than it actually is.

a(ii) If the air bubble stays in the buret during the entire titratioon, then there is no effect. However, if it is replaced by the $KMnO_4$ solution during the titration, then less solution leaves the buret than you are reporting, leading to a higher than correct percentage of the iron (II).

a(iii) If more water is added to the flask that contains the iron salt, it will not affect the number of moles of iron (II) and therefore not affect the reported result.

a(iv) If the same point on the meniscus is used as a reference, there will not be any effect on the percentage of iron reported.

(b) % Error
= (your lab. results – accepted value) / accepted value × 100%
$$(6.896 - 7.77) / 7.77 \times 100\% = -11\%$$
Note that the answer is negative, and known only to two sig.fig.

(c) $5Fe^{2+} + MnO_4{}^- + 8H^{1+} \rightarrow 5Fe^{3+} + Mn^{2+} + 4H_2O$.
$Fe^{2+}$ is undergoing oxidation (the loss of electrons) and so it is the reducing agent.

(*Chemistry* 6th ed. pages 161–162, 168–177 / 7th ed. pages 151–152, 158–167)

**Question 5: Answers**

(a) The thermal energy added is used to overcome the forces of attraction (IMF) in the solid (not to make the particles move faster, which is an increase in kinetic energy, thereby temperature.)

(b) The distance between $K^+$ and $Cl^-$ is greater than the distance between $Na^+$ and $F^-$, and so the lattice energy is smaller.

(c) $Br_2$ (i) is nonpolar and has lower IMF than polar ICl.

(d) The bonding electrons are shared much more unevenly in HF, giving rise to a very polar molecule. This leads to stronger dipole–dipole forces in HF (called hydrogen 'bonding,' of course).

(e) An endothermic process causes this. The alcohol evaporates. This requires energy which is lost by your skin, so the temperature of your skin decreases.

(*Chemistry* 6th ed. pages 488–490, 360–365, 459–460, 354–356, 243–249 / 7th ed. pages 463–465, 340–345, 435–436, 335–337, 231–236)

**Question 6: Answers**

(a) By running a series of reactions while changing the concentration of only substance X, or of only substance Y, and noting how this affects the reaction rate.

(b) It is first order with respect to X, it is first order with respect to Y, and second order overall.

(c) The rate constant may be changed by changing the temperature (but not by changing the concentration of X or of Y). Recall the Arrhenius equation.

(d) For zero-order reactions, the rate is constant {Rate $= k[A]^0 = k(1) = k$}. This means that the rate constant does not change with concentration as it does with a first-order reaction {Rate $= k[A]^1$}.

(e) All reactants in the rate determining step must be part of the rate law expression because each of those reactants controls the reaction rate.

(*Chemistry* 6th ed. pages 563–565, 568–570, 576, 579–582, Table 12.6 on page 578 / 7th ed. pages 534–536, 538–540, 546, 549–552, Table 12.6 on page 548)

## SCORING THE DIAGNOSTIC TEST FREE-RESPONSE QUESTIONS

It is difficult to come up with an exact score for this section of test. However, if you compare your answers to the answers in this book, remembering that each part of the test you answer correctly is worth points even if the other parts of the answer are incorrect (see the section titled "Scoring for the Free-Response Questions" on page 12 of this book), you can get a general idea of the percentage of the questions for which you would get credit. If you believe that you got at least one-third of the possible credit, you would probably receive a 3 on this part of the test. If you believe that you would receive close to half or more of the available credit, your score would more likely be a 4 or a 5.

# Part II

## A Review of AP Chemistry

# 1

# CHEMICAL FOUNDATIONS

In this chapter, you will perform the basic calculations used in chemistry. Chemists use equipment with varying degrees of precision. The measurements recorded in an experiment must reflect the precision of the equipment used. The results of the calculations in an experiment also must reflect the precision of the equipment.

Many of the problems in chemistry use dimensional analysis to convert from one unit to another or to solve stoichiometry problems. Other basic calculations that will be discussed are density and temperature conversions.

You should be able to

- Identify the number of significant figures in a given measurement.
- Perform calculations involving significant figures.
  - ☐ Memorize the rules for counting and performing operations with significant figures.
- Differentiate between measurements which are accurate and precise.
- Determine the density of solids and liquids and calculate volumes or masses using the given density.
- Convert between units of temperature: degrees Celsius and Fahrenheit, and Kelvin.
- Identify the characteristics of the states of matter: solids, liquids, and gases.
- Identify substances such as elements, compounds, or mixtures.
- Identify methods of separation of mixtures.
- Identify changes as being physical or chemical.

## AP tips

Significant figures are important in calculations. One point out of the nine possible in the free-response sections will be deducted for errors in significant figures that are off by ± 1 significant figure.

Units are the key to problem solving. Use conversion factors and cancel out units or problems involving conversions or stochiometry. Every measured number or number calculated from measurements has a unit except for equilibrium constants, which by convention are reported without units.

# UNCERTAINTY IN MEASUREMENT

(*Chemistry* 6th ed. pages 11–14 / 7th ed. pages 10–13)

## SIGNIFICANT FIGURES

The significant figures of a measurement are all of the certain digits in a measurement and the first uncertain digit (estimated number). Students should be able to read measurements to the proper number of significant figures.

EXAMPLE: Figure 1.9 on page 11 in the Zumdahl 6th edition text and page 10 in the 7th edition shows the measurement of a volume of liquid using a buret. The *certain digits* in the measurement are the three numbers 20.1. The digit to the right of the one must be estimated by interpolating between the 0.1 mL marks. The measurement with *uncertainty* can be reported as 20.15 mL

## PRECISION AND ACCURACY

Accuracy refers to the agreement of a particular value with a true value.
Precision refers to the degree of agreement among several measurements of the same quantity. The degree of precision refers to the number of digits that a measuring device permits one to measure. In a measuring device, all except the last digit, which is estimated, are certain. For example, a balance which measures to the nearest 0.0001 g is more precise than one that measures to the nearest 0.01 g.

$$\text{Percent Error} = \frac{\text{Experimental Value} - \text{Actual Value}}{\text{Actual Value}} \times 100\%$$

EXAMPLE: In an experiment, the density of aluminum is to be determined. Two students perform the experiment three times and obtain the following results.

| Trial | Student A | Student B |
|---|---|---|
| 1 | 2.45 g/mL | 2.69 g/mL |
| 2 | 2.43 g/mL | 2.70 g/mL |
| 3 | 2.44 g/mL | 2.71 g/mL |

Describe the accuracy and precision of each student's results. For Student A, calculate the mean and the percent error if the actual value is 2.70 g / mL.

SOLUTION: The values for Student A are precise, but inaccurate. The values for student B are both precise and accurate.

Random error (indeterminate error) means that a measurement has an equal probability of being high or low.

Systematic error (determinate error) occurs in the same direction each time; it is either always high or always low.

EXAMPLE: A balance could have a defect causing it to give a result that is consistently 1.000 g too high.

## SIGNIFICANT FIGURES AND CALCULATIONS

It is important to know the uncertainty in the final result in an experiment The final reported result cannot have more certainty than the least accurate measurement. The number of significant figures in a single value will be determined. Memorize the rules below and use them to answer the examples that follow.

---

### Rules for Counting Significant Figures

1.  Nonzero integers always count as significant figures.

2.  Zeros: There are three classes of zeros.

    a.  Leading zeros precede all the nonzero digits and do not count as significant figures. Example: 0.0025 has 2 significant figures.

    b.  Captive zeros are zeros between nonzero numbers. These always count as significant figures. Example: 1.008 has 4 significant figures.

    c.  Trailing zeros are zeros at the right end of the number.

        Trailing zeros are only significant if the number contains a decimal point. Example: $1.00 \times 10^2$ has three significant figures.

        Trailing zeros are not significant if the number does not contain a decimal point. Example: 100 has one significant figure.

3.  Exact numbers, which can arise from counting or definitions such as 1 in = 2.54 cm, never limit the number of significant figures in a calculation.

---

EXAMPLE: How many significant figures are in each of the following?

100 L

SOLUTION: There is 1 significant figure. Trailing zeros do not count. Only the "1" is significant.

0.001010 L

SOLUTION: There are 4 significant figures. Leading zeros do not count. The numbers "1010" are significant because one zero is captive and the other zero is trailing, but a decimal is present.

---

### Rules for Significant Figures in Calculations

1. For multiplication and division, the number of significant figures in the result is the same as the number with the least number of significant figures in the calculation.

2. For addition and subtraction, the result has the same number of decimal places as the number with the least number of decimal places in the calculation.

3. Rules for rounding:

   In a series of calculations, carry the extra digits to the final result, then round.


   If the digit to be removed

   > is less than 5, the preceding digit stays the same. For example, 2.44 rounds to 2.4.

   > is greater than or equal to 5, the preceding digit is increased by 1. For example, 2.45 rounds to 2.5.

---

It is important to calculate the results of mathematical expressions to the proper number of significant figures. Memorize the rules above and apply them to the examples that follow.

EXAMPLE: Perform the following calculations to the correct number of significant figures.

1) 16.8 g + 3.2557 g

SOLUTION: The calculator answer is 20.0557. The correct answer is 20.1g. The answer should have one decimal place.

2) 27 g / 4.148 mL

SOLUTION: The calculator answer is 6.509161041. The correct answer is 6.5 g/ mL. The answer should have two significant figures.

## DIMENSIONAL ANALYSIS

(*Chemistry* 6th ed. pages 18–21 / 7th ed. pages 16–19)

Dimensional analysis is used to convert from one unit to another. It is the single most valuable mathematical technique that you will use in general chemistry. The method involves conversion factors to cancel units until you have the proper unit in the proper place. When you are setting up problems using dimensional analysis, you should be more concerned with units than numbers.

EXAMPLE: The density of mercury is 13.6 g/cm³. How many pounds would 1.00 liter of mercury weigh?

SOLUTION:

$$1.00L \times \frac{1000 \text{ cm}^3}{1L} \times \frac{13.6 \text{ g}}{1 \text{ cm}^3} \times \frac{1 \text{ pound}}{453.6 \text{ g}} = 30.0 \text{ pounds}$$

Double check that all of your units cancel properly. If they do, your numerical answer is probably correct. If they don't, your answer is definitely wrong.

# DENSITY

(*Chemistry* 6th ed. pages 25–26 / 7th ed. pages 24–25)

Density is the mass of substance per unit volume of substance. Density = mass/volume.

The density of an object can be determined through the *water displacement method*. The object is massed and then submerged in a measured amount of water in a graduated cylinder. The final volume in the graduated cylinder is read. The volume of water displaced by the object is the volume of the object.

EXAMPLE: A sample containing 33.42 g of metal pellets is poured into a graduated cylinder containing 12.7 mL of water, causing the water level in the cylinder to rise to 21.6 mL. Calculate the density of the metal.

SOLUTION:

Volume of metal = (Volume H₂O metal) – Volume H₂O

= 21.6 mL – 12.7 mL = 8.9 mL

Density of metal = 33.42 g / 8.9 mL = 3.8 g/mL

Note: The answer has 2 significant figures.

# TEMPERATURE

(*Chemistry* 6th ed. pages 22–24 / 7th ed. pages 19–22)

You should be able to interconvert among Fahrenheit, Celsius and Kelvin. You should also know the freezing and boiling points of water on each scale.

$$°C \times (9/5) + 32 = °F$$

$$T_K = T_C + 273.15$$

EXAMPLE: The boiling point of water on top of Long's Peak in Colorado (14,255 feet above sea level) is about 86.0°C. What is the boiling point in Kelvins and degrees Fahrenheit?

SOLUTION:

$$T_K = T_C + 273.15 = 86.0°C + 273.15 = 359.2\ K$$

$$T_F = T_C \times (9/5) + 32.0°F = 86.0 \times (9/5) + 32 = 187°F$$

## CLASSIFICATION OF MATTER

(*Chemistry* 6th ed. pages 26–31 / 7th ed. pages 25–28)

Matter exists in three states: solid, liquid, and gas. Properties of these three states of matter are listed below.

| State of Matter | Shape | Volume |
|---|---|---|
| Solid | Fixed | Fixed |
| Liquid | Not definite | Fixed |
| Gas | Not fixed; shape of container | Not fixed; volume of container |

## MIXTURES AND PURE SUBSTANCES

(*Chemistry* 6th ed. page 30 / 7th ed. page 28)

Pure substances, such as elements or compounds, make up mixtures. The following chart summarizes mixtures and the methods for separating them.

# CHANGES IN MATTER

Matter can undergo physical or chemical changes.

Physical changes in matter do not change the original composition of the substance. Changes in state such as boiling or melting are physical changes. Changes involving an alteration in the form of the substance such as grinding or tearing are physical. Physical properties are properties of a substance that can be observed without changing the composition of the substance. During a physical change, bonds are not broken and no reaction between atoms occurs. For example, density, color, and boiling point are physical properties.

Chemical changes in matter change the composition of the original substance by breaking and making bonds between atoms. A new substance is produced when a chemical change occurs. Evidence that a chemical change has occurred includes change in color or odor or the production of a gas or a solid (precipitate). Some examples of chemical properties include flammability and reactivity to air.

# MULTIPLE-CHOICE QUESTIONS

No calculators are to be used in this section.

When choosing your answers, it is helpful to make the general observation that numerical answers tend to fall into one of two areas. Quite often the answers will vary greatly from each other. For example, you usually know if an answer of 3.0 grams or 300. grams is more reasonable. Secondly, consider how many significant figures an answer should have when examining all of the answers.

1.  Measurements indicate a charge of 0.444 C passes a point in 0.12 seconds. The current (i.e., the rate of charge flow, in C/s) is best expressed as
    (A) 0.27 C/s.
    (B) 0.270 C/s.
    (C) 3.7 C/s.
    (D) 3.70 C/s.
    (E) 3.700 C/s.

ANSWER: C Even if you do not understand what is meant by a term (in this case, current) you can often successfully attack a problem by careful attention to units. In the problem, you are seeking an answer in C/s, so divide the 0.444 C by 0.12 s (and keep your answer to two significant figures (*Chemistry* 6th ed. pages 14–15 / 7th ed. pages 13–14).

2. A given sample contains 2.0 g of hydrogen, 33.1 g of sulfur, and 75.01 g of oxygen. What is the total mass of the sample?
   (A) 110.12 g
   (B) 110.1 g
   (C) 110. g
   (D) $1.1 \times 10^2$ g
   (E) $1.1 \times 10^{-2}$ g

ANSWER: **B** Each of these values is known to one-tenth of a gram (0.1 g) or more, so the total may be known to no more than the nearest 0.1 g. Remember, when adding or subtracting, you do not count the number of significant figures but instead look at the position of the decimal point in each of the data used in the calculation (*Chemistry* 6th ed. pages 14–15 / 7th ed. pages 13–14).

3. What is the percent mass of sulfur in the above mixture (see number 2)?
   (A) 3.01%
   (B) 3.0%
   (C) 3.02%
   (D) 30.1%
   (E) 30%

ANSWER: **D** Now you are dividing, so the answer is known to the number of significant figures which is the least number of significant figures in data used to obtain the answer (here, three significant figures in the mass of sulfur limits the answer to three significant figures even though you know the total mass to four). Also note that the 100% is known by definition, that is, it is considered an "exact number" (*Chemistry* 6th ed. pages 14–15 / 7th ed. pages 13–14).

4. The correct name for $Li_3N$ is
   (A) lithium nitride.
   (B) lithium nitrate.
   (C) trilithium nitride.
   (D) monolithium mononitride.
   (E) lithium nitrogen.

ANSWER: **A** When naming binary ionic compounds, you are to use the names of the cation followed by the anion. Prefixes are used when considering nonmetal–nonmetal compounds (*Chemistry* 6th ed. pages 62–65 / 7th ed. pages 57–61).

5. The correct name for $Mg(OH)_2$ is
   (A) magnesium hydroxide.
   (B) magnesium(I) hydroxide.
   (C) magnesium(II) hydroxide.
   (D) magnesium(II) hydroxide(I).
   (E) magnesium hydrogen oxide.

ANSWER: **A** Roman numerals are used with transition metal compounds to indicate the charge or oxidation number of the metal ion in this compound. Since transition metal ions can often have more than one charge (oxidation number), this identification is necessary. There is no need for such nomenclature when the magnesium ion used for it can have only a 2+ charge. Note also that even the transition metal ions $Ag^+$, $Zn^{2+}$, and $Cd^{2+}$ generally are found with only the oxidation number indicated, and since there is only one form of ion, Roman numerals are not used with these three transition metal ions (*Chemistry* 6th ed. pages 62–65, esp. Table 2.4 / 7th ed. pages 57–61, esp. Table 2.4).

6. The correct name for $Co_2O_3$ is
   (A) cobalt oxide.
   (B) cobalt(II) oxide.
   (C) cobalt oxide(III).
   (D) cobalt(III) oxide.
   (E) dicobalt trioxide.

ANSWER: D As indicated in the discussion for question #5, it is necessary to identify the oxidation number of the cobalt in this oxide since it is not always three (*Chemistry* 6th ed. pages 62–65 / 7th ed. pages 57–61).

7. The correct formula for copper(II) phosphate is
   (A) $CuPO_4$.
   (B) $CuP$.
   (C) $Cu_3(PO_4)_2$.
   (D) $Cu_2P$.
   (E) $Cu_2PO_4$.

ANSWER: C Both copper with a 3+ and a 1+ charge are common ions. It is necessary to assign copper the 2+ charge in this case and then determine the ratio of $Cu^{2+}$ and $PO_4^{3-}$ to make the unit neutral; hence $Cu_3(PO_4)_2$ (*Chemistry* 6th ed. page 67, esp. Table 2.5 / 7th ed. page 62, esp. Table 2.5).

8. Element 104, rutherfordium, has a half-life of 65 seconds. How much of a 1.00 g sample of element 104 will remain at the end of 2.1667 minutes?
   (A) 0.25 g
   (B) 0.250 g
   (C) 0.50 g
   (D) 0.500 g
   (E) 0.75 g

ANSWER: A Consider that 2.1667 minutes × 60. s/min = 130. s, and that is equal to two half-lives. So 1.00 g × 1/2 × 1/2 = 0.25 g. Note also that 65s for one half-life, given in the problem, is known only to two significant figures and that is what controls the number of significant figures in this problem (*Chemistry* 6th ed. pages 14–18 / 7th ed. pages 13–16).

9. The density of copper is 8.96 g/cm³. What is the mass of 18.88 cm³ of pure copper?
   (A) 1.69 g
   (B) 16.9 g
   (C) 169 g
   (D) 169.0 g
   (E) 1690 g

ANSWER: C From density = $m/v$, solving for mass and using mass = ($d$) × ($v$)= 8.96 g/cm³ × 18.88 cm³. A 'quick and dirty' estimate using 9 × 20 would given an answer of about 180 g; only three significant figures are allowed by the density value since it is limited to three significant figures, so there is only one possible selection from the possible answers (*Chemistry* 6th ed. pages 25–26 / 7th ed. pages 24–25).

10. The density of a piece of metal can be determined from mass and water displacement data. A piece of metal with a mass of 15.54 g is placed in a flask with a volume of 50.00 cm³. It is found that 40.54 g of water (*d* = 0.9971 g/cm³) is needed to fill the flask with the metal in it. The density of the metal is most nearly (all answers in g/cm³):
    (A) 1.66.
    (B) 1.7.
    (C) 9.46.
    (D) 9.5.
    (E) 40.7.

ANSWER: A  The volume of water displaced by the piece of metal is 50.00 cm³ - 40.66 = 9.34 cm³. Using mass / volume = 15.54 g / 9.34 cm³ = 1.66 g / cm³. Note that 40.54 g of water with *d* = 0.9971 g/cm³ has a volume of 40.66 cm³ (*Chemistry* 6th ed. pages 25–26 / 7th ed. pages 24–25).

## FREE-RESPONSE QUESTIONS

1(a)  Describe how, in the laboratory, you might determine experimentally the density of a solid, such as a sugar cube, which is water soluble. Indicate what equipment you might best use in the process.

1(b)  Then describe a second experimental method for determining the density of this same object, so that you might verify the results of the first method.

ANSWER: 1(a)  Since density is a ratio of the mass of the object compared to the volume of the object, you might first determine the mass of the sugar cube using a balance. Greater precision is possible with an analytical balance. Then determine the volume of this regular cube by measuring its length, width, and depth (they should all be the same, of course!) and calculating the volume [$V = (l) \times (w) \times (d) = (\text{side})^3$]. It is then a simple step to divide the mass by the volume just calculated (*Chemistry* 6th ed. pages 25–26 / 7th ed. pages 24–25).

ANSWER: 1(b):  The second method suggests a liquid displacement method for determining volume. However the liquid used may not be water since the object is water soluble. Select a nonpolar liquid to keep this polar solid from dissolving, perhaps, 1,1,1-trichloroethane (*Chemistry* 6th ed. pages 25–26 and 134–136 / 7th ed. pages 24–25 and 127–129).

2.  Discuss the terms *precision* and *accuracy*. Describe what each term is used to indicate.

ANSWER: 2  Precision indicates how reproducible a value is (i.e., agreement with other measured data). Accuracy indicates the extent of agreement with an accepted value (i.e., with a standard value) (*Chemistry* 6th ed. pages 11–14 / 7th ed. pages 10–13).

# 2

# ATOMIC THEORY AND ATOMIC STRUCTURE

This chapter will help you to understand the nature of atoms and how they combine. You will also review the rules for naming compounds.

You should be able to

- Discuss the development of the atom from its earliest model to the modern day atom.
- Identify the correct number of subatomic particles for atoms, ions, and isotopes.
- Calculate the average atomic mass of an atom from isotopic data.
- Name compounds and write chemical formulas for binary compounds, ternary compounds (those with polyatomic ions), and acids.
  - ☐ Memorize the chemical formulas and charges of the polyatomic ions and the most common transition metal ions.

## AP tips

Mastery of chemical nomenclature is essential in writing net-ionic equations in the free response section on the AP test. Students must have memorized the formulas and charges of monoatomic ions, including the transition metals, and also polyatomic ions.

# FUNDAMENTAL CHEMICAL LAWS

(*Chemistry* 6th ed. pages 43–46 / 7th ed. pages 41–43)

## THE LAW OF DEFINITE PROPORTION

The law of definite proportion states that a given compound always contains exactly the same proportions of elements by mass.

**EXAMPLE:** A sample of $H_2SO_4$ contains 2.02 g hydrogen, 32.07 g sulfur, and 64.00 g oxygen. How many grams of sulfur and grams of oxygen are present in a second sample of $H_2SO_4$ containing 7.27 g of hydrogen?

**SOLUTION:** Hydrogen is increased by a factor of 3.60 (7.27/2.02). Therefore

g sulfur = 32.07 × 3.60 = 115.45 g sulfur = 115 g sulfur

g oxygen = 64.00 g × 3.60 = 230.4 g oxygen = 230 g oxygen

## THE LAW OF MULTIPLE PROPORTIONS

The law of multiple proportions states that when two elements form a series of compounds, the ratio of masses of the second element that combine with 1 g of the first element can always be reduced to the smallest whole numbers.

**EXAMPLE:** Sulfur and oxygen can react to form both sulfur dioxide and sulfur trioxide. In sulfur dioxide there are 32.06 g sulfur and 32.00 g oxygen. In sulfur trioxide there are 32.06 g sulfur combined with 48.00 g oxygen. What is the ratio of the weights of oxygen that combine with 32.06 g sulfur?

**SOLUTION:** 48.00/32.00 = 1.5 or 3:2, demonstrating the law of multiple proportions.

# HISTORY OF THE ATOM

(*Chemistry* 6th ed. pages 46–53 / 7th ed. pages 43–49)

You should know the historical developments leading to the modern day atom, including the works of Dalton, Avogadro, Thomson, and Rutherford.

# MODERN VIEW OF ATOMIC STRUCTURE

(*Chemistry* 6th ed. pages 53-54, 56 / 7th ed. pages 49–52)

## ASSIGNMENT OF SUBATOMIC PARTICLES TO ATOMS

The atomic number equals the number of protons in a neutral atom. In an atom, the number of protons equals the number of electrons.

The mass number of an isotope equals the total number of protons and neutrons in an atom which can be obtained by rounding the atomic mass in the periodic table to the nearest whole number. This is not the same as the atomic mass, which is an average of the masses of the isotopes of the particular element.

## WRITING SYMBOLS

(*Chemistry* 6th ed. page 54 / 7th ed. pages 50–51)

Mass number → $^{23}_{11}$Na
Atomic Mass → 

EXAMPLE: How many protons, neutrons and electrons are in an atom of sodium – 23?

SOLUTION: There are 11 protons since the atomic number is 11. There are also 11 electrons since atoms are neutral. There are 12 neutrons (mass number – atomic number = 23 – 11).

## ISOTOPES

(*Chemistry* 6th ed. pages 81–85 / 7th ed. pages 78–81)

Isotopes are atoms of the same element with different numbers of neutrons and therefore different atomic masses.

## CALCULATION OF AVERAGE ATOMIC MASS FROM ISOTOPIC DATA

(*Chemistry* 6th ed. pages 81–85 / 7th ed. pages 78–81)

The average atomic mass of an element can be calculated from the percent abundance and mass of each isotope for that element.

Avg Atomic Mass = $\Sigma$(Percent abundance of isotope × mass value )

EXAMPLE: Element "E" is present with the following mass values and natural abundances.

| Isotope | Mass Value (amu) | Percent Abundance |
|---------|------------------|-------------------|
| 10 E | 10.01 | 19.78 % |
| 11 E | 11.01 | 80.22 % |

What is the average atomic mass of the element, E? What is the element?

(0.1978) 10.01 + (0.8022) 11.01 = 10.812 amu.

The element is boron.

## IONS

(*Chemistry* 6th ed. pages 57, 62, 63, 67 / 7th ed. pages 53, 57–58, 62)

An ion is an atom that has lost or gained electrons. A polyatomic ion is a group of atoms bonded together as a single unit that carries a charge.

EXAMPLE: Aluminum forms a cation, a positive ion, by losing three electrons:

$$Al \rightarrow Al^{3+} + 3e^-$$

Oxygen forms an anion, a negative ion, by gaining three electrons:

$$O + 2e^- \rightarrow O^{2-}$$

# NOMENCLATURE

(*Chemistry* 6th ed. pages 62–72 / 7th ed. pages 57–67)

When you finish this section you will be able to name, or give formulas for, the following classes of compounds: binary salts, salts with polyatomic ions, binary covalent compounds and acids.

Use the flowcharts that follow to answer the questions below.

**Naming Compounds from the Given Formula**

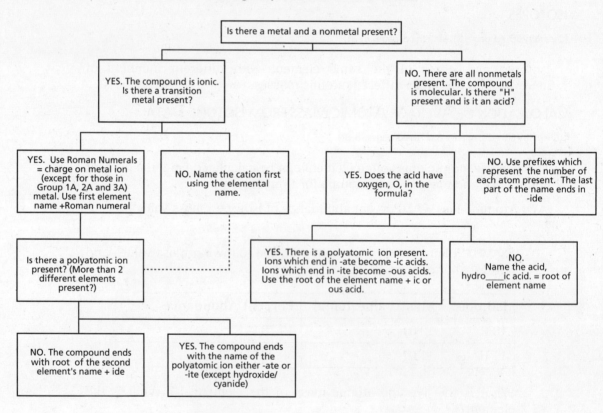

**Writing Formulas from the Given Name**

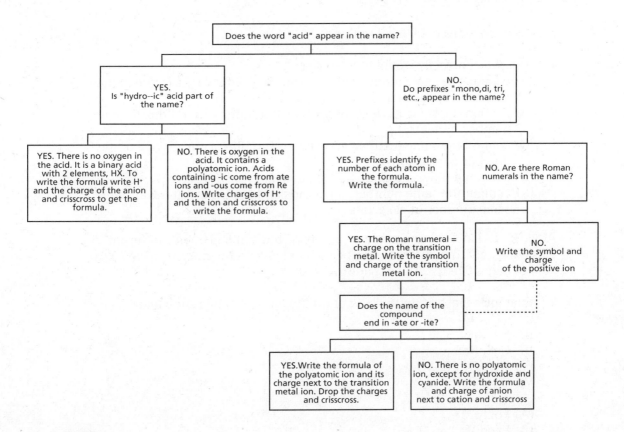

EXAMPLES:

1.  Name each of the following compounds.

    a. $CaF_2$                    d. $KIO_3$

    b. $Cl_2O_7$                  e. HF (aq)

    c. CuO                        f. $HNO_2$

2. Write the formulas for each of the following compounds.

    a. Tin (IV) oxide             d. Sulfurous acid

    b. Hydrosulfuric acid         e. Potassium chloride

    c. Nickel (II) nitrate        f. Iodine pentafluoride

SOLUTIONS:

1.  a. Calcium fluoride          d. Potassium iodate

    b. Dichlorine heptoxide      e. Hydrofluoric acid

    c. Copper (II) oxide         f. Nitrous acid

2.  a. $SnO_2$                    d. $H_2SO_3$

    b. $H_2S$                     e. KCl

    c. $Ni(NO_3)_2$               f. $IF_5$

## MULTIPLE-CHOICE QUESTIONS

No calculators are to be used in this section.

1. It may be correctly stated that isotopes
   (A) contain the same number of electrons but differ in mass due to the number of neutrons.
   (B) contain the same number of protons but differ in mass due to the number of neutrons.
   (C) contain the same number of neutrons but differ in mass due to the number of electrons.
   (D) contain a different number of protons.
   (E) contain the same number of protons and neutrons but never the same number of electrons.

ANSWER: B Isotopes are made of atoms of the same element, differing in mass due only to a different number of neutrons in the nucleus (*Chemistry* 6th ed. pages 81–87 / 7th ed. pages 78–83).

2. Examples illustrating the Law of Multiple Proportions shown are
   I.   $CO$, $CO_2$.
   II.  $Ca$, $BaO$.
   III. $CaS$, $BaS$.
   IV.  $Na_2CO_3$, $Na_2SO_4$.
   V.   $O_2$, $O_3$.

   (A) I only
   (B) I and V
   (C) III and IV
   (D) III and V
   (E) I and III

ANSWER: A The Law of Multiple Proportions deals with the relative masses of two elements that combine in more than one ratio with each other (*Chemistry* 6th ed. pages 46–48 / 7th ed. pages 43–45).

3. The chemist who is given credit for being the first to use symbols for elements and who developed a system of formulas for compounds was
   (A) Berzelius.
   (B) Thomson.
   (C) Lavoisier.
   (D) LeChatelier.
   (E) Arrhenius.

ANSWER: A (*Chemistry* 6th ed. pages 50–51 / 7th ed. pages 46–47).

4. The chemist given credit for first developing a valid model of the nucleus as a result of experiments with gold foil and alpha particle scattering was
   (A) Boyle.
   (B) Thomson.
   (C) Rutherford.
   (D) Berzelius.
   (E) Arrhenius.

ANSWER: C The summary of the work of many of the early chemists is worth reviewing (*Chemistry* 6th ed. pages 43–52 / 7th ed. pages 41–48).

5. The correct name of $Ca_3(PO_4)_2$ is
   (A) tricalcium diphosphate.
   (B) calcium diphosphate.
   (C) calcium (II) phosphate.
   (D) calcium phosphate.
   (E) calcium phosphide.

ANSWER: D This is a compound made up of a nontransition metal and a polyatomic ion, hence it is named by noting the metal followed by the polyatomic ion (*Chemistry* 6th ed. pages 62–63 / 7th ed. pages 57–58).

6. The correct number of atomic particles for $^{51}_{23}V$ is

|       | electrons | protons | neutrons |
|-------|-----------|---------|----------|
| (A)   | 28        | 28      | 23       |
| (B)   | 22        | 22      | 29       |
| (C)   | 23        | 28      | 23       |
| (D)   | 28        | 23      | 23       |
| (E)   | 23        | 23      | 28       |

ANSWER: E The atomic number, 23, indicates the number of protons and if this is not an ion, the number of electrons in the atom. The mass number, here 51, gives the total number of protons plus neutrons (51 – 23 = 28) (*Chemistry* 6th ed. pages 53–56 / 7th ed. pages 49–52).

7. In which of the following is a pair of isotopes represented?

|         | Atomic Number | Mass Number |
|---------|---------------|-------------|
| (A) X   | 24            | 52          |
| Y       | 25            | 52          |
| (B) X   | 26            | 56          |
| Y       | 27            | 59          |
| C) X    | 30            | 66          |
| Y       | 30            | 66          |
| (D) X   | 28            | 58          |
| Y       | 28            | 59          |
| (E) X   | 16            | 20          |
| Y       | 8             | 10          |

ANSWER: D Isotopes must have the same atomic number since they have the same number of protons. The mass number differs because of a difference in the number of neutrons (*Chemistry* 6th ed. pages 81–90 / 7th ed. pages 78–85).

8. The correct systematic name for SnI₄ is
    (A) silver iodide.
    (B) tin iodide.
    (C) tin iodate.
    (D) tin (IV) iodide.
    (E) tin iodide (IV).

Answer: **D** Tin, symbol Sn, commonly exists in ionic form with either a 2+ or a 4+ charge. It is necessary to indicate which ion is involved in this transition metal–nonmetal compound (*Chemistry* 6th ed. pages 63–65 / 7th ed. pages 58–61).

9. The correct systematic name for N₂O₄ is
    (A) dinitrogen trioxide.
    (B) dinitrogen tetroxide.
    (C) nitrogen (II) oxide.
    (D) nitrogen (IV) oxide.
    (E) nitrogen (II) oxide (IV).

Answer: **B** This is a good example of a binary covalent compound of Type III (*Chemistry* 6th ed. pages 62–71 / 7th ed. pages 57–66).

10. The correct formula for silver chlorate is
    (A) AgCl.
    (B) Ag₂Cl.
    (C) S₂ClO₃.
    (D) Ag₂ClO₃.
    (E) AgClO₃.

Answer: **E** Even though it is a transition metal ion, silver ions almost always have a 1+ charge. Therefore there is no need to indicate the charge on this ion with Roman numerals. You do need to know the charge on the silver ion as well as the formula and charge of the chlorate ion (*Chemistry* 6th ed. pages 63–68 / 7th ed. pages 58–63).

Note: Students who expect to do well on the AP Chemistry Examination should have gotten all, or nearly all, of these ten questions correct.

## FREE-RESPONSE QUESTIONS

1. In the late 1880s, J.J. Thomson experimented with cathode rays by bending them in an applied electromagnetic field. He discovered that the charge / mass (e/m) ratio was the same no matter what the cathode was made of.
    (a) What did this suggest?
    (b) What were these particles called?
    (c) What model of the atom's structure did he propose?

Answer: **1** J.J. Thomson was able to determine the ratio of the charge to mass of the electron, showing that cathode rays consist of discrete particles. He also showed that cathode rays always had the same kind of properties. He concluded that atoms of all substances contain the same kind of negative particles, called electrons. From this he developed the plum pudding model (electrons were dispersed in an atom like raisins in a pudding) (*Chemistry* 6th ed. pages 49–53 / 7th ed. pages 45–49).

2. (a) Draw and label the basic parts of Rutherford's famous gold foil experiment.
   (b) Describe his results and the conclusions he reached.
   (c) What would be the results of Rutherford's experiment if J.J. Thomson's model had been correct?

ANSWER: 2 In your diagram, you must show at least (1) a source of alpha particles, (2) a thin sheet of gold foil, and (3) a detecting screen. It is important to indicate that most of the alpha particles were deflected, a few through very large angles, and that this suggested to Rutherford that most of the atom is empty space and that the smaller number of particles experiencing the wide angle deflections suggest a very small and very dense nucleus. If Thomson's model had been correct, none of the alpha particles would have been deflected through wide angles (*Chemistry* 6th ed. pages 52–53 / 7th ed. pages 48–49).

3. Chemist Lawrence Strong once said that the model "must not be confused with the reality that it mirrors." Explain what he meant.

ANSWER: 3 A model, as scientists use the term, is neither correct or incorrect, but <u>successful or unsuccessful</u> in matching observations, giving understanding to relationships, and allowing for useful predictions (*Chemistry* 6th ed. pages 368–370 / 7th ed. pages 348–350).

Note: It is doubtful that a question this philosophical will appear on the AP Chemistry Examination.

# 3

# STOICHIOMETRY

Stoichiometry includes calculations of average atomic mass, the mole, molar mass, percent composition, molarity and empirical formula. The quantities of materials consumed and produced in chemical reactions are also considered. The multiple-choice section of the AP Exam may contain questions that require computations without calculators. The free response portion of the exam may contain one multistep question on stoichiometry.

You should understand

- Moles, mass, representative particles (atoms, molecules, formula units), molar mass, and Avogadro's number.
- Molarity; preparation of solutions.
- The percent composition of an element in a compound.
- Balanced chemical equations; for example, for a given mass of reactant, calculate the amount of product produced.
- Limiting reactants: calculate the amount of product formed when given the amounts of all of the reactants present.
- Reactions in solution: given the molarity and the volume of the reactants, calculate the amount of product produced or the amount of reactant required to react.
- The percent yield of a reaction.

## THE MOLE

(*Chemistry* 6th ed. pages 86–90 / 7th ed. pages 82–85)

The mole is defined in the table below. Calculations involving moles utilize the conversions in the table below.

| |
|---|
| 1 mol of a monoatomic element = $6.02 \times 10^{23}$ atoms |
| 1 mol of a molecular compound or diatomic element = $6.02 \times 10^{23}$ molecules |
| 1 mol of an ionic compound = $6.02 \times 10^{23}$ formula units |

## MOLAR MASS

(*Chemistry* 6th ed. pages 90–93 / 7th ed. pages 86–88)

Molar mass is the mass of 1 mole of an element or compound.

EXAMPLE: The principal ore in the production of aluminum cans has a molecular formula of $Al_2O_3 \cdot 2H_2O$. What is the mass in grams of $2.10 \times 10^{24}$ formula units (f.u.) of $Al_2O_3 \cdot 2H_2O$?

$$2.10 \times 10^{24} \text{ formula units} \times \frac{1 \text{ mol } Al_2O_3 \cdot 2H_2O}{6.02 \times 10^{23} \text{ f.u.}} \times \frac{138 \text{ g}}{1 \text{ mol } Al_2O_3 \cdot 2 H_2O} = 481 \text{ g}$$

## MOLARITY

(*Chemistry* 6th ed. pages 140–146 / 7th ed. pages 133–139)

The number of moles of solute in 1 liter of solution is a measure of its molarity or solution concentration.

EXAMPLE: Prepare 2.00 L of 0.250 M NaOH from solid NaOH.

$$2.00 \text{ L} \times \frac{0.250 \text{ mol NaOH}}{\text{L}} \times \frac{40.00 \text{ g NaOH}}{1 \text{ mol}} = \textbf{20.0 g NaOH}$$

Place 20.0 g NaOH in a 2-L volumetric flask; add water to dissolve the NaOH, and fill to the mark with water, mixing several times along the way.

EXAMPLE: Prepare 2.00 L of 0.250 $M$ NaOH from 1.00 $M$ NaOH.

$$M_1V_1 = M_2V_2$$

$$1.00 \ M \ V_1 = 0.250 \ M \times 2.00 \text{ L} \qquad V_1 = \textbf{0.500 L}$$

Add 500. mL of 1.00 $M$ NaOH stock solution to a 2-L volumetric flask; add deionized water in several increments, mixing until the flask is filled to the mark on the neck of the flask.

# PERCENT COMPOSITION

(*Chemistry* 6th ed. pages 93–96 / 7th ed. pages 89–91)

The mass percentages of elements in a compound can be obtained by comparing the mass of each element present in 1 mole of the compound to the total mass of 1 mole of compound.

$$\frac{\text{Mass of element in 1 mol of compound}}{\text{Mass of 1 mol of compound}} \times 100\%$$

EXAMPLE: Calculate the percent of oxygen in $Mg(NO_3)_2$
$$= (16 \times 6/148) \times 100\% = \mathbf{65\%}$$

# DETERMINATION OF EMPIRICAL FORMULA BY COMBUSTION ANALYSIS

(*Chemistry* 6th ed. pages 96–98 / 7th ed. pages 91–93)

The empirical formula is the simplest whole number ratio of atoms in a compound.

(*The following problem involves two steps. First, the percent composition of the compound is determined from combustion data. Second, the empirical formula of the compound is determined from the percent composition.*)

EXAMPLE: A compound contains only carbon, hydrogen, and oxygen. Combustion of 10.68 mg of the compound yields 16.01 mg $CO_2$ and 4.37 mg of $H_2O$. What is the percent composition of the compound?

SOLUTION: Assume that all the carbon in the compound is converted to $CO_2$ and determine the mass of carbon present in the 10.68-mg sample.

$$.01601 \text{ g } CO_2 \times \frac{12.01 \text{ g C}}{44.01 \text{ g } CO_2} = 4.369 \text{ mg C}$$

The mass percent of C in this compound is
$$\frac{4.369 \text{ mg C}}{10.68 \text{ mg}} \times 100\% = 40.91\% \text{ C}$$

The same procedure can be used to find the mass percent of hydrogen in the unknown compound. We assume that all the hydrogen present in 10.68 mg of compound was converted to $H_2O$.

$$0.0437 \text{ g } H_2O \times \frac{2.016 \text{ g H}}{18.02 \text{ g } H_2O} = 0.489 \text{ mg H}$$

The mass percent of H in the compound is

$$\frac{0.489 \text{ mg}}{10.68 \text{ mg}} \times 100\% = 4.58\% \text{H}$$

The unknown compound contains only carbon, hydrogen and oxygen. The remainder must be oxygen.

$$100.00 \% - (40.91\% \text{ C} + 4.58\% \text{ H}) = 54.51\% \text{ O}$$

## EMPIRICAL FORMULA

(*Chemistry* 6th ed. pages 98–101 / 7th ed. pages 93–95)

What is the empirical formula of a compound that contains 40.91% carbon, 4.58% hydrogen and 54.51 % oxygen?

$$40.91 \text{ g C} \times \frac{1 \text{ mol C}}{12.01 \text{ g}} = 3.406 \text{ mol C}$$

$$4.58 \text{ g H} \times \frac{1 \text{ mol H}}{1.008 \text{ g}} = 4.54 \text{ mol H}$$

$$54.51 \text{ g O} \times \frac{1 \text{ mol O}}{16.00 \text{ g}} = 3.407 \text{ mol O}$$

*Divide by the smallest number*

$$\frac{4.54 \text{ mol H}}{3.406} = 1.33 \text{ mol H} \times 3 = 3.99 \sim 4.00$$

$$C_{\frac{3.406}{3.406}} H_{\frac{4.54}{3.406}} O_{\frac{3.407}{3.406}} = (C_{1.00} H_{1.33} O_{1.00})3 = C_3 H_4 O_3$$

Empirical Formula

Since a whole number ratio is required, we multiply all subscripts by the lowest factor (3) to obtain the whole number. Rounding can only occur when the subscripts are within 0.1 of the nearest whole number.

## CONVERSIONS INVOLVED IN STOICHIOMETRIC CALCULATIONS

| Type of Conversion | Example | Reference |
|---|---|---|
| **g → mol** (using molar mass) | $\dfrac{18 \text{ g}}{1 \text{ mol } H_2O}$ | *Chemistry* 6th ed. pages 90–91 / 7th ed. page 86 |
| **mol → mol** (using mole ratio – coefficients from balanced chemical equation) | $\dfrac{1 \text{ mol } O_2}{2 \text{ mol } H_2O}$ | *Chemistry* 6th ed. pages 108–110 / 7th ed. pages 102–104 |
| **mol → L** (using molarity which is a solution's concentration equal to the number of mols of solute in 1 liter of solution) | $\dfrac{6 \text{ mol HCl}}{1L}$ | *Chemistry* 6th ed. pages 140–148 / 7th ed. pages 133–140 |

## SOLUTION STOICHIOMETRY

(*Chemistry* 6th ed. pages 161–164 / 7th ed. pages 151–154)

Solution stochiometry involves calculations for reactions that occur in aqueous solution.

EXAMPLE: The compound with the empirical formula, $C_3H_4O_3$, is a monoprotic acid. A 2.20-g sample of the unknown monoprotic acid is dissolved in 1.0 L of water. A titration required 25.0 mL of 0.500 M NaOH to react completely with all the acid present. What is the molecular formula of the acid? First, find the molar mass of the acid.

SOLUTION: To find the molar mass of the acid, let HA = unknown acid.

Write a balanced molecular equation.
$$HA + NaOH \rightarrow NaA + H_2O$$

Find the moles of HA present from the volume and molarity of NaOH.

$$0.250 \text{ L NaOH} \times \frac{0.500 \text{ mol NaOH}}{1 \text{ L}} \times 1 \text{ mol HA} = 0.0125 \text{ mol HA}$$

To find the molar mass of HA, divide mass (g) of HA which is given by mol of HA.

$$\frac{2.20 \text{ g HA}}{0.0125 \text{ mol HA}} = 176 \text{ g/mol.}$$

To find the molecular formula of the compound if the empirical formula and molar mass are known, use the following equation (*Chemistry* 6th ed. pages 100-102 / 7th ed. pages 95–96):

| $\dfrac{\text{Molar mass}}{\text{Empirical formula mass}} = \text{Number}$ | *Multiply each subscript in the empirical formula by the number* |
|---|---|

Determine the empirical formula mass = 3 (12) + 4(1) + 3(16) = 88 g/mol

Since 176/88 = 2.0, the **molecular formula** is $(C_3H_4O_3)_2 = \mathbf{C_6H_8O_6}$

## LIMITING REACTANT

(*Chemistry* 6th ed. pages 113–118 / 7th ed. pages 106–111)

In a limiting reactant problem, the amounts of all reactants are given and the amount of product is to be determined. The limiting reactant is completely consumed when the reaction goes to completion. It determines how much product is formed.

EXAMPLE: What mass of precipitate can be produced when 50.0 mL of 0.200 $M$ Al $(NO_3)_3$ is added to 200.0 mL of 0.100 $M$ KOH?

■ Always begin with a balanced molecular equation.

$$Al(NO_3)_3 \text{ (aq)} + 3KOH(aq) \rightarrow Al(OH)_3 \text{ (s)} + 3KNO_3(aq)$$

■ For each reactant, determine the amount of product produced (in mol or g).

$$0.0500 \text{ L} \times \frac{0.200 \text{ mol Al(NO}_3)_3}{\text{L}} \times \frac{1 \text{ mol Al(OH)}_3}{1 \text{ mol Al(NO}_3)_3}$$

$$= 0.0100 \text{ mol Al(OH)}_3$$

$$0.200 \text{ L} \times \frac{0.100 \text{ mol KOH}}{\text{L}} \times \frac{1 \text{ mol Al(OH)}_3}{3 \text{ mol KOH}}$$

$$= 0.00667 \text{ mol Al(OH)}_3$$

■ *The limiting reactant produces the least amount of product.* The limiting reactant in this case is KOH. It produces only 0.00667 mol of product which is less than 0.0100 mol.

The mass of precipitate produced is:

$$0.00667 \text{ mol Al(OH)}_3 \times \frac{78.00 \text{ g Al(OH)}_3}{1 \text{ mol Al(OH)}_3} = \textbf{0.520 g Al(OH)}_3$$

## PERCENT YIELD

(*Chemistry* 6th ed. pages 118–120 / 7th ed. pages 111–113)

The *percent yield* of a reaction is the actual yield of a product as a percentage of the theoretical yield. The *actual yield* is the amount of product obtained in an experiment. The *theoretical yield* is the amount of product calculated from the amounts of reactants used.

EXAMPLE: If the reaction above has an 85.3% yield of precipitate, how much aluminum hydroxide is produced?

$$\text{Percent Yield} = \frac{\text{Actual Yield}}{\text{Theoretical Yield}} \times 100\%$$

$$\text{Actual Yield} = \frac{85.3\% \times 0.520 \text{ g}}{100\%} = \textbf{0.444 g Al(OH)}_3$$

# MULTIPLE-CHOICE QUESTIONS

No calculators are to be used in this section.

1. There are three naturally occurring isotopes of oxygen: oxygen-16, oxygen-17, oxygen-18. The atomic mass of oxygen is 16.00 amu. From these data it may be concluded that
   (A) most of the oxygen atoms are $^{17}O$.
   (B) most of the oxygen atoms are $^{18}O$.
   (C) very few of the atoms are $^{17}O$ and $^{18}O$.
   (D) there is equal abundance of all three isotopes in nature.
   (E) more than half of the atoms are $^{18}O$.

ANSWER: C   In order for the average mass of naturally occurring oxygen to be 16.00 amu, there cannot be a significant contribution from either oxygen-17 or oxygen-18 (*Chemistry* 6th ed. pages 81–101 / 7th ed. pages 78–95).

2. Analysis of a sample of an oxide of chromium is reported as 26 g of chromium and 12 g of oxygen. From these data determine the empirical formula of this compound.
   (A) CrO
   (B) $Cr_2O_3$
   (C) $CrO_3$
   (D) $CrO_2$
   (E) $Cr_4O_6$

ANSWER: B   You have 0.5 mol of Cr (26 g × 1 mol/52 g) and 0.75 mol of O (12 g × 1 mol/16 g). The ratio of mol Cr: mol O is 1/2:3/4 = (1/2 × 4/3) = 2/3 or 2 mol Cr: 3 mol O. While answer E gives the same ratio, it is not the smallest whole number ratio (*Chemistry* 6th ed. pages 98–101 / 7th ed. pages 93–95).

3. Aluminum reacts with sulfuric acid, $H_2SO_4$, to form aluminum sulfate, $Al_2(SO_4)_3$, and hydrogen gas. Give the sum of all coefficients (all reactants and all products) for this balanced chemical expression.
   (A) 4
   (B) 5
   (C) 6
   (D) 9
   (E) 12

ANSWER: D   The equation is $2Al + 3H_2SO_4 \rightarrow Al_2(SO_4)_3 + 3H_2$. Note that hydrogen is diatomic. Be sure to count the coefficients for the whole formula; it is understood that the coefficient in front of the aluminum sulfate is one (*Chemistry* 6th ed. pages 102–108 / 7th ed. pages 96–102).

4. Methane reacts with oxygen to form carbon dioxide and water as shown in the following chemical equation: $CH_4 + 2O_2 \rightarrow CO_2 + 2 H_2O$. If 72 g of water form, how much methane must have reacted?
   (A) 16 g
   (B) 32 g
   (C) 36 g
   (D) 84 g
   (E) 72 g

ANSWER: B The equation indicates a mole ratio of water to methane of 2:1. The 72 grams of water is 4.0 moles of water (72 g / 18 g/mol = 4.0 moles), so 2.0 moles of methane are required. Two moles of methane have a mass of 32 grams (16 g/mol × 2) (*Chemistry* 6th ed. pages 108–112 / 7th ed. pages 102–106).

5. Hydrogen reacts with oxygen to form only water. If 16 grams of hydrogen is mixed with 16 grams of oxygen, how much water can form?
   (A) 0.50 grams
   (B) 8.0 grams
   (C) 18 grams
   (D) 72 grams
   (E) 144 grams

ANSWER: C Write an equation for the reaction: $2H_2 + O_2 \rightarrow 2H_2O$. The mole ratio can then be seen as 2 mol $H_2$ for 1 mole $O_2$. Determine the number of moles of each reactant: 16 g $H_2$ × 1 mol $H_2$/2.0 g = 8 mol $H_2$ and 16 g $O_2$ × 1 mol $O_2$/32 g = 0.50 mol $O_2$.

Since $O_2$ is the limiting reagent, it determines the amount of water formed. From the mole ratio in the equation, it follows that 0.50 mol $O_2$ forms two times that amount of water or 1 mol of water. The molar mass of water is 18 (*Chemistry* 6th ed. pages 113-120 / 7th ed. pages 106–113).

6. How many moles of barium sulfide, BaS, will form when 60.0 mL of 1.0 M $Ba(NO_3)_2$ are mixed with 25.0 mL of 0.80 $M$ $K_2S$ solution to form barium sulfide solid?
   (A) 0.020 mol
   (B) 0.040 mol
   (C) 0.060 mol
   (D) 0.080 mol
   (E) 0.10 mol

ANSWER: A The equation, $Ba(NO_3)_2 + K_2S \rightarrow BaS + 2KNO_3$, indicates that with 0.060 mole $Ba(NO_3)_2$ and 0.020 mole $K_2S$, the potassium sulfide is the limiting reactant (they react in a one-to-one ratio). The equation also shows that for one mole of potassium sulfide, one mole of barium sulfide forms, hence 0.020 mol of potassium sulfide will allow for the formation of 0.020 mol of the solid BaS (*Chemistry* 6th ed. pages 146–153, 156–158 / 7th ed. pages 139–144, 147–148).

7. Consider the neutralization reaction $Be(OH)_2 + 2HCl \rightarrow BeCl_2 + 2H_2O$. What volume of 5.00 $M$ HCl is required to react completely with 4.30 g of $Be(OH)_2$? (Molar mass of $Be(OH)_2 = 3.0$ g/mol.)
   (A) 10.0 mL
   (B) 30.0 mL
   (C) 40.0 mL
   (D) 50.0 mL
   (E) 80.0 mL

ANSWER: C The mass of 4.3 g of $Be(OH)_2$ represents 0.10 mol of this base; the equation indicates that twice that amount (0.20 mol) of acid is needed for neutralization to be complete. Since both the concentration and number of moles needed are known for the acid, the volume of the acid can be calculated (0.20 mol / 5.0 mol/L = 0.040 L = 40. mL) (*Chemistry* 6th ed. pages 163–164 / 7th ed. pages 153–154).

8. Sucrose, $C_{12}H_{22}O_{11}$, has a molar mass of 342 g/mol. How many moles of molecules of this white crystal are there in 684 g of sucrose?
   (A) 2.00 mole of molecules
   (B) 21.0 mole of molecules
   (C) 42.0 mole of molecules
   (D) $6.02 \times 10^{23}$ mole of molecules
   (E) $3.01 \times 10^{23}$ mole of molecules

ANSWER: A It takes 342 g to make one mole of the compound and you have twice that amount. Therefore you have 2.00 mole of this substance (684 g / 342 g/mol) (*Chemistry* 6th ed. pages 86–93 / 7th ed. pages 82–88).

9. Nitric acid reacts with silver metal: $4HNO_3 + 3Ag \rightarrow NO + 2H_2O + 3AgNO_3$. Calculate the number of grams of NO formed when 10.8 g of Ag reacts with 12.6 g of $HNO_3$.
   (A) 0.999 g
   (B) 9.0 g
   (C) 12.0g
   (D) 18.0 g
   (E) 30.0 g

ANSWER: A First determine the number of moles of each reactant to determine which is in excess and which is the limiting reactant: 10.8 g Ag / 108 g/mol = 0.100 mol of Ag; 12.6 g $HNO_3$ / 63.0 g /mol = 0.40 mol of $HNO_3$. Since the Ag / $HNO_3$ ratio (from the equation) is 3 / 4, Ag is the limiting reactant. 0.100 mol of Ag will form one third as much NO (0.100 × 1NO / 3 Ag) or 0.0333 mol of NO. Finally, 0.0333 mol NO × 30.0 g NO / mol = 0.999 g NO (*Chemistry* 6th ed. pages 113–120 / 7th ed. pages 106–113).

10. How many grams of chromium are in 58.5 g of $K_2Cr_2O_7$? (Molar mass = 294 g/mol.)
    (A) 200 g of Cr
    (B) 10.4g of Cr
    (C) 15.6 g of Cr
    (D) 20.8 g of Cr
    (E) 208. g of Cr

Answer: **D** First determine the number of moles of the salt: 58.5 g / 294 g/mol = 0.200 mol $K_2Cr_2O_7$. Since Cr has an atomic mass of 52.0 and there are two Cr per each unit of $K_2Cr_2O_7$, 0.200 mol × 2 Cr / $K_2Cr_2O_7$ × 52.0 g / mol = 20.8 g of Cr (*Chemistry* 6th ed. pages 87–89 / 7th ed. pages 83–85).

## Free-Response Questions

Calculators may be used for this section.

1. A test tube containing $CaCO_3$ is heated until all of the compound decomposes. If the test tube plus calcium carbonate originally weighed 30.08 grams and the loss of mass during the experiment is 4.40 grams, what was the mass of the empty test tube?

Answer: **20.1 g** Since 0.100 mol of $CO_2$ was lost (4.40g / 44.0 /mol = 0.100 mol), this means that the same number of moles of $CaCO_3$ decomposed. (0.100 mol of $CaCO_3$ is 10.0 g (100 g/mol × 0.100 mol = 10.0 g of $CaCO_3$.) If the test tube and the salt weighed 30.08 g and 10.0 g was the calcium carbonate, then the mass of the test tube must have been 20.1 g (*Chemistry* 6th ed. pages 108–112 / 7th ed. pages 102–106).

2. A 10.0 g sample of an oxide of copper is heated in a stream of pure hydrogen, forming 1.26 g of water.
   (a) Determine the percentage of copper in the compound.
   (b) Determine the empirical formula of the copper oxide. Name it.

Answer:

(a) **88 % copper**

$$1.26 \text{ g } H_2O \times \frac{1 \text{ mol}}{18.0 \text{ g } H_2O} = 0.0700 \text{ mol } H_2O \times \frac{1 \text{ mol O}}{1 \text{ mol } H_2O}$$
$$= 0.0700 \text{ mol O atoms}$$

0.0700 mol × 16.0 g/mol = 1.12 g of oxygen in the oxide

10.0 g – 1.12 g = 8.8 g of copper; 8.8 g/10.0 g × 100% = 88% Copper

(b) **$Cu_2O$ copper (I) oxide**

Consider the mole ratio of Cu to O:

[8.8 g / 63.55 g/mol = 0.14 mol Cu]

$$\frac{0.14 \text{ mol Cu}}{0.070 \text{ mol O}} = \frac{2 \text{ mol Cu}}{1 \text{ mol O}}$$ $Cu_2O$ is the empirical formula.

This is called copper(I) oxide since the Cu has a charge of +1 in this compound (*Chemistry* 6th ed. pages 108–112, 98–102, 92–95 / 7th ed. pages 102–106, 93–96, 88–91).

## ESSAY FOR STOICHIOMETERY

The percentage yield for some reactions is low. Discuss several reasons why this might be so. [Note: While this is not specifically discussed in the chapter, it should direct your attention to practical limitations of chemical reactions often found in your own work in the laboratory.]

ANSWER:

1.  The possibility of mechanical loss. Things like spillage, products getting stuck to equipment and not weighed can often occur in the separation and purification processes which are usually necessary to give a pure product.

2.  The system may react in such a way that it reaches equilibrium far to the left, i.e., with more reactant than product present at equilibrium. [This is the case with systems that have a low value for the equilibrium constant.]

3.  There can also be side reactions that lower the yield by reacting with reactants of the original process thereby preventing them from forming the desired products.

For these reasons the actual yield of products is almost always less than the theoretical yield.

# 4

# CHEMICAL REACTIONS

Chemical equations represent chemical changes in which atoms in one or more substances are reorganized. Skills involved in this section include balancing chemical reactions, classifying reactions by type, predicting products of chemical reactions, and writing complete and net-ionic equations.

The multiple choice section of the AP Exam will include questions on these topics.

The free-response portion of the exam includes a separate reactions section in which the student is given the names of the reactants combined in eight different laboratory situations and asked to provide the net-ionic equation for five of the situations. *Balancing is not required on this portion of the free-response exam. The physical states of the reactants and products do not need to be indicated in this section,* although they are extremely helpful in writing the net-ionic equation. If you do not know the products which are worth 2 points, at least write down the correct reactants to obtain 1 of the 3 points for each reaction.

You should be able to
- Classify reactions by type.
- Write a balanced molecular equation, complete ionic equation, and a net ionic equation.
- Balance oxidation-reduction reactions.
- Predict if a precipitate will form using the solubility rules.
- Predict products of reactions given the chemical names of the reactants.
  - ☐ Memorize the list of common oxidizers and reducers given in this chapter. Know the chemical formulas of the products they form.
  - ☐ This is a separate section in the AP Exam.

## DESCRIBING REACTIONS IN AQUEOUS SOLUTION

(*Chemistry* 6th ed. pages 154–155 / 7th ed. pages 145–146)

### MOLECULAR EQUATION

A molecular equation gives the overall reaction. Although it gives information on stoichiometry, it gives no information on whether or not compounds really exist as ions in solution. For example, $2KCl \ (aq) + Pb(NO_3)_2 \ (aq) \rightarrow PbCl_2 \ (s) + 2KNO_3 \ (aq)$.

### COMPLETE IONIC EQUATION

The complete ionic equation gives the equation including all the ions in solution. For example, $2K^+ + 2Cl^- + Pb^{2+} + 2NO_3^- \rightarrow PbCl_2 \ (s) + 2K^+ + 2NO_3^-$ The states are aqueous , unless otherwise indicated.

### NET-IONIC EQUATION

The net-ionic equation only gives information on those species that undergo a chemical change. The spectator ions are omitted from the complete ionic equation.

$Pb^{2+} + 2 \ Cl^- \rightarrow PbCl_2 \ (s)$ This is the answer for the AP Exam, except that it does not need to be balanced.

## TYPES OF REACTIONS

Classification of chemical reactions by type allows the prediction of products if given the reactants.

| Main Reaction Types |
| --- |
| Combustion, Combination, Complex ion formation |
| Acid-base |
| Redox (oxidation–reduction) |
| Precipitation |

# COMBUSTION

(*Chemistry* 6th ed. page 1050 / 7th ed. page 1003)

Combustion is the complete oxidation of any organic compound containing C, H and/or O to yield $CO_2$ and $H_2O$.

Ethene is completely burned in oxygen.
$C_2H_4 (g) + 3O_2 (g) \rightarrow 2CO_2 (g) + 2H_2O (g)$

Nonmetallic hydrides combine with oxygen to form oxides and water.

Gaseous silane is burned in oxygen.
$SiH_4 (g) + 2O_2 (g) \rightarrow SiO_2 (s) + 2H_2O (g)$

Nonmetallic sulfides combine with oxygen to form sulfur dioxide and oxides.

Carbon disulfide vapor is burned in excess oxygen.
$CS_2 (s) + 3O_2 (g) \rightarrow CO_2 (g) + 2SO_2 (g)$

# COMBINATION

Two or more elements $\rightarrow$ 1 compound

Magnesium ribbon is burned in pure nitrogen gas.
$3Mg (s) + N_2 (g) \rightarrow Mg_3N_2 (s)$

Two or more compounds $\rightarrow$ 1 compound
Metal oxide + water $\rightarrow$ base

Solid lithium oxide is added to water.
$Li_2O (s) + H_2O (l) \rightarrow 2 Li^+ (aq) + 2OH^- (aq)$

Nonmetal oxide + water $\rightarrow$ acid (no change in oxidation state)

Sulfur trioxide is bubbled into water.
$SO_3 (g) + H_2O (l) \rightarrow 2H^+ (aq) + SO_4^{2-} (aq)$

Metal oxide + nonmetal oxide $\rightarrow$ salt

A mixture of solid calcium oxide and solid tetraphosphorus decaoxide is heated.
$6CaO (s) + P_4O_{10} (s) \rightarrow 2Ca_3(PO_4)_2 (s)$

# DECOMPOSITION

Decomposition = 1 compound $\rightarrow$ 2 or more elements or compounds (reverse of combination)

Magnesium carbonate is heated strongly in a crucible.
$MgCO_3 (s) \rightarrow MgO (s) + CO_2 (g)$

## COMPLEX-ION FORMATION

(*Chemistry* 6th ed. pages 988, 998, 1001–1003 / 7th ed. pages 944, 955, 958–960)

Transiton metal ions such as $Cu^{2+}$, $Ag^+$, and $Ni^+$, react with excess ligands to form coordination complexes.

| Examples of Ligands |
|---|
| $OH^-$, hydroxo |
| $CN^-$, cyano |
| $NH_3$, amine |
| $H^-$, halo (halides) |

Sometimes, # ligands = 2 (charge of transition metal ion)
Charge on complex ion = total charge of all ligands + T.M. Ion

Solid zinc nitrate is treated with excess sodium hydroxide solution.
$Zn(NO_3)_2(s) + 4OH^-(aq) \rightarrow Zn(OH)_4^{2-}(aq) + 2NO_3^-(aq)$

Complex ions are broken down by the addition of a strong acid which reacts with basic ligands.

An aqueous solution of diamminesilver chloride is treated with dilute nitric acid.
$Ag(NH_3)_2^+(aq) + Cl^-(aq) + 2H^+(aq) \rightarrow AgCl(s) + 2NH_4^+(aq)$

## ACID-BASE REACTIONS

(*Chemistry* 6th ed. pages 158–164 / 7th ed. pages 149–154)

Acid + Base $\rightarrow$ Salt + $H_2O$

Memorize the list of strong acids and strong bases below. All others are weak acids or bases.

| Strong Acids | Strong Bases |
|---|---|
| HCl, HBr, HI | Group I Hydroxides, (LiOH, NaOH . . .) |
| $H_2SO_4$ ($HSO_4^-$ = weak) | Group II Hydroxides of Ca, Sr and Ba |
| $HNO_3$ | |
| $HClO_4$ | |

## STRONG ACID–STRONG BASE REACTIONS

(*Chemistry* 6th ed. page 158 / 7th ed. page 149)

Strong acids and bases are assumed to be 100% ionized in aqueous solution; write all reactants in ionic form. Use solubility rules for the salt formed. (See end of this chapter for solubility rules.)

For the examples that follow, all states are aqueous, unless otherwise indicated.

> EXAMPLE: Equimolar solutions of hydrochloric acid and sodium hydroxide are mixed.

$$H^+ + Cl^- + Na^+ + OH^- \rightarrow Na^+ + Cl^- + HOH(l)$$

Omit spectator ions to get the net-ionic equation:

$$H^+ + OH^- \rightarrow HOH$$

*Note: For most strong acid–strong base reactions, this will be the net-ionic equation, but make sure that the salt formed is soluble.*

EXAMPLE: $Ba(OH)_2 + H_2SO_4 \rightarrow BaSO_4 (s) + H_2O(l)$

The net-ionic equation for the reaction above is

$Ba^{2+} + 2OH^- + 2H^+ + SO_4^{2-} \rightarrow BaSO_4(s) + H_2O (l)$

## WEAK ACID–STRONG BASE REACTIONS (OR STRONG ACID–WEAK BASE)

(*Chemistry* 6th ed. page 159 / 7th ed. page 150)

Weak acids and bases ionize only slightly. Write the formula for the weak acid or weak base in molecular form (do not separate into ions!).

> EXAMPLE: Equimolar solutions of acetic acid and sodium hydroxide are mixed.
>
> $HC_2H_3O_2(aq) + Na^+ + OH^- \rightarrow Na^+ + C_2H_3O_2^- + HOH(l)$
>
> *Omit spectator ions to get:* $HC_2H_3O_2(aq) + OH^- \rightarrow C_2H_3O_2^- + HOH(l)$
>
> All states are aqueous, unless otherwise indicated.

## OXIDATION–REDUCTION REACTIONS

(*Chemistry* 6th ed. pages 164–179 / 7th ed. pages 154–168)

Oxidation reactions involve the transfer of electrons. Skills required for this section include the assignment of oxidation numbers and the identification of oxidation–reduction reactions. In an oxidation–reduction reaction, the atoms oxidized will need to be identified as well as the atoms reduced, the oxidizing agent, and the reducing agent.

| **Rules for Assigning Oxidation States** (*Chemistry* 6th ed. pages 167–168 / 7th ed. pages 157–158) | |
|---|---|
| The oxidation state of . . . | Examples |
| An atom in element is zero. | Na(s), $O_2$(g) |
| A monoatomic ion is the same as its charge. | $Na^+$ |
| Oxygen is usually –2 in its compounds. Exception: peroxides (containing $O_2^{2-}$) in which oxygen is 1–. | $H_2O$, $CO_2$ |
| Hydrogen is +1 in its covalent compounds. (hydrogen is –1 in binary hydrides, ex. NaH). | $H_2O$, $NH_3$ |
| For an electrically neutral compound, the sum of the oxidation states must be zero. | $KMnO_4$ <br> Solution: K = +1, O = -2 <br> +1 + Mn + 4(–2) = 0 <br> –7 + Mn = 0 ; Mn = +7 |
| For an ionic species, the sum of the oxidation states must equal the overall charge. | Ex: $Cr_2O_7^{2-}$ <br> Solution: O = –2 , <br> 2 Cr + 7(-2) = –2 ; 2Cr = +12; <br> Cr = +6 |

EXAMPLE: Is the following reaction an oxidation–reduction reaction? If it is an oxidation–reduction, identify which atoms are oxidized, which atoms are reduced, the oxidizing agent, and the reducing agent (Chemistry 6th ed. pages 168–170 / 7th ed. pages 158–161).

$$Zn + Cu(NO_3)_2 \rightarrow Cu + Zn(NO_3)_2$$

SOLUTION: The nitrate ion appears unchanged on both sides of the reaction, so neither element in the ion is oxidized or reduced. *Copper* goes from $Cu^{+2}$ to $Cu^0$ (It is reduced; its oxidation # goes down. It gains 2 electrons.) *Zinc* goes from $Zn^0$ to $Zn^{2+}$ (it is oxidized, its oxidation goes up, and it loses 2 electrons.)

*Oxidizing agents* are reduced and cause something else to be oxidized. $Cu^{2+}$ is the oxidizing agent. *Reducing agents* are oxidized and cause something else to be reduced. Zn is the reducing agent.

NOTE: If there is no change in oxidation numbers in the reaction, then the reaction is not an oxidation–reduction reaction.

## STEPS FOR BALANCING OXIDATION–REDUCTION REACTIONS

### BALANCE OXIDATION–REDUCTION REACTIONS IN ACIDIC SOLUTION USING THE HALF-REACTION METHOD

(*Chemistry* 6th ed. pages 172–176 / 7th ed. pages 162–166)

EXAMPLE: $Cr_2O_7^{2-} + Cl^- \rightarrow Cr^{3+} + Cl_2$

STEP 1: *Write separate half-reactions*:

$$Cr_2O_7^{2-} \rightarrow Cr^{3+} \qquad\qquad Cl^- \rightarrow Cl_2$$

**STEP 2**: *Balance all atoms except H and O*:

$$Cr_2O_7{}^{2-} \rightarrow 2Cr^{3+} \qquad\qquad 2Cl^- \rightarrow Cl_2$$

**STEP 3**: *Balance oxygen using water*:

$$Cr_2O_7{}^{2-} \rightarrow 2Cr^{3+} + 7H_2O \qquad\qquad 2Cl^- \rightarrow Cl_2$$

**STEP 4**: *Balance hydrogen with $H^+$*:

$$14H^+ + Cr_2O_7{}^{2-} \rightarrow 2Cr^{3+} + 7H_2O \qquad\qquad 2Cl^- \rightarrow Cl_2$$

**STEP 5**: *Balance charge using electrons*:

$$6e^- + 14H^+ + Cr_2O_7{}^{2-} \rightarrow 2Cr^{3+} + 7H_2O \qquad 2Cl^- \rightarrow Cl_2 + 2e\text{-}$$

**STEP 6**: *Equalize electron transfer*. Multiply each reaction by numbers that will allow both reactions to have the same number of electrons exchanged:

$$6e^- + 14H^+ + Cr_2O_7{}^{2-} \rightarrow 2Cr^{3+} + 7H_2O \qquad 6Cl^- \rightarrow 3\,Cl_2 + 6e^-$$

NOTE: 2nd reaction is multiplied by 3.

Add the two half reactions canceling out all the electrons and the formulas which appear on both sides of the equation.

$$\mathbf{14H^+ + Cr_2O_7{}^{2-} + 6Cl^- \rightarrow 2Cr^{3+} + 7H_2O + 3Cl_2}$$

Sum of Charges:   $+14 - 2 - 6 = +6 \rightarrow +6 + 0 + 0 = +6$

**STEP 7**: *Double check that* there is the same number of each kind of atom on both sides and that the sums of all charges are the same on both sides.

## BALANCE OXIDATION–REDUCTION REACTIONS IN BASIC SOLUTION USING THE HALF-REACTION METHOD

(Chemistry 6th ed. pages 177–179 / 7th ed. pages 167–168)

Repeat steps 1–5 above. After Step 5: Add $OH^-$ to both sides of the equation (equal to $H^+$). Form $H_2O$ on the side containing $H^+$ and $OH^-$ ions. Eliminate number of $H_2O$ appearing on both sides. Continue with steps 6 and 7.

**EXAMPLE**: $NO_2{}^- + Al \rightarrow NH_3 + AlO_2{}^-$

**STEP 1**: Write separate half-reactions.

$$Al \rightarrow AlO_2\text{-} \qquad\qquad NO_2{}^- \rightarrow NH_3$$

**STEP 2**: All atoms except H and O are already balanced.

STEPS 3 AND 4: Balance O with $H_2O$ and H with $H^+$.

$$2H_2O + Al \rightarrow AlO_2^- + 4H^+ \qquad 7H^+ + NO_2^- \rightarrow NH_3 + 2H_2O$$

STEP 5: Balance charge using electrons.

$$6\ e^- + 7H^+ + NO_2^- \rightarrow NH_3 + 2H_2O$$

$$\underline{(2H_2O + Al \rightarrow AlO_2^- + 4H^+ + 3e^-) \times 2}$$

$$7H^+ + NO_2^- + 4H_2O + 2Al \rightarrow 2AlO_2^- + 8H^+ + NH_3 + 2H_2O$$

CANCEL: $NO_2^- + 2H_2O + 2Al \rightarrow 2AlO_2^- + H^+ + NH_3$

New step for basic solution: Add $OH^-$ to both sides equal to the number of $H^+$ ions:

$$OH^- + NO_2^- + 2H_2O + 2Al \rightarrow 2AlO_2^- + H^+ + NH_3 + OH^-$$

CANCEL: $OH^- + NO_2^- + 2H_2O + 2Al \rightarrow 2AlO_2^- + NH_3 + H_2O$

THE ANSWER: $OH^- + NO_2^- + H_2O + 2Al \rightarrow 2AlO_2^- + NH_3$

## REDOX REACTIONS IN THE AP CHEMISTRY FREE-RESPONSE SECTION

| Oxidizing Agent | | Reducing Agent | |
|---|---|---|---|
| Appears in reactants | Appears in products | Appears in reactants | Appears in products |
| $MnO_4^-$, in acid | $Mn^{2+}$ | Halide ions | Halogens, diatomic |
| $MnO_4^-$ in neutral or basic | $MnO_2$ | Metal element | Metallic ion |
| $MnO_2$, in acid | $Mn^{2+}$ | Sulfite | Sulfate |
| $Cr_2O_7^{2-}$, in acid | $Cr^{3+}$ | Nitrite | Nitrate |
| $HNO_3$, concentrated | $NO_2$ | Halogens, dilute base | Hypohalite ion, $ClO^-$ |
| $HNO_3$, dilute | $NO$ | Halogens, conc. base | Halite ion. $ClO_2^-$ |
| $H_2SO_4$, hot , conc. | $SO_2$ | Metalous ion | Metalic ion |
| Metalic ions (ex. $Pb^{+4}$) | Metalous ions ($Pb^{+2}$) | $H_2O_2$ | $O_2$ |
| Halogens, diatomic | Halide ions | $C_2O_4^{2-}$ | $CO_2$ |
| $Na_2O_2$ | NaOH | | |
| $HClO_4$ | $Cl^-$ | | |
| $H_2O_2$ | $H_2O$ | | |

It is important to be very familiar with the list of common oxidizing agents and reducing agents above.

EXAMPLE: A solution of sodium dichromate is added to an acidified solution of sodium iodide.

$$Cr_2O_7{}^{2-} + H^+ + I^- \rightarrow Cr^{3+} + I_2 + H_2O$$

Oxidizing Agent        Reducing Agent

In this reaction, $Cr_2O_7{}^{2-}$ is the oxidizing agent which gets reduced to $Cr^{3+}$. The reducing agent, $I^-$, gets oxidized to $I_2$. $H_2O$ must be added to the products to balance the $H^+$ in the reactants.

EXAMPLE: Dilute potassium permanganate solution is added to an oxalic acid solution which was acidified with a few drops of sulfuric acid.

$$MnO_4{}^- + C_2O_4{}^{2-} + H^+ \rightarrow Mn^{2+} + CO_2 + H_2O$$

# PRECIPITATION REACTIONS

## PRECIPITATION REACTIONS INVOLVE THE FORMATION OF A SOLID WHEN TWO SOLUTIONS ARE MIXED

(*Chemistry* 6th ed. 158–168 / 7th ed. pages 149–158)

| Solubility Rules |
| --- |
| 1. Most nitrate salts are soluble. |
| 2. Most salts containing Group I ion and ammonium ion, $NH_4{}^+$, are soluble. |
| 3. Most chloride, bromide, and iodide salts are soluble, except $Ag^+$, $Pb^{2+}$ and $Hg_2{}^{2+}$. |
| 4. Most sulfate salts are soluble, except $BaSO_4$, $PbSO_4$, $Hg_2SO_4$, and $CaSO_4$. |
| 5. Most hydroxides except Group I and $Ba(OH)_2$, $Sr(OH)_2$, and $Ca(OH)_2$ are only slightly soluble. |
| 6. Most sulfides, carbonates, chromates, and phosphates are only slightly soluble. |

# PREDICTING PRECIPITATES

When two solutions of ionic compounds are mixed, exchange anions to form products, remembering that each product formed must be neutral. Use the solubility rules above.

EXAMPLE: Solutions of potassium chloride and lead (II) nitrate are mixed:

MOLECULAR EQUATION:

$$2KCl\ (aq) + Pb(NO_3)_2\ (aq) \rightarrow PbCl_2\ (s)\ +\ 2KNO_3\ (aq)$$

chlorides of Pb insoluble     all nitrates soluble

COMPLETE IONIC EQUATION:

$$2K^+ + 2Cl^- + Pb^{2+} + 2NO_3^- \rightarrow PbCl_2\ (s) + 2K^+ + 2NO_3^-$$

NET-IONIC EQUATION:  **$Pb^{2+} + 2Cl^- \rightarrow PbCl_2$ (s)**  This is the answer for the AP Exam, except that it does not need to be balanced.

# MULTIPLE-CHOICE QUESTIONS

No calculators are to be used in this section.

1.  Consider the equation: $Cl_2(g) + 2KI(aq) \rightarrow I_2(s) + 2KCl(aq)$. Which species is oxidized?
    (A) $Cl_2$
    (B) $K^+$
    (C) $I^-$
    (D) $I_2$
    (E) $Cl^-$

ANSWER:  **C**  We need only look at reactants when asked to identify the species oxidized or reduced. KI is an ionic compound and we can observe that the oxidation number of $K^+$ is the same in both KI and KCl. Whenever an element is found in a compound on one side of the equation and as a free element on the other side of the equation, there has to be a change in oxidation number for that species.

By definition, oxidation is the loss of electrons. Writing the two half-reactions [$Cl_2 + 2e^- \rightarrow 2Cl^-$ and $2I^- -2e^- \rightarrow I_2$] shows that it is the two iodide ions that lose two electrons to become diatomic iodine (*Chemistry* 6th ed. pages 164–171 / 7th ed. pages 154–162).

2.  A solid dissolves easily in benzene ($C_6H_6$) but it does not dissolve in water. The molecules of the solid can best be described as
    (A) polar molecules.
    (B) nonpolar molecules.
    (C) either polar or nonpolar.
    (D) small with much charge separation.
    (E) large with a high dipole moment.

ANSWER:  **B**  The phrase "like dissolves like" refers to the polarity of the solute and of the solvent. Benzene is a nonpolar solvent (from the formula it can be observed that only nonpolar C–H bonds are involved), so the solid is most likely nonpolar. Water is a polar solvent (*Chemistry* 6th ed. pages 134–136 / 7th ed. pages 127–129).

3. How many grams of NaOH are contained in 200. mL of a 0.400 $M$ solution?
   (A) 3.20 grams of NaOH
   (B) 8.00 grams of NaOH
   (C) 16.0 grams of NaOH
   (D) 40.0 grams of NaOH
   (E) 80.0 grams of NaOH

ANSWER: A There are 0.400 moles in each liter of this solution, and you are using 0.200 liter of the solution; there are 40.0 grams in each mol of NaOH, hence 0.400 mol/L NaOH × 0.200 L × 40.0 g NaOH / mol NaOH = 3.20 grams of NaOH (*Chemistry* 6th ed. pages 140–142 / 7th ed. pages 133–134).

4. Which of the following solutions contains the largest number of ions?
   (A) 500. mL of 0.10 $M$ $CaCl_2$
   (B) 500. mL of 0.10 $M$ $FeCl_3$
   (C) 700. mL of 0.20 $M$ NaOH
   (D) 400. mL of 0.100 $M$ $Al(NO_3)_3$
   (E) 600. mL of 0.200 $M$ $AlCl_3$

ANSWER: E There are three factors which must be considered in this problem, the number of ions per formula unit, the concentration of the solution, and the volume of the solution used. For example, in the first solution there are 0.15 moles of ions present: 0.500L × 0.100 mol of formula units of $CaCl_2$ / L × 3 ions/ $CaCl_2$ = 0.15 moles of ions. In response E: 0.600L × 0.200 mol of formula units of $AlCl_3$ × 4 ions/ $AlCl_3$ = 0.48 mol. of ions (*Chemistry* 6th ed. pages 140–143 / 7th ed. pages 133–135).

5. There are six strong acids. Which of the following is NOT a strong acid?
   (A) HCl
   (B) HF
   (C) HBr
   (D) HI
   (E) $HNO_3$

ANSWER: B Even though hydrofluoric acid will dissolve glass (!), it does not ionize significantly. It is important for you to know the six strong acids (then you know that all others are weak!). The six strong acids are $HClO_4$, HCl, HBr, HI, $H_2SO_4$, and $HNO_3$. These are the acids that are 100% ionized in water. You can assume that all other acids are weak (*Chemistry* 6th ed. pages 136–138, 657–658, Appendix 5.1 / 7th ed. pages 129–131, 626–627, Appendix 5.1).

6. Spectator ions are those which are in solution but do not react. Identify any spectator ions for the reaction of sodium phosphate with calcium bromide.
(A) only $Na^+(aq)$
(B) only $PO_4^{3-}(aq)$
(C) $Na^+(aq)$ and $PO_4^{3-}(aq)$
(D) $Na^+(aq)$ and $Br^-(aq)$
(E) $Ca^{2+}(aq)$ and $PO_4^{3-}(aq)$

ANSWER: D The total unbalanced expression for this reaction is: $Na^+ + PO_4^{3-} + Ca^{2+} + Br^- \rightarrow Ca_3(PO_4)_2 + Na^+ + Br^-$. In this reaction, the only insoluble compound is calcium phosphate, leaving the other two ions in solution unreacted (i.e., NaBr is soluble in water). You must know the solubility rules to write equations successfully for reactions (*Chemistry* 6th ed. pages 154–155, 158 / 7th ed. pages 145–146, 149).

7. A precipitate forms when which of the following solutions are added to a solution of NaCl?
I   $AgNO_3$
II  $Pb(NO_3)_2$
III $Cr(NO_3)_2$

(A) I only
(B) I and II
(C) I and III
(D) II and III
(E) I, II, and III

ANSWER: B The solubility rules indicate that all chlorides are soluble except for AgCl, $PbCl_2$ and $Hg_2Cl_2$. Again, you must know the solubility rules (*Chemistry* 6th ed. pages 151–152 / 7th ed. pages 142–143).

8. Consider the following three equations for chemical reactions:

$2Na(s) + Cl_2(g) \rightarrow 2NaCl(s)$
$2NaCl(aq) + Pb(NO_3)_2(aq) \rightarrow PbCl_2(s) + 2NaNO_3(aq)$
$NaOH(aq) + HBr(aq) \rightarrow H_2O(l) + NaBr(aq)$

These are examples of:
(A) three precipitation reactions.
(B) three acid–base reactions.
(C) a redox reaction, a precipitation reaction, then an acid–base reaction.
(D) three redox reactions.
(E) a neutralization reaction, then two precipitation reactions.

ANSWER: C Being able to type reactions will help in predicting products. In the first reaction, Na is oxidized and $Cl_2$ is reduced (a metal–nonmetal reaction can always be assumed to be a redox reaction); in the second reaction, insoluble lead(II) chloride forms; in the third reaction, a base and an acid form water and a salt (*Chemistry* 6th ed. pages 156–160, 168–171 / 7th ed. pages 147–151, 158–162).

9. Which of the following pairs of ions would <u>not</u> form a solid in aqueous solution?
   (A) $Ba^{2+}$ and $SO_4^{2-}$
   (B) $Pb^{2+}$ and $Br^-$
   (C) $Mg^{2+}$ and $SO_4^{2-}$
   (D) $Pb^{2+}$ and $S^{2-}$
   (E) $Ca^{2+}$ and $CO_3^{2-}$

ANSWER: **C** Most sulfates are soluble. Exceptions are barium, lead(II), mercury(I), and calcium sulfates (*Chemistry* 6th ed. pages 151–152 / 7th ed. pages 142–143).

10. Which of the following ions are likely to form a soluble sulfate in aqueous solution?
   (A) $Ba^{2+}$
   (B) $Pb^{2+}$
   (C) $Ca^{2+}$
   (D) $NH_4^{1+}$
   (E) $Hg_2^{2+}$

ANSWER: **D** Salts containing the ammonium ion are soluble (*Chemistry* 6th ed. pages 151–152 / 7th ed. pages 142–143).

# FREE-RESPONSE QUESTIONS

Every AP Chemistry exam has a question that requires prediction of the product(s). The directions indicate that in all cases a reaction will occur and that you are to write the net-ionic equation for each of the reactions indicated in words. Do not show any spectator ions. Do not balance. Showing physical state is not required. On the actual exam, you are asked to choose 5 out of 8 for scoring. The scoring is 1 point for the correct formula of the reactants and 2 points for the products; points are subtracted for improper formulas and spectator ions.

1. Write net-ionic equations for each of the following reactions indicated in words. Remember not to show spectator ions. Do not balance. Showing physical state is not required.
   (a) Solutions of barium chloride and silver nitrate are mixed.
   (b) Hydrochloric acid is added to a solution of mercury(I) nitrate.
   (c) Hydrogen sulfide gas is bubbled through potassium hydroxide solution.
   (d) Solutions of sodium iodide and lead(II) nitrate are mixed.
   (e) Excess ammonia is added to a solution of copper(II) chloride.

ANSWER: 1(a)

$$Cl^- + Ag^+ \rightarrow AgCl$$

Know solubility rules, watch for formation of solids.

ANSWER: 1(b)

$$Cl^- + Hg_2^{2+} \rightarrow Hg_2Cl_2$$

Do not assume that an acid in a reaction always means an acid–base reaction. Here the hydrochloric acid, a strong acid, is the source of the chloride ion. Mercury(I) chloride is insoluble (watch how you write the formula for that solid!).

ANSWER: 1(c)

$$H_2S + OH^- \rightarrow S^{2-} + H_2O$$

This is an acid–base reaction. Since hydrogen sulfide is a weak acid, it must be written as a molecular species.

ANSWER: 1(d)

$$Pb^{2+} + I^- \rightarrow PbI_2$$

Lead(II) iodide is insoluble. The spectator ions, $Na^+$ and $NO_3^-$, are not shown.

ANSWER: 1(e)

$$NH_3 + Cu^{2+} \rightarrow Cu(NH_3)_4^{2+}$$

This is a complex ion formation reaction. Know that ammonia complexes with Cd, Cu, and Zn ions in a 4 to 1 ratio, and with $Ag^+$ in a 2 to 1 ratio. Also watch for the formation of $Al(OH)^{4-}$ (*Chemistry* 6th ed. pages 148–155, 998–999 / 7th ed. pages 140–146, 955–956).

2. Consider the reaction $3O_2(g) + 2PbS(s) \rightarrow 2PbO(s) + 2SO_2(g)$. Show the two half-reactions and identify the reducing agent for this reaction.

ANSWER: 2

$$O_2 \rightarrow 2O^{2-}$$

Since the oxidation state of each oxygen decreases from zero to 2-, the oxygen is reduced.

$$S^{2-} \rightarrow S^{4+}$$

Sulfur is oxidized (electrons are lost as the substance becomes more positive). The reducing agent for this reaction is therefore sulfur in lead(II) sulfide (*Chemistry* 6th ed. pages 164–171 / 7th ed. pages 154–162).

# 5

# GASES

In this section, the laws describing gas behavior are explored. Mathematical laws relating the properties of pressure, volume, temperature, and moles of gas are considered. Densities and molar masses of ideal gases will be determined from the ideal gas law. Stoichiometric calculations will be performed for reactions involving gases. Pressures of gases in a mixture will be examined. The behavior of ideal gases will be explained by the kinetic molecular theory. Real gases will be compared to ideal gases.

Questions about gas behavior and gas laws can be either qualitative or quantitative on the multiple choice section.

The free-response portion of the exam can include questions about an experiment involving gases such as the determination of the molar mass or molar volume of a gas. Essay questions about ideal or real gas behavior or the kinetic molecular theory may also appear.

You should be able to

- Perform calculations with gas laws: Boyle's, Charles', Avogadro's, Combined, and Ideal.
- Perform calculations with the ideal gas law to find the density or molar mass of the gas.
- Interpret or draw graphical relationships between gas variables.
- Perform stoichiometric calculations for reactions which produce gases.
- Perform calculations with molar volume.
- Perform calculations with Dalton's Law for a mixture of gases.
- Perform calculations for gases collected over water.
- Perform calculations with rates of effusion to find the molar mass.
- Perform calculations with root mean square velocity.
- Compare real gases to ideal gases.

---

## AP tip

Mathematical equations will be provided in the free-response section of the exam, but not for the multiple-choice part. It is best to memorize the equations or know how to derive the equations from one equation.

---

## GAS LAWS

(*Chemistry* 6th ed. pages 192–202 / 7th ed. pages 181–190)

The table below summarizes the gas laws that can be derived from the ideal gas law. A typical problem involves solving for the missing variable when given a set of initial and final conditions for a gas sample. Perform all gas calculations involving temperature using the Kelvin scale, not degrees Celsius. Temperature (K) = °C + 273.15.

| Gas Law | Equations | Definition |
|---|---|---|
| Boyle's law | $P = (nRT)\,1/V$ <br><br> $P = \text{(constant)}\,1/V$ <br><br> $\boxed{P_1V_1 = P_2V_2}$ | The product of the pressure and the volume is a constant, $k$, for a trapped sample of gas at constant temperature |
| Charles' law | $V = (nR/P)\,T$ <br><br> $V = \text{(constant )}T$ <br><br> $\boxed{\dfrac{V_1}{T_1} = \dfrac{V_2}{T_2}}$ | The volume of a gas at constant pressure increases linearly with the temperature of the gas. |
| Avogadro's law | $V = (RT/P)\,n$ <br><br> $V = \text{(constant)}\,n$ <br><br> $\boxed{\dfrac{V_1}{n_1} = \dfrac{V_2}{n_2}}$ | Equal volumes of gases at the same temperature and pressure contain the same number of moles of gas. |
| Ideal Gas law | $\boxed{PV = nRT}$ <br><br> $R = $ universal gas constant <br><br> $= \ 0.08206 \ \text{L} \times \text{atm}/(\text{K} \times \text{mol})$ | An equation of state for a gas, where the state of a gas is its condition at a given time. |
| Dalton's law | $\boxed{P_{total} = P_1 + P_2 + P_3 + \dots.}$ <br><br> $\boxed{P_{total} = n_{tot}RT/V}$ | For a mixture of gases, the sum of the pressures of the individual gases is equal to the total pressure. The total pressure depends on the total moles of gas present. |

## GRAPHICAL RELATIONSHIPS FOR GAS LAWS

(*Chemistry* 6th ed. pages. 192–195 / 7th ed. pages 181–184)

BOYLE'S LAW Pressure is inversely related to volume. A graph of $P$ vs. $1/V$ is a straight line.

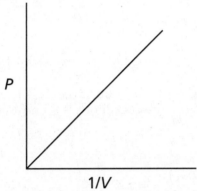

CHARLES' LAW Volume is directly proportional to temperature. A graph of $T$ vs. $V$ is linear. Extrapolation of this graph for all gases to zero volume results in the same temperature, –273.2°C (0 K), which is absolute zero.

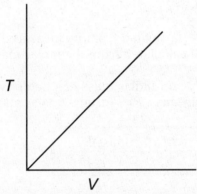

EXAMPLE: A sample tube containing 103.6 mL of CO gas at 20.6 torr is connected to an evacuated 1.13 liter flask. What will the pressure be when the CO is allowed into the flask?

$$P_1V_1 = P_2V_2; \ P_2 = \frac{P_1V_1}{V_2} = \frac{(20.6 \text{ torr})(0.1036 \text{ L})}{1.47 \text{ L}} = 1.45 \text{ torr}$$

Note: Milliliters need to be converted to liters so that the units of volume cancel out.

EXAMPLE: A quantity of gas at 27.0°C is heated in a closed vessel until the pressure is doubled. To what temperature is the gas heated?

Pressure is directly proportional to temperature. If the pressure is doubled, the temperature will also double. 27.0°C + 273.2 = 300.2 K. The answer is 600.4 K (or 327.2°C).

## GAS STOICHIOMETRY

(*Chemistry* 6th ed. pages 203–206 / 7th ed. pages 190–194)

If the pressure, volume, and temperature of ideal gases are known, stochiometric calculations can be performed.

## MOLAR VOLUME

(*Chemistry* 6th ed. page 203 / 7th ed. pages 190–191)

One mole of an ideal gas at 0°C and 1 atm occupies 22.4 L. Standard temperature and pressure (STP) are the conditions 0°C and 1 atm.

EXAMPLE: What mass of helium is required to fill a 1.5-L balloon at STP?

$$1.5 \text{ L} \times \frac{1 \text{ mol He}}{22.4 \text{ L}} \times \frac{4.0 \text{ g He}}{1 \text{ mol}} = 0.27 \text{ g He}$$

## MOLAR MASS OF A GAS

(*Chemistry* 6th ed. pages 205–206 / 7th ed. pages 193–194)

The molar mass of a gas can be calculated from its measured density. The density of a gas can also be calculated from the molar mass of the gas.

The equation relating density and molar mass can be derived from $PV = nRT$. Since $n$ = grams/molar mass, $PV$= (grams /molar mass) $RT$.

$$\text{Molar Mass} = \frac{dRT}{P} = \frac{(\text{mass})RT}{V \times P}$$

EXAMPLE: A sample of gas weighing 0.800 g occupies a 256 mL flask at 100°C and 750.0 torr. Determine the molar mass of the gas.

$$\text{Molar mass} = \frac{0.800 \text{ g} \times \frac{0.08206 \text{ L atm}}{\text{mol K}} \times 373 \text{ K}}{0.256 \text{ L} \times 0.986 \text{ atm}} = 97.0 \text{ g/mol}$$

## DALTON'S LAW

(*Chemistry* 6th ed. pages 206–211 / 7th ed. pages 194–199)

For a mixture of gases in a container, the total pressure is the sum of the pressures that each gas would exert if it were alone.

$$P_{\text{total}} = P_1 + P_2 + P_3 + \ldots$$

The partial pressure of a gas, $P_{gas}$, is the pressure that a particular gas would exert if it were alone in the container. The subscripts refer to the individual gases (gas$_1$, gas$_2$, and so on). The partial pressure of each gas can be calculated from the ideal gas law:

$$P_1 = n_1 RT/V$$

The total pressure of the mixture, $P$, can be represented as

$$P_{total} = n_{tot} RT/V$$

The mole fraction is the ratio of the number of moles of a given component in a mixture to the total number of moles in the mixture.

$$X_1 = n_1/n_{tot}$$

EXAMPLE: A mixture of 1.00 g of $H_2$ and 1.00 g of He is placed in a 1.00 L container at 27°C. Calculate the mole fraction and partial pressure of each gas. Calculate the total pressure in the container.

$$n_{H_2} = 1.00 \text{ g } H_2 \times \frac{1 \text{ mol } H_2}{2.02 \text{ g } H_2} = 0.496 \text{ mol } H_2$$

$$n_{He} = 1.00 \text{ g He} \times \frac{1 \text{ mol He}}{4.00 \text{ g}} = 0.250 \text{ mol He}$$

$$X_{H2} = \frac{0.496 \text{ mol}}{0.496 \text{ mol} + 0.250 \text{ mol}} = 0.665; \quad X_{He} = \frac{0.250 \text{ mol}}{0.496 \text{ mol} + 0.250 \text{ mol}} = 0.335$$

$$P_{H_2} = \frac{n_{H_2} RT}{V} = \frac{0.496 \text{ mol} \times \dfrac{0.08206 \text{ L} \times \text{atm}}{\text{mol} \times \text{K}} \times 300 \text{ K}}{1.00 \text{ l}} = 12.2 \text{ atm}$$

$$P_{He} = \frac{n_{He} RT}{V} = \frac{0.250 \text{ mol} \times \dfrac{0.08206 \text{ L} \times \text{atm}}{\text{mol} \times \text{K}} \times 300 \text{ K}}{1.0 \text{ L}} = 6.15 \text{ atm}$$

$$P_{total} = P_{He} + P_{H_2} = 12.2 \text{ atm} + 6.15 \text{ atm} = 18.4 \text{ atm}$$

## GAS COLLECTION OVER WATER

(*Chemistry* 6th ed. pages 210–211 / 7th ed. pages 197–199)

A gas can be collected by displacement of water. A mixture of gases results due to a mixture of water vapor and the gas being collected.

EXAMPLE: A sample weighing 0.986 g contains zinc and some impurities. Excess hydrochloric acid is added and reacts with the zinc but not the impurities. Determine the percentage of zinc in the sample if 240.0 mL hydrogen gas are collected over water at 30.0°C and 1.032 atm.

Begin with the balanced equation: $Zn + 2\,HCl \rightarrow ZnCl_2 + H_2$.

Using Dalton's law and the vapor pressure of water at the specified temperature:

$$P_{total} = P_{atm} = P_{H_2} + P_{H_2O} \; ; \; P_{H_2} = P_{atm} - P_{H_2O}$$

$$P_{H_2} = 1.032 \text{ atm} - 0.042 \text{ atm} = 0.990 \text{ atm}$$

$$n_{H_2} = \frac{P_{H_2} \times V}{RT} = \frac{0.990 \text{ atm} \times 0.2400 \text{ L}}{\dfrac{0.08206 \text{ L atm} \times 303.2 \text{ K}}{\text{mol K}}} = 0.00955 \text{ mol H}_2$$

$$0.00955 \text{ mol H}_2 \times \frac{1 \text{ mol Zn}}{1 \text{ mol H}_2} \times \frac{65.4 \text{ g Zn}}{1 \text{ mol Zn}} = 0.625 \text{ g Zn}$$

$$\%Zn = \frac{g\,Zn \times 100\%}{g\,sample} = \frac{0.625 \text{ g}}{0.986 \text{ g}} \times 100\% = 63.3\% Zn$$

# KINETIC MOLECULAR THEORY

(*Chemistry* 6th ed. pages 212–219 / 7th ed. pages 199–206)

The kinetic molecular theory is a model that attempts to explain the properties of an ideal gas. The postulates of the kinetic molecular theory are
1. The volume of the individual particles of a gas can be assumed to be negligible.
2. The gas particles are in constant motion. The pressure exerted by the gas is due to the collisions of the gases with the walls of the container.
3. The gases are not attracted to one another.
4. The average kinetic energy of a gas is directly proportional to the Kelvin temperature.
$$(KE)_{average} = 3/2\,RT = \tfrac{1}{2}\,mv_{avg}^2$$

EXAMPLE: Three identical flasks are filled with three different gases.

Flask A: CO at 760 torr and 0°C

Flask B: $N_2$ at 250 torr and 0°C

Flask C: $H_2$ at 100 torr and 0°C

In which flask will the molecules have the greatest average kinetic energy? In which flask will the molecules have the greatest average velocity?

All molecules will have the same average kinetic energy since they are all at the same temperature. Flask C will have the greatest average velocity since hydrogen has the lowest molar mass. At constant $T$, the lightest molecules are fastest, on average.

## ROOT-MEAN-SQUARE VELOCITY

(*Chemistry* 6th ed. pages 217–218 / 7th ed. pages 204–205)

The root-mean-square velocity is the square root of the squares of the individual velocities of the gas particles.

$$u_{rms} = (3RT/M)^{1/2}$$

8.3145 J/K × mol

$R$ = Ideal gas constant

$I$ = Kelvin tempature

$M$ = Molar mass of the gas

**EXAMPLE:** Which one of the following molecules has the largest root-mean-square velocity? $H_2$, $N_2$, $CH_4$, $C_2H_6$

Hydrogen, the substance with the lowest molar mass, has the largest root-mean-square velocity.

## EFFUSION AND DIFFUSION

(*Chemistry* 6th ed. pages 219–222 / 7th ed. pages 206–208)

Diffusion is the mixing of gases. Effusion is the passage of a gas through a small opening into an evacuated chamber.

$$\text{For gases at the same temperature,}$$
$$\frac{\text{Rate of effusion for gas}_1}{\text{Rate of effusion for gas}_2} = \frac{(M_2)^{1/2}}{(M_1)^{1/2}}$$

**EXAMPLE:** The rate of effusion of a particular gas was measured and found to be 24.0 mL/min. Under the same conditions, the rate of pure methane ($CH_4$) gas is 47.8 mL/min. What is the molar mass of the unknown gas?

$$\frac{24.0}{47.8} = \frac{(16.04)^{1/2}}{(M1)^{1/2}} = 0.502$$

$16.04 = (0.502)^2 \times M_1$; $M_1 = 16.04 / 0.252 = 63.7$ g/mol

## Real Gases

<small>(*Chemistry* 6th ed. pages 222–224 / 7th ed. pages 208–210)</small>

The ideal gas model fails at high pressure and low temperature. Real gas molecules take up space and experience attractive forces between molecules. At high pressure there is less empty space between molecules, and the volume of molecules becomes more significant. An ideal gas could be compressed to zero volume, but for a real gas, as the pressure doubles, the volume of empty space cannot continue to be halved. As the temperature decreases, the molecules have less kinetic energy to overcome the attractive forces between gas molecules; these attractive forces may cause the gas to condense.

The van der Waals equation modifies the assumptions of the kinetic molecular theory to fit the behavior of real gases.

$$[P_{obs} + a(n/V)^2] \times (V - nb) = nRT$$

EXAMPLE: Which of the following gases would you expect to have the largest value of the van der Waals constant, *b*: $H_2$, $N_2$, $CH_4$, $C_2H_6$, or $C_3H_8$?

The van der Waals constant, *b*, is a measure of the size of the molecule. $C_3H_8$ has the largest molar volume and should have the largest value of *b*.

EXAMPLE: Which of the following gases would you expect to have the largest value of the van der Waals constant, *a*: $H_2$, $CO_2$, $CH_4$, or $N_2$?

$CO_2$ has the largest value for a, which measures intermolecular attractions. All the molecules are nonpolar so the only force present is an induced dipole or London force which increases as the number of electrons and protons in the molecule increases. (This will be discussed in Chapter 8.)

## Multiple-Choice Questions

No calculators are to be used in this section.

1. Ideal gases vary from real gases at conditions of
   (A) high temperature and low pressure.
   (B) high temperature and high pressure.
   (C) low temperature and low pressure.
   (D) low temperature and high pressure.
   (E) both high density and low pressure.

ANSWER: **D** At low temperature and high pressure, the molecules are closer together and therefore the forces between molecules become more important as they are stronger. The actual (finite) volume of individual molecules also becomes more important as more of the total space is actually occupied by finite molecular volume (*Chemistry* 6th ed. pages 222–224 / 7th ed. pages 208–210).

2. What is the volume of 3.00 mol of gas @ STP?
   (A) 22.4 L
   (B) 3 × 22.4 L
   (C) 3 × 22.4 L × 760
   (D) 3 × 22.4 LL × 273 / 760
   (E) It cannot be determined without knowing which gas is involved.

**ANSWER: B** The molar volume of all gases at STP is about 22.4 L, so three moles would occupy 22.4 L /mol × 3 moles. (*Chemistry* 6th ed. pages 222–224 / 7th ed. pages 208–210).

3. An ideal gas of volume 189. mL is collected over water at 30°C and 777 torr. The vapor pressure of water is 32 torr @ 30°C. What pressure is exerted by the dry gas under these conditions?
   (A) 320 torr
   (B) 745 torr
   (C) 777 torr
   (D) 32 / 77 torr
   (E) 32 x 777 torr

**ANSWER: B** The total pressure = 777 torr. Of this, 32 torr is due to the water vapor, hence 777–32 = 745 torr of pressure are allocated to the dry gas (*Chemistry* 6th ed. pages 206–211 and 483–492 / 7th ed. pages 194–199 and 459–466).

4. A 14.0-L cylinder contains 5.60 g $N_2$, 79.9 g Ar and 6.40 g $O_2$. What is the total pressure in atm at 27°C? ($R$ = the ideal gas constant.)
   (A) 20 $R$
   (B) 26 $R$
   (C) 30 $R$
   (D) 60 $R$
   (E) 120 $R$

**ANSWER: C** Use $P_{Total} = N_{Total}RT / V$ to determine the pressure. From the mass of each of the gases you can find 0.200 mol of $N_2$, 1.00 mol of Ar, and 0.200 mol of $O_2$ to give a total number of moles ($n$) of 1.4 mol. Therefore $P_{Total} = R \times 1.40$ mol × 300 K / 14.0 L = 30 $R$. Note: You do not need a calculator to divide 1.4 by 14 and then multiply by 300! (*Chemistry* 6th ed. pages 198–202 / 7th ed. pages 186–190).

5. In a closed inflexible system, 7.0 mol $CO_2$, 7.0 mol Ar, 7.0 mol $N_2$, and 4.0 mol Ne are trapped, with a total pressure of 10.0 atm. What is the partial pressure exerted by the neon gas?
   (A) 1.6 atm
   (B) 4.0 atm
   (C) 10.0 atm
   (D) 21.0 atm
   (E) 29.0 atm

**ANSWER: A** The total number of moles is 25 (7.0 + 7.0 + 7.0 + 4.0 = 25), hence the Ne is 4 /25 of the total amount of gas and exerts 4 /25 of the total pressure, or 4/25 x 10 atm = 1.6 atm (*Chemistry* 6th ed. pages 206–211 / 7th ed. pages 194–199).

6.  Consider the reaction: $C_2H_6(g) + 7/2\ O_2(g) \rightarrow 2CO_2(g) + 3H_2O(g)$. If 6.0 g ethane, $C_2H_6(g)$ burn (as shown above), what volume of $CO_2(g)$ will be formed at STP?
    (A) 0.20 L
    (B) 0.40 L
    (C) 2.2 L
    (D) 9.0 L
    (E) 22.4 L

ANSWER: **D**  6.0  g of ethane / 30. g/mol yields 0.20 mol of ethane, which forms twice that number of moles of carbon dioxide. Since one mol of gas occupies 22.4 L at STP, 22.4 L/mol × 0.20 mol × 2 $CO_2/C_2H_6$ = 8.96 L of carbon dioxide (rounded to two significant figures = 9.0 L) (*Chemistry* 6th ed. pages 203–206 / 7th ed. pages 190–194).

7.  $Cl_2$ and $F_2$ combine to form a gaseous product; one volume of $Cl_2$ reacts with three volumes of $F_2$ yielding two volumes of product. Assuming constant conditions of temperature and pressure, what is the formula of the product?
    (A) ClF
    (B) $Cl_2F_2$
    (C) $ClF_2$
    (D) $Cl_2F$
    (E) $ClF_3$

ANSWER: **E**  This problem assumes you understand Avogadro's Law, "At constant temperature and pressure, the volume of a gas is directly proportional to the number of moles of the gas". To apply that to this problem, 1 volume $Cl_2$ + 3 volumes $F_2$ → 2 volumes $Cl_xF_{3x}$. The simplest formula for the product becomes $ClF_3$ (*Chemistry* 6th ed. pages 192–198 / 7th ed. pages 181–186).

8.  Decreasing the temperature of an ideal gas from 80°C to 40°C causes the average kinetic energy to
    (A) decrease by a factor of two.
    (B) decrease by a  factor of four.
    (C) increase by a  factor of two.
    (D) increase by a  factor of four.
    (E) decrease by less than a factor of two.

ANSWER: **E**  Be careful here to note the difference between the kinds of temperature scales and what they mean. Even though the Celsius temperature is half as much, the average kinetic energy is proportional to the Kelvin temperature. In this case that ratio is only 353 / 313 = 1.13 so the average kinetic energy decreases only by a factor of 1.13 (*Chemistry* 6th ed. pages 211–219 / 7th ed. pages 198–206).

9. The effusion rate of helium gas, He, compared with that of methane gas, $CH_4$, is
   (A) four times greater for helium.
   (B) four times less for helium.
   (C) twice as great for helium.
   (D) twice as great for methane.
   (E) sixteen times as great for helium.

ANSWER: C Effusion rates are inversely proportional to the square root of the molecular masses of the gases (*Chemistry* 6th ed. pages 219–222 / 7th ed. pages 206–208).

10. The average speed of the molecules of a gas is proportional to the
    (A) volume of the container.
    (B) reciprocal of absolute temperature, $(1 / T)$.
    (C) absolute temperature.
    (D) square root of the absolute temperature.
    (E) square of the absolute temperature.

ANSWER: D In this case the question is about the average speed (not the energy) of the molecules. Review root-mean-square velocity (*Chemistry* 6th ed. pages 217–219 / 7th ed. pages 204–206).

## FREE-RESPONSE QUESTIONS

Calculators may be used for this section.

1. (a) Compare the temperature of freshly made coffee made at lower altitudes to coffee made at higher altitudes. Support your ANSWER with discussion.
   (b) The van der Waals equation, $(P + a/V_2)(V - b) = nrt$, is a necessary form of $PV = nRT$ under certain conditions.
      i. What are these conditions? Explain your ANSWER.
      ii. What are the physical properties which the constants "*a*" and "*b*" are designed to describe? Explain your ANSWER.
   (c) Explain on a molecular level why gases exert pressure.
   (d) How does gaseous pressure relate to changes in volume? Explain.

ANSWER: 1(a) At low altitudes, the amount of air above the surface of the earth, and therefore the total atmospheric pressure, would be greater. Water boils at a higher temperature under such conditions since a higher vapor pressure is required for the liquid to become a gas (*Chemistry* 6th ed. pages 495–497 / 7th ed. pages 469–470).

ANSWER: 1(b) In this question, be careful not to confuse "conditions" with "physical properties" or "characteristics of the gaseous molecules." The conditions under which real gases do not act like ideal gases and therefore do not obey the simpler $PV = nRT$ are those of high pressure and low temperature. This is because when the gas molecules are closer together, the forces of attraction between the molecules (accounted for by the constant "*a*") become more important. Also, the finite volume of individual molecules (accounted for by constant "*b*") becomes a larger part of the total volume occupied by the gas (*Chemistry* 6th ed. pages 198–202 / 7th ed. pages 186–190).

**ANSWER: 1(c)** Gases exert pressure because the molecules are moving and strike the sides of the container exerting force on it; this force per unit of area is what we call pressure (*Chemistry* 6th ed. pages 190–191 / 7th ed. pages 179–180).

**ANSWER: 1(d)** Gaseous pressure is inversely proportional to the volume of the container. If only the volume of the container is less, for example, the molecules have less room to move around before striking the sides and therefore strike the sides of the container more often. This assumes that the temperature is held constant (*Chemistry* 6th ed. pages 192–195 / 7th ed. pages 181–184).

Note: What assumption is made in your ANSWER to D?

2. Assume that two cylinders at 27°C are connected by a closed stopcock (valve) system. The right-hand cylinder contains 2.4 L of hydrogen at 0.600 atm; the left cylinder is larger and contains 6.8 L of helium at 1.40 atm.
   (a) How many moles of each gas are present?
   (b) What is the total pressure when the valve is open?
   (c) Determine the partial pressure of these two gases at 27°C when the stopcock is opened.

**ANSWER: 2** Solving this problem has three steps. (1) To determine the pressure exerted by each gas, first calculate the number of moles of each ($n = PV / RT$), (2) and then use the total number of moles of gas and $p = nRT/V$ to calculate the total pressure when the total volume is 9.20 L. (3) Finally, the partial pressure of each gas is the total pressure × the mol fraction of that gas (*Chemistry* 6th ed. pages 198–202 and 206–207 / 7th ed. pages 186–190 and 194–195).

For hydrogen: (0.600 atm × 2.40 L) / ( R × 300 K) = 0.0585 mol of $H_2$

For helium: (1.40 atm × 6.80 L) / (R × 300 K) = 0.397 mol of He

Total number of moles = 0.0585 + 0.397 mol = 0.456 mol

Total pressure is then (0.456 mol × R × 300 K) / (9.20 L) = 1.22 atm

Use the mol fractions: (0.0585 / 0.456) × 1.22 atm = 0.157 atm $H_2$

(0.397 / 0.456) × 1.22 atm = 1.06 atm He

# 6

# THERMOCHEMISTRY AND THERMODYNAMICS

Thermochemistry describes the heat flow of a chemical reaction or physical change. Changes in enthalpy are calculated through calorimetry, Hess's Law, and standard heats of formation. The study of thermodynamics allows one to predict whether a process will occur or not. The dependence of Gibbs free energy on temperature, entropy, and enthalpy will be identified. The laws of thermodynamics will be explored.

In the free-response portion of the exam, one of the questions in Part B usually involves entropy, enthalpy, and free energy. A lab question regarding the use of a calorimeter to determine the heat of a reaction may be asked in the essay portion of the test.

You should be able to
  - Perform stochiometric calculations with the enthalpy of the reaction.
  - Perform calculations with specific heat.
  - Discuss how a calorimeter is used and perform related calculations.
  - Draw, label, and perform associated calculations for heating curves involving specific heat and changes in enthalpy for phase changes.
  - Perform calculations with Hess' Law.
  - Perform calculations with standard heats of formation, $\Delta H^\circ_f$.
  - Write reactions representing standard heats of formation, $\Delta H^\circ_f$.
  - Compare the absolute entropies, $S^\circ$, of elements and compounds.
  - Perform calculations with entropy $S^\circ$ and $\Delta S^\circ$.
  - Perform calculations with free energy, $\Delta G^\circ_f$.

## FIRST LAW OF THERMODYNAMICS

(*Chemistry* 6th ed. page 244 / 7th ed. pages 231–232)

The first law of thermodynamics states that energy can be converted from one form to another, but can neither be created nor destroyed: The energy of the universe is constant.

## ENTHALPY

(*Chemistry* 6th ed. pages 248–250 / 7th ed. pages 235–236)

At constant pressure, the change in enthalpy ($\Delta H^\circ$) of a system is equal to the energy flow as heat. Enthalpy is a state function, so the change in $H$ is independent of the pathway.

### EXOTHERMIC

A reaction that is exothermic gives off heat and the change in enthalpy is negative ($\Delta H^\circ < 0$). The temperature of the surroundings increases during an exothermic process.

### ENDOTHERMIC

A reaction that is endothermic absorbs heat and the change in enthalpy is positive ($\Delta H^\circ > 0$). The temperature of the surroundings decreases in an endothermic process.

EXAMPLE: How much heat is released when 4.03 g of hydrogen is reacted with excess oxygen?

$$2H_2(g) + O_2(g) \rightarrow 2H_2O(l) \qquad\qquad \Delta H^\circ = -572 \text{ kJ}$$

$$4.03 \text{ g H}_2 \times \frac{1 \text{ mol H}_2}{2.02 \text{ g}} \times \frac{-572 \text{ kJ}}{2 \text{ mol H}_2} = -572 \text{ kJ}$$

Note that conversion factors can be written using the relationship between the enthalpy, $\Delta H^\circ$, and the moles of reactants and products. 572 kJ of heat is produced when 1 mole of $O_2$ reacts with 2 moles of $H_2$ to produce 2 mol of $H_2O$.

## CALORIMETER

(*Chemistry* 6th ed. pages 250–256 / 7th ed. pages 237–242)

A calorimeter is a device used to experimentally determine the heat energy change of a chemical reaction. Calorimetry can be carried out under constant pressure (measures $\Delta H$) or constant volume (measures $\Delta E$).

## SPECIFIC HEAT

(*Chemistry* 6th ed. pages 250–251 / 7th ed. pages 237–238)

The amount of heat lost or gained in a reaction depends on
1.   The change in temperature during the reaction.
2.   The amount of substance present.
3.   The heat capacity (*C*) of a substance which is the amount of energy required to change the temperature of an object by 1°C.
     $C$ = heat absorbed / increase in temperature = $J$ / °C
     a.   Specific heat capacity, *s*, is the amount of energy required to change the temperature of 1.0 g of a substance by 1°C (J/ g °C).
     b.   Molar heat capacity is the amount of energy required to change the temperature of 1 mole of a substance by 1°C.

EXAMPLE: The same amount of heat energy is applied to 1.0 g of each of solid aluminum, solid iron, mercury liquid, and carbon graphite. Which one of these substances will reach the highest temperature?

The answer is Hg(l). Mercury has the lowest specific heat of the four mentioned substances (see table in the textbook), which means it takes a lesser amount of heat to raise 1 gram of mercury 1°C.

The heat released by a reaction can be determined by

$$q = m \times s \times \Delta T$$

EXAMPLE: A 110.g sample of copper (specific heat capacity 0.20 J / (g° C) is heated to 82.4°C and then placed in a container of water at 22.3° C. The final temperature of the water and the copper is 24.9° C. What was the mass of the water in the original container, assuming that all the heat lost by the copper is gained by the water?

The heat lost by copper is equal to:

$$110. \text{ g Cu} \times \frac{0.20 \text{ J}}{\text{g°C}} \times (82.4°C - 24.9°C) = 1265 \text{ J} = 1.3\text{kJ}$$

The heat lost by the copper is equal to the heat gained by the water.

−(heat lost by copper) = (heat gained by water)

$$\text{Mass of water} = \frac{q}{s \times \Delta T} = \frac{1265 \text{ J}}{\dfrac{4.184 \text{ J} \times 2.6°\text{C}}{\text{g}°\text{C}}} = 120 \text{ g water}$$

## AP tips

Students sometimes have trouble with the sign of $\Delta T$. Change in temperature, $\Delta T$, is usually calculated $T_{\text{final}} - T_{\text{initial}}$. When the heat lost by one substance equals the heat gained by the other substance, $\Delta T$ for the substance losing heat needs to be calculated $T_{\text{initial}} - T_{\text{final}}$ in order for the heat gained to equal the heat lost. (Otherwise you will end up with $-q = q$ .)

## HEATING CURVES

(*Chemistry* 6th ed. pages 488–489 / 7th ed. pages 463–464)

When a system is heated, energy is transferred into it. In response to the energy it receives, the system changes, for example, by increasing its temperature. A plot of the temperature versus time is called **the heating curve**. One such heating curve is shown above. The physical states of the substance and the phase transitions are identified along the curve.

The temperature of the system usually increases when energy is applied. However, when the energy supplied is used for phase transition, a change in the physical state, the temperature (average kinetic energy) remains constant because the potential energy of the system is being increased as the molecules are rearranged in the phase change.

## HEAT OF FUSION

(*Chemistry* 6th ed. page 489 / 7th ed. page 464)

The heat of fusion, $\Delta H_{fus}$ is the enthalpy change that occurs in melting a solid at its melting point.

> EXAMPLE: What quantity of heat is required to melt 1.00 kg of ice at its melting point? For ice, $\Delta H_{fus} = 6.0$ kJ/mol
>
> $$1.00 \text{ kg} \times \frac{1000 \text{ g}}{1 \text{ kg}} \times \frac{1 \text{ mol}}{18.0 \text{ g}} \times \frac{6.0 \text{ kJ}}{1 \text{ mol}} = 333 \text{ kJ}$$

## HEAT OF VAPORIZATION

(*Chemistry* 6th ed. page 483 / 7th ed. page 458)

The heat of vaporization, $\Delta H_{vap}$ , is the energy required to vaporize one mole of a liquid at a pressure of one atmosphere.

> EXAMPLE: What quantity of heat is required to vaporize 130 g of water? For liquid, water, $\Delta H_{vap} = 43.9$ kJ/mol.
>
> $$130 \text{ g} \times \frac{1 \text{ mol}}{18.0 \text{ g}} \times \frac{43.9 \text{ kJ}}{1 \text{ mol}} = 317 \text{ kJ}$$

> EXAMPLE: The substance, X, has the following properties:

|  | Specific Heat |
| --- | --- |
| $\Delta H_{vap}$ =20. kJ/mol | Solid = 3.0 J/ (g °C) |
| $\Delta H_{fus}$ = 5.0 kJ/mol | Liquid 2.5 J/ (g °C) |
| Boiling point 75°C | Gas 1.0 J/ (g °C) |
| Melting point –15°C | |

Calculate the energy required to convert 250. g of substance X from a solid at –50.0°C to a gas at 100°C. Assume that X has a molar mass of 75.00 g/mol.

There are 5 steps involved.

1.  Heating the solid from –50.0°C to –15°C.

2.  Melting the solid at –15°C.

3.  Heating the liquid to its boiling point from –15°C to 75°C.

4.  Boiling the liquid at 75°C.

5.  Heating the gas to 100°C

The energy used = sum of energies from individual steps.

There are 3.33 mol of X in 250. g of X.

Step 1: $q = m \times s \times \Delta T$; 250. g × [3.0 J/(g °C)] × 35°C = 26 kJ

Step 2: mol × $\Delta H_{fus}$ = 3.33 mol × 5.0 kJ/mol = 17 kJ

Step 3: $q = m \times s \times \Delta T$; 250.0g × [2.5 J/(g °C)] × 90°C = 56 kJ

Step 4: mol × $\Delta H_{vap}$ = 3.33 mol × 20. kJ/mol = 67 kJ

Step 5: $q = m \times s \times \Delta T$; 250.0g × [1.0 J/(g °C)] × 25 °C = 6.2 kJ

172 kJ

The total energy required is the sum of the energies of the individual steps.

# HESS'S LAW

(*Chemistry* 6th ed. pages 256–260 / 7th ed. pages 242–246)

Hess's Law states that enthalpy change, in going from a particular set of reactants to a particular set of products, is the same whether the reaction takes place in one step or in a series of steps. Enthalpy is a state function.

**EXAMPLE:** Given the following reactions and $\Delta H°$ values,

$\Delta H°$

$Si(s) + 2H_2(g) \rightarrow SiH_4(g)$     34 kJ/mol

$Si(s) + O_2(g) \rightarrow SiO_2(s)$     –911 kJ/mol

$H_2(g) + \frac{1}{2} O_2(g) \rightarrow H_2O(g)$     –242 kJ/mol

Calculate $\Delta H$ for

$SiH_4(g) + 2O_2(g) \rightarrow SiO_2(s) + 2 H_2O(g)$     $\Delta H° = ?$

The idea is to manipulate the three equations so that they add up to the overall reaction. There are two ways that the equations can be manipulated.

1. Reverse the reaction. The sign of $\Delta H$ is also reversed.

2. Multiply the coefficients in a balanced reaction by an integer. The value of $\Delta H$ is multiplied by the same integer.

**SOLUTION:** Start by finding a substance that appears only once in all of the reactions, $SiH_4$ or $H_2O$. Look to see where that substance is in the final reaction and in what amount. Modify the reaction so that the substance appears where it should be and in the correct amount. $SiH_4$ needs to be in the reactants, so the first reaction is reversed. $H_2O$ stays in the products, but there are two of them in the final reaction, so, the third equation needs to be multiplied by two.

$SiH_4(g) \rightarrow \text{S̶i̶(s)} + 2H_2(g)$   $-$ (34 kJ/mol)  (Note: This was reversed.)

$\text{S̶i̶(s)} + O_2(g) \rightarrow SiO_2(s)$   $-$ 911 kJ/mol

(Note: The 2nd reaction stays the same so that Si can cancel out.)

$\underline{2H_2(g) + O_2(g) \rightarrow 2H_2O(g)}$        $\underline{2(-242kJ/mol)}$

$SiH_4(g) + 2O_2(g) \rightarrow SiO_2(s) + 2\ H_2O(g)$

$$\Delta H^\circ = -1429 \text{ kJ}$$

## STANDARD ENTHALPIES OF FORMATION, $\Delta H^\circ_F$

(*Chemistry* 6th ed. pages 260–267 / 7th ed. pages 246–252)

The standard enthalpy of formation of a compound is the change in enthalpy that accompanies the formation of one mole of a compound from its elements with all substances in their standard states.

The degree symbol on $\Delta H^\circ$ indicates that the process occurred under standard conditions of 25°C , 1 atm and $1M$ solutions.

By definition, the standard heat of formation for elements in their standard states equals zero.

> EXAMPLE: Which of the following will have standard heats of formation equal to zero?
>
> $H_2(g)$, $Hg(s)$, $CO_2(g)$, $H_2O(l)$, $Br_2(l)$
>
> The only elements present in their standard states are $H_2(g)$ and $Br_2(l)$. Mercury is a liquid under standard conditions. Carbon dioxide and water are compounds.
>
> EXAMPLE: Write the balanced molecular equation representing the $\Delta H_f^0$, standard heat of formation reaction of ethanol, $C_2H_5OH(l)$.
>
> $2C(s) + 3H_2(g) + 1/2\ O_2(g) \rightarrow C_2H_5OH(l)$

Note that 1 mole of product is produced according to the definition of the standard heat of formation. All of the reactants are elements in their standard states.

To calculate the enthalpy change for a given reaction, the enthalpies of formation of the reactants are subtracted from the enthalpies of formation of the products.

$$\Delta H^\circ_{reaction} = \Sigma \Delta H^\circ_{f(products)} - \Sigma \Delta H^\circ_{f(reactants)}$$

EXAMPLE: Using the standard heats of formation, $\Delta H^{\circ}_f$ ,calculate the change in enthalpy, $\Delta H^0$, for the following reaction. The standard heats of formation for hydrogen peroxide and water are –187 kJ/mol and –285 kJ/mol, respectively.

$$2H_2O_2(l) \rightarrow 2\ H_2O(l) + O_2(g)$$

$$\Delta H^{\circ}_{reaction} = [2\ (-285\ kJ) + 0] - 2\ (-187\ kJ) = -196\ kJ$$

EXAMPLE: Calculate the heat of formation of gaseous carbon monoxide based on the reaction below.

$$2CO(g) + C(s) \rightarrow C_3O_2(g) \qquad \Delta H^{\circ} = 127.3\ kJ$$

$\Delta H^{\circ}_f$ for $C_3O_2(g)$ is –93.7 kJ/mol .

$$\Delta H^{\circ}_{reaction} = \Sigma\ \Delta H^{\circ}_{f(products)} - \Sigma\ \Delta H^{\circ}_{f(reactants)}$$

$$\Delta H^{\circ}_{reaction} = \Delta H^{\circ}_f \text{ for } C_3O_2(g) - [2\ \Delta H^{\circ}_f \text{ for } CO(g) - \Delta H^{\circ}_f \text{ for } C(s)]$$

Rearrange the equation and solve for missing variable; note $\Delta H^{\circ}_f$ for C(s) is zero.

$$\Delta H^{\circ}_f \text{ for } CO(g) = [\Delta H^{\circ}_f \text{ for } C_3O_2(g) - \Delta H^{\circ}_{reaction}] / 2$$

$$= (-93.7\ kJ - 127.3\ kJ) / 2 = -110.5\ kJ\ /mol$$

# ENTROPY

(*Chemistry* 6th ed. pages 786–790 / 7th ed. pages 752–755)

Entropy, *S,* is the driving force for a spontaneous process which readily occurs without intervention. Entropy measures molecular randomness or disorder.

Positional entropy describes the number of arrangements or positions available to a system in a given state. Positional entropy increases in going from a solid to liquid to gas. In the solid state, molecules are much closer together than in the gaseous state, with very few positions available for them.

Entropy during a process is said to be increasing when the value of $\Delta S > 0$; The value of $\Delta S$ is positive.

Entropy during a process is said to be decreasing when the value of $\Delta S < 0$; The value of $\Delta S$ is negative.

EXAMPLE: Which of the following pairs is likely to have the higher positional entropy per mole at a given temperature?

1)   Solid $CO_2$ or gaseous $CO_2$ ?

2)   $N_2$ gas at 1.0 atm and $N_2$ gas at 0.001 atm?

1)   Gaseous $CO_2$ has more positional entropy than solid $CO_2$ since there are more positions for the molecules the gaseous state to move to than in the solid state.

2) $N_2$ gas at 0.001 atm has more positional entropy than $N_2$ gas at 1.0 atm, because at a lower pressure, there is more volume for the molecules to move than at a higher pressure.

EXAMPLE: Describe the change in entropy when solid salt is added to water.

The entropy increases when salt in its solid state is dissolved in water. The ions fixed in a crystal lattice in the solid are free to move about in the water due to solute–solvent interactions.

## SECOND LAW OF THERMODYNAMICS

(*Chemistry* 6th ed. page 790 / 7th ed. page 755)

The second law of thermodynamics states that in any spontaneous process there is always an increase in the entropy of the universe. The change in entropy of the universe is equal to the change in the entropy of the system and the change in entropy of the surroundings.

$$\Delta S_{univ} = \Delta S_{sys} + \Delta S_{surr}$$

$$\Delta S_{surr} = -\Delta H/T$$

The sign of $\Delta S_{surr}$ depends on the direction of the heat flow. $\Delta S_{surr}$ is positive at constant temperature when the reaction is exothermic, since heat flows to the surroundings increasing the random motions and the entropy of the surroundings. The opposite is true for an endothermic reaction at constant pressure.

The magnitude of $\Delta S_{surr}$ depends on temperature. At a low temperature, the production of heat effects a much greater percent change in the randomness of the surroundings than it does at high temperature.

## ENTROPY CHANGES IN A CHEMICAL REACTION

(*Chemistry* 6th ed. pages 798–801 / 7th ed. pages 762–766)

The change in entropy, $\Delta S$, for a chemical reaction can be predicted without calculation.

For a chemical reaction involving only gaseous reactants and products, entropy is related to the total number of moles of gas on either side of the equation. If the moles of gas increase from reactants to products in a chemical reaction, the entropy is increasing. If the moles of gas decrease in a chemical reaction, the entropy is decreasing.

For a chemical reaction involving solid, liquids and gases, the production of a gas will, in general, increase the entropy of the reaction much more than an increase in the number of moles of liquids or solids.

EXAMPLE: Predict the sign of $\Delta S°$ for the following reaction:

$N_2(g) + 3H_2(g) \rightarrow 2NH_3(g)$

The entropy decreases as the reaction proceeds from reactants to products because the number of moles of gas decreases from four total moles to two moles. The sign of $\Delta S^\circ$ is negative.

The change in entropy for a reaction, $\Delta S^\circ$, can also be calculated using tabulated thermodynamic values in the appendix of the textbook.

$$\Delta S^\circ_{reaction} = \Sigma\, S^\circ_{(products)} - \Sigma\, S^\circ_{(reactants)}$$

**Example:** Calculate $\Delta S^\circ$ for the following reaction:

$N_2(g) + 3H_2(g) \rightarrow 2NH_3(g)$

$\Delta S^\circ = 2(193) - [1(192) + 3\,(131)] = -199$ J/K.

The sign of $\Delta S^\circ$ is negative, confirming the prediction in the previous example.

## FREE ENERGY

*(Chemistry 6th ed. pages 802–809 / 7th ed. pages 766–773)*

Free energy, $G$, is a thermodynamic function whose value describes whether or not a process is spontaneous. Gibbs free energy is dependent on the change in enthalpy, change in entropy, and temperature of the system.

$$\Delta G = \Delta H - T\,\Delta S; \; T \text{ is the Kelvin temperature}$$

Using the chart below, you can predict if a reaction will occur without the exact value for $\Delta H$ and $\Delta S$.

| The Dependence of Spontaneity on Temperature | | |
|---|---|---|
| $\Delta S$ | $\Delta H$ | $\Delta G$ |
| + | – | Spontaneous at all temperatures |
| + | + | Spontaneous at high temperatures |
| – | – | Spontaneous at low temperatures |
| – | + | Process not spontaneous at any temperature |

## FREE ENERGY AND CHEMICAL REACTIONS

*(Chemistry 6th ed. pages 802–806 / 7th ed. pages 766–770)*

The standard free energy change, $\Delta G^\circ$, is the change in free energy that will occur if the reactants in their standard states are converted to the products in their standard states. This value cannot be measured directly, but it can be calculated from other measured quantities such as the equilibrium constant and the standard cell potential.

The standard free energy change, $\Delta G$, can be calculated from the changes in enthalpy and entropy.

$\Delta G° = \Delta H° - T \Delta S°$

**EXAMPLE:** Consider the reaction $2\ POCl_3(g) \rightarrow 2PCl_3(g) + O_2(g)$.

The value of $\Delta S°$ is 179 J/K. The value of $\Delta H°$ is 542 kJ. At what temperature is this reaction spontaneous? Assume that $\Delta H°$ and $\Delta S°$ do not depend on temperature.

When the process is spontaneous, solve for the temperature when $\Delta G° = 0$.

$\Delta G° = \Delta H° - T \Delta S°$; $T = \Delta H° / \Delta S° = 542$ kJ/ (0.179 kJ/K) = 3030 K.

The standard free energy change, $\Delta G$, is a state function and can be calculated by

$$\Delta G°_{reaction} = \Sigma \Delta G°_{f(products)} - \Sigma \Delta G°_{f(reactants)}$$

# THE DEPENDENCE OF FREE ENERGY ON PRESSURE

(*Chemistry* 6th ed. pages 806–808 / 7th ed. pages 770–772)

The free energy of a reaction is dependent on the pressure of a gas or the concentration of species in solution.

Consider the reaction: $N_2(g) + 3H_2(g) \rightarrow 2NH_3(g)$

$$\Delta G° = \Delta G°_{reaction} + RT \ln Q$$

$$\Delta G = \Delta G°_{reaction} + RT \ln \left[ \frac{(P_{NH_3})^2}{(P_{N_2})(P_{H_2})^3} \right]$$

$R$ = Gas law constant = 8.3145 J/K

$\Delta G°$ = the free energy change with all reactants and products at a pressure of 1 atm and 25°C.

$\Delta G$ = the free energy change for the reaction for the specified pressures of reactants and products.

$Q$ = the reaction quotient and will be discussed in the equilibrium chapter in this manual.

# FREE ENERGY AND EQUILIBRIUM

(*Chemistry* 6th ed. pages 810–814 / 7th ed. pages 774–778)

$$\Delta G° = -RT \ln K; K = \text{equilibrium constant}$$

This will be discussed in the chapter on equilibrium in this manual.

## FREE ENERGY AND CELL POTENTIAL

(*Chemistry* 6th ed. pages 837–840 / 7th ed. pages 800–803)

$$\Delta G^\circ = -n\, FE^\circ; \; E^\circ \text{ is the standard cell potential}$$

This will be discussed in the chapter on electrochemistry in this manual.

## MULTIPLE-CHOICE QUESTIONS

No calculators are to be used in this section.

1. The standard enthalpy of formation for nitrogen dioxide is the enthalpy change of the reaction
   (A) $1/2\ N_2O_4(g) \rightarrow NO_2(g)$.
   (B) $1/2\ N_2(g) + O_2(g) \rightarrow NO_2(g)$.
   (C) $N_2(g) + 2O_2(g) \rightarrow N_2O_4(g)$.
   (D) $N_2(g) + 2O_2(g) \rightarrow 2NO_2(g)$.
   (E) $NO(g) + 1/2\ O_2(g) \rightarrow NO_2(g)$.

ANSWER: **B** This question should determine if you understand how to apply the definition of the term "standard enthalpy of formation." Like may terms in chemistry, this has a very specific meaning: It is the energy involved in forming one mole of a compound from its elements in their standard or free state at 25°C and 1 atm. pressure. As applied here, you must write the equation showing the formation of $NO_2(g)$ as the product from the elements nitrogen and oxygen (both are diatomic elements under these conditions), and balance it so that only one mole of the product compound is formed (*Chemistry* 6th ed. pages 260–266 / 7th ed. pages 246–251).

2. Which of the following has a standard enthalpy of formation which is not zero?
   (A) Na(s)
   (B) Hg(l)
   (C) $H_2O(l)$
   (D) $N_2(g)$
   (E) C(s)

ANSWER: **C** By definition, standard enthalpy of formation is the energy involved when one mole of a compound is formed from its elements in their standard states at 25°C and 1 atm. In this question, only water is a compound. (*Chemistry* 6th ed. pages 260–261 / 7th ed. pages 246–247).

THERMOCHEMISTRY AND THERMODYNAMICS ❖ 121

3.  For endothermic reactions at constant pressure
    (A)  $\Delta H < 0$.
    (B)  $\Delta H > 0$.
    (C)  $\Delta G < 0$.
    (D)  $\Delta S > 0$.
    (E)  $\Delta S < 0$.

ANSWER: B In an endothermic reaction, heat is gained by the system. The convention to show this is to indicate that the enthalpy change is positive (*Chemistry* 6th ed. pages 256–258 / 7th ed. pages 242–244).

4.  At a certain temperature $C_{(s)} + O_{2(g)} \rightarrow CO_{2(g)}$ has a $\Delta G$ of –339.4 kJ/mol. This means that at this temperature
    (A)  the system is at equilibrium.
    (B)  gaseous carbon dioxide is unstable.
    (C)  gaseous carbon dioxide spontaneously forms.
    (D)  this system has a high reaction rate.
    (E)  the system will not react.

ANSWER: C A negative Gibbs free energy value indicates a spontaneous reaction; $CO_2$ forms. Note that it says nothing about the rate of reaction. Kinetics will be the topic of a later chapter (*Chemistry* 6th ed. pages 795–798, 809–813 / 7th ed. pages 759–762, 773–776).

For questions 5, 6, and 7, consider four separate situations:

I with $\Delta H = +$ and $\Delta S = +$

II with $\Delta H = +$ and $\Delta S = -$

III with $\Delta H = -$ and $\Delta S = -$

IV with $\Delta H = -$ and $\Delta S = +$

Assume that both $\Delta H$ and $\Delta S$ are temperature independent.

5.  Which processes are spontaneous at all temperatures?
    (A) I only
    (B) II only
    (C) III only
    (D) IV only
    (E) I and IV

ANSWER: D From $\Delta G = \Delta H - T\Delta S$. For example, if $\Delta H$ is negative and $\Delta S$ is positive the reaction must be spontaneous because $\Delta G$ is negative in all such cases (*Chemistry* 6th ed. pages 795–804 / 7th ed. pages 759–768).

6. The process which must be nonspontaneous at all values of temperature is
   (A) I.
   (B) II.
   (C) III.
   (D) IV.
   (E) Impossible to identify without more data.

ANSWER: B From $\Delta H - T\Delta S = \Delta G$. If $\Delta H$ is positive and $T\Delta S$ is negative, then $\Delta G$ is positive in all cases, so the reaction is always nonspontaneous (*Chemistry* 6th ed. pages 795–804 / 7th ed. pages 759–768).

7. Which of these four processes is improbable at a low temperature but becomes more probable as the temperature rises?
   (A) I
   (B) II
   (C) III
   (D) IV
   (E) Impossible to identify without more data

ANSWER: A From $\Delta H - T\Delta S = \Delta G$, if the temperature is high, the $T\Delta S$ factor is large and "overcomes" the influence of $\Delta H$ (+) to make $\Delta G$ negative; hence it becomes spontaneous at the higher temperatures (*Chemistry* 6th ed. pages 795–804 / 7th ed. pages 759–768).

8. In which of the following four processes is there an increase in entropy?
   I   $2SO_2(g) + O_2(g) \rightarrow SO_3(g)$
   II  $H_2O(g) \rightarrow H_2O(s)$
   III $Hg(l) \rightarrow Hg(g)$
   IV  $H_2O_2(l) \rightarrow H_2O(l) + 1/2\ O_2(g)$

   (A) all of the above
   (B) I, II
   (C) I, IV
   (D) III, IV
   (E) II, III, IV

ANSWER: D If there is an increase in entropy, then the products must be more disordered than the reactants. When a liquid becomes a gas (process III) the disorder increases; when one mole of a liquid becomes one mole of a different liquid and half a mole of gas (process IV), the entropy increases (*Chemistry* 6th ed. pages 798–801 / 7th ed. pages 762–766).

For questions 9 and 10 refer to the following exothermic reactions involving gases.

$$\text{I} \qquad CH_4 + 2O_2 \rightarrow CO_2 + 2H_2O$$

$$\text{II} \qquad 1/2\ C + 1/2\ O_2 \rightarrow 1/2\ CO_2$$

$$\text{III} \qquad CH_4 + O_2 \rightarrow C + 2H_2O$$

$$\text{IV} \qquad C + O_2 \rightarrow CO_2$$

9. According to the data given, which reaction liberates the most energy?
(A) I
(B) II
(C) III
(D) IV
(E) This cannot be determined without more data.

ANSWER: A Since II is exactly half of IV, it must be less exothermic. Note also that if you add III to IV you get I, hence I must be the most exothermic of these four processes (each of the parts must be less than the whole (*Chemistry* 6th ed. pages 256–260 / 7th ed. pages 242–246).

10. According to the data given, which reaction liberates the least energy?
(A) I
(B) II
(C) III
(D) IV
(E) This cannot be determined without more data.

ANSWER: E From these data there is no certain way to determine if II or III is less energetic (*Chemistry* 6th ed. pages 256–260 / 7th ed. pages 242–246).

## FREE-RESPONSE QUESTIONS

1. Ozone can be prepared from molecular oxygen by subjecting oxygen to an electrical discharge. The standard free energy of formation of ozone is +163.4 kJ/mol.
   (a) Write the chemical equation for the formation of ozone.
   (b) What is the sign of the entropy change for this reaction? Justify your answer.
   (c) Comment on the stability of ozone based on these data.
   (d) As temperature increases, would you expect ozone to become more or less stable? Justify your answer.

ANSWER:

(a) $\quad 3O_2(g) \rightarrow 2O_3(g)$

(b) $\Delta S$ is (–) since order is increasing (this is to say, entropy is decreasing) as 3 moles of $O_2$ form 2 mol of $O_3$.

(c) Note that the reaction in (A) is for the formation of ozone and that $\Delta G$ is a large positive value. This means that the reverse reaction must have a large negative $\Delta G$ and that ozone will spontaneously decompose, hence ozone is less stable than oxygen (*Chemistry* 6th ed. pages 786–790, 798–801 / 7th ed. pages 752–755, 762–766).

(d) As temperature increases, it is expected that $O_3$ is less stable. From $\Delta G = \Delta H - T \Delta S$. As $T$ increases, the value of $\Delta G$ becomes more positive and therefore the reaction is less likely to occur.

2. The molar heats of fusion and vaporization of benzene are 10.9 kJ/mol and 31.0 kJ/mol, respectively. The melting temperature of benzene is 5.5° C and it boils at 80.1°C.
   (a) Calculate the entropy changes for solid → liquid, and for liquid → vapor for benzene.
   (b) Would you expect the $\Delta S$ for these two changes to be about the same?
   (c) Comment on the physical significance of the difference in these two values.
   (d) Why are the values for heat of vaporization usually so much greater than the heats of fusion?

ANSWER: When a liquid boils at its boiling temperature or freezes at its freezing temperature, no useful work can be done by the process, i.e., $\Delta G$ is zero. Therefore, under these conditions,

$$\Delta S = \Delta H/T.$$

In melting, $\Delta S = 10.9 \times 10^3$ J/mol /(5.5 + 273 K) = +39.1 J/K•mol.

In boiling, $\Delta S = +87.8$ J/K•mol.

Vaporization involves a much greater change in disorder than melting (gases are very disordered compared to liquids, whereas liquids and solids vary less in disorder, distance between molecules, and number of possible positions for molecules); hence $\Delta S_{vap} > \Delta S_{fus}$ (*Chemistry* 6th ed. pages 798–801 / 7th ed. pages 762–766).

# 7

# ATOMIC STRUCTURE AND PERIODICITY

In this chapter, coverage of the electromagnetic spectrum will lead to the electronic structure of atoms. Properties such as the sizes of atoms and ionization energy can be predicted from electronic structures. Bond formation between atoms can also be rationalized from the electronic structure of atoms in the next chapter.

Questions on electronic structure, periodicity, and bonding are often found in the free-response essay section as well as in the multiple-choice section of the AP Chemistry exam.

You should be able to:

- Identify characteristics of and perform calculations with frequency and wavelength.
- Know the relationship between types of electromagnetic radiation and energy; for example, gamma rays are the most damaging.
- Know what exhibits continuous and line spectra.
- Know what each of the four quantum numbers $n$, $\ell$, $m_\ell$ and $m_s$ represents.
- Identify the four quantum numbers for an electron in an atom.
- Write complete and abbreviated electron configurations as well as orbital diagrams for an atom or ion of an element.
- Identify the number and location of the valence electrons in an atom.
- Apply the trends in atomic properties such as atomic radii, ionization energy, electronegativity, electron affinity, and ionic size.

## ELECTROMAGNETIC RADIATION

(*Chemistry* 6th ed. pages 290–291 / 7th ed. pages 275–276)

Electromagnetic radiation provides an important means of energy transfer. Light from the sun reaches the earth in the form of radiation that includes ultraviolet and visible radiation. Electromagnetic radiation, from the shortest to the longest wavelength, includes gamma rays, x-rays, ultraviolet, visible, infrared, microwaves, and radio waves. All types of electromagnetic radiation exhibit both wavelike and particle-like behavior and travel at the speed of light in a vacuum.

Chemists have used the interaction of light with matter to study the structure of the atom.

## CHARACTERISTICS OF WAVES

(*Chemistry* 6th ed. page 290 / 7th ed. pages 275–276)

Wavelength, $\lambda$, is the distance between two consecutive peaks or troughs in a wave.

Frequency, $\upsilon$, is the number of waves (cycles) per second that passes a given point in space.

Wavelength and frequency are related by the equation:

$$\upsilon = c / \lambda$$

in which $c$ is the speed of light in a vacuum, $2.9979 \times 10^8$ m/s. Wavelength is measured in meters, and frequency is measured in 1/s, or hertz (Hz).

## ENERGY OF ELECTROMAGNETIC RADIATION

(*Chemistry* 6th ed. pages 290–292 / 7th ed. pages 275–277)

Energy is quantized and can occur only in whole number multiples of $h\upsilon$. Each small packet of energy is called a quantum. A system can transfer energy only in discrete quanta. The change in energy for a system can be represented by

$$\Delta E = h\upsilon$$

where $n$ is an integer (1,2,3,...), $h$ is Planck's constant which equals $6.626 \times 10^{-34}$ Js, and $\upsilon$ is the frequency.

Einstein viewed electromagnetic energy as a stream of particles called photons. And the energy of each photon can be given by

$$E_{photon} = hc/\lambda$$

EXAMPLE: The laser in an audio compact disc player uses light with a wavelength of 7.80 x $10^2$ nm. Calculate the frequency of this radiation. What is the energy of this radiation per photon?

$$v = \frac{c}{\lambda} = \frac{2.998 \times 10^8 \text{ m/s}}{7.80 \times 10^{-7} \text{ m}} = 3.84 \times 10^{14} \text{ s}^{-1}$$

$E_{\text{photon}} = hc / \lambda$

= (6.626 × $10^{-34}$ Js)(2.998 × $10^8$ m/s) / 7.80 × $10^{-7}$ m

= 25.5 J × $10^{-19}$

# ATOMIC SPECTRUM

*(Chemistry* 6th ed. pages 297–298 / 7th ed. pages 284–285)

A continuous spectrum contains all wavelengths of visible light. A line spectrum contains only a few lines, each corresponding to a discrete wavelength. When atoms are excited by adding energy, the excess energy is given off by emitting light of various wavelengths to produce an emission spectrum or line spectrum.

# BOHR MODEL

*(Chemistry* 6th ed. page 299 / 7th ed. page 285)

Bohr proposed that the electron in the hydrogen atom moves around the nucleus only in certain allowed circular orbits and the line spectra of elements are the result of electrons moving between these allowed orbits. It has often been described as a solar system model. Since a moving electron radiates energy, this model was unstable using classical mechanics and the model was shown to be incorrect.

# QUANTUM MECHANICAL MODEL

*(Chemistry* 6th ed. pages 304–307 / 7th ed. pages 290–293)

Our current model of the atom is the quantum mechanical model. In this model, we treat the electrons as a wave and do not believe it is possible to describe the path of an electron. An orbital is a three-dimensional electron density map in which there is a 90% probability of finding the electron. The Heisenberg Uncertainty Principle is an important part of the model. It states that we cannot know both the momentum and location of an electron.

# QUANTUM NUMBERS

*(Chemistry* 6th ed. pages 307–309 / 7th ed. pages 293–294)

Quantum numbers characterize various properties of the orbitals. There are four quantum numbers. The Pauli Exclusion Principle states that no two electrons can have the same set of four quantum numbers.

The principal quantum number, n, is related to the size and energy of the orbital. As n increases, the orbital becomes larger, which also means it is higher in energy.

The angular momentum quantum number, ℓ, is related to the shape of atomic orbitals.

The magnetic quantum number, m is related to the orientation of the orbital in space relative to the other orbitals in the atom.

The electron spin quantum number, ms, has only two values: +1/2 and –1/2. This quantum number is necessary to describe the fact that the electron has two possible orientations, or spins, when placed in an external magnetic field.

The four quantum numbers are summarized in the following table.

## Quantum Numbers

| Name | Designation | Possible range of values | Property of the orbital |
|---|---|---|---|
| Principal quantum number | $n$ | Integers > 0 (1,2,3,...) | Related to size and energy of the orbital |
| Angular momentum quantum number | $\ell$ | Integers from 0 to $n-1$ | Related to the shape of the orbital |
| Magnetic quantum number | $m_\ell$ | Integers from $-\ell$ to $+\ell$ | Related to the position of the orbital in space in relation to other orbitals |
| Electron spin quantum number | $m_s$ | $+\frac{1}{2}$ or $-\frac{1}{2}$ | Related to the spin of the electron, which can be only one of two values |

For each value of the angular quantum number, ℓ, the designated atomic orbitals are as follows:

## The Angular Momentum Quantum Number, ℓ, and the Letters Used to Designate Atomic Orbitals

| Values of $\ell$ | 0 | 1 | 2 | 3 | 4 |
|---|---|---|---|---|---|
| Letter used | s | p | d | f | g |
| Number of orbitals | 1 | 3 | 5 | 7 | 9 |
| Total number of electrons in sublevel | 2 | 6 | 10 | 14 | 18 |

The principal quantum number, n, is the number which appears in front of the types of orbital s, p, d or f. In general, for periods (rows) 1, 2 and 3, the principal quantum number, n, is the same as the row number. For periods 3, 4, 5, 6, and 7, the principal quantum number written before s and p orbitals is the same as the row number for the main group of elements, those in columns 1A through 8A. For the transition elements, the value of n which is written in front of the d orbital is 1 number less than the row number. For example, scandium, in period 4, has a value of $n$ equal to 3 for its d electron. In the sixth row of the periodic table, the f-transition elements occur after lanthanum. For these elements, the value of $n$ which is placed before the f is 2 less than the row in which they occur.

## ORBITAL SHAPES AND ENERGIES

(*Chemistry* 6th ed. pages 309–312 / 7th ed. pages 295–298)

### s ORBITALS

The s orbitals have a spherical distribution of electron density. The s orbitals become larger as the value of $n$ increases.

Nodes are areas in which there is zero probability of finding an electron. The 2s has one node which separates areas of high probability, the 3s has two nodes, and so forth.

### p ORBITALS

There are three p orbitals for each value of $n > 1$. The p orbitals have two lobes which are separated by a node at the nucleus. The p orbitals are labeled $n_{px}$, $n_{py}$, and $n_{pz}$, according to the axis along which the lobe lies.

Degenerate orbitals are orbitals with the same value of $n$ which have the same energy. The three p orbitals in the same energy level are degenerate or equal in energy.

### d AND f ORBITALS

The d orbitals first occur when $n = 3$ ($\ell = 2$), but in the fourth period of the periodic table. The f orbitals first occur when $n = 4$ and in the sixth period.

> **EXAMPLE:** Which of the following orbital designations is incorrect? 1s, 1p, 2d, 4f?
>
> 1p is incorrect: $n = 1$, for a p orbital, $\ell$ must equal 1. If $n = 1$, $\ell = n - 1$ or 0 is the only possible value.
>
> 2d is incorrect: $n = 2$, for a d orbital, $\ell$ must equal 2. If $n = 2$, the largest value possible of $\ell$ is $n - 1$ or 1.

> **EXAMPLE:** Give the maximum number of electrons in an atom that can have the quantum number $n = 3$.
>
> $\ell$ can be 0, 1, or 2. Thus we have s ($2^{e-}$) p ($6^{e-}$), and d ($10^{e-}$). The total number of electrons to fill these orbitals is 18.
>
> For any energy level, $2n^2 = $ max number electrons for any energy level. For the above example, $2(3)^2 = 18$.

> **EXAMPLE:** Give the maximum number of electrons in an atom that can have the quantum numbers $n = 2$ and $\ell = 1$.

If $\ell = 1$, then it is a p-type sublevel, so there are three orbitals in that sublevel, each of which can hold two electrons or a total of six electrons is possible in the sublevel.

## AUFBAU PRINCIPLE AND ELECTRON CONFIGURATIONS

(*Chemistry* 6th ed. pages 317–324 / 7th ed. pages 302–309)

The Aufbau principle is a scheme used to reproduce the electron configuration of the ground states of atoms by successively filling sublevels with electrons in a specific order.

The electron configuration for an atom of an element represents all of the electrons in the atom and it shows in which energy levels and orbitals the electrons reside.

EXAMPLE: Write an electron configuration for oxygen.

The electron configuration for oxygen is $1s^2 2s^2 2p^4$.

To write an electron configuration for an atom of an element, follow the Aufbau principle and add electrons to the lowest energy orbital, the 1s. Using the Pauli Exclusion Principle, each orbital holds two electrons, we fill up the 1s with two electrons of opposite spin, then the 2s, to get $1s^2 2s^2$. We have used four electrons so far and oxygen has eight total electrons. The remaining four electrons go into the three 2p orbitals. Hund's Rule requires that we place one electron in each separate p orbital with parallel spin before pairing electrons. Write the electron configuration as $1s^2 2s^2 2p^4$.

EXAMPLE: Write the abbreviated electron configuration for oxygen.

The abbreviated electron configuration for oxygen is [He] $2s^2 2p^4$.

To write the abbreviated electron configuration, write the noble gas symbol which comes before the element in brackets followed by the remaining electrons.

## VALENCE ELECTRONS

(*Chemistry* 6th ed. page 319 / 7th ed. page 304)

Valence electrons are the electrons in the outermost principal quantum level of an atom. In the example with oxygen, the valence electrons are in the second quantum (energy) level. In the second energy level, there are four valence electrons, two in the 2s and two in the 2p. Coincidentally, the valence electrons of a main group element are the same as the group number, except for helium which has two valence electrons. The two electrons in the 1s orbital are referred to as the core (inner) electrons. Pages 321 and 322 of the 6th edition of *Chemistry* and pages 306 and 307 of the 7th edition display helpful diagrams which show how the orbitals are filled in various parts of the periodic table.

> ## AP tip
>
> Questions about trends in the periodic table often appear in the essay portion of the free-response section. It is important to be clear and concise in your writing. Different essay writing styles are discussed in this book in the test preparation chapter.

There are exceptions to electron configurations. One exception is chromium. Instead of having the expected electron configuration $[Ar]4s^23d^4$, the configuration is $[Ar]4s^13d^5$. This configuration is energetically more favorable (lower in energy) than the configuration with only four d electrons. This also occurs for molybdenum. A similar situation occurs for Cu and all of the transition elements beneath it. Instead of having the electron configuration $[Ar]4s^23d^9$, the configuration for copper is $[Ar]4s^13d^{10}$. The filled d sublevel is lower in energy than the partially filled sublevel.

## ORBITAL DIAGRAMS

(*Chemistry* 6th ed. pages 317–319 / 7th ed. pages 302–304)

The orbital diagram displays the same information as the electron configuration. In addition, the spin of the electron is also represented.

**EXAMPLE:** Give the orbital diagram for oxygen.

$$\underline{\uparrow\downarrow}\quad\underline{\uparrow\downarrow}\quad\underline{\uparrow\downarrow}\;\underline{\uparrow}\;\underline{\uparrow}$$
$$\;\;\;1s\quad\;\;2s\quad\quad\;\;2p$$

The first arrow in the 1s orbital represents an electron spinning in a particular direction. The second electron occupying the same orbital must spin in the opposite direction since, according to the Pauli exclusion principle, only two electrons with opposite spin can occupy the same orbital. When the 1s orbital is filled, the 2s orbital fills in the same manner as the 1s. There are three degenerate 2p orbitals, so one electron must go into each of the 2p orbitals. The fourth 2p electron becomes paired with the first 2p electron.

**EXAMPLE:** Give an acceptable set of four quantum numbers for each electron in oxygen:

|  |  ↑↓ | ↑↓ | ↑↓ | ↑ | ↑ |
|---|---|---|---|---|---|
|  | 1s | 2s |  | 2p |  |
| $n$ | 11 | 22 | 22 | 2 | 2 |
| $\ell$ | 00 | 00 | 11 | 1 | 1 |
| $m_\ell$ | 00 | 00 | −1−1 | 0 | +1 |
| $m_s$ | +½ −½ | +½ −½ | +½ −½ | +½ −½ | +½ −½ |

Begin with the orbital diagram for oxygen. Write the set of four quantum numbers under each electron (arrow). The value of $n$ is obtained from the integers in front of the type of orbital, s or p. For the s orbitals, $\ell = 0$ and for p $\ell = 1$.

The values for $m_\ell$ can be from $+\ell$ to $-\ell$. For $\ell = 0$, $m_l = 0$. For l $\ell = 1$, $m_l$ can be −1, 0, +1. Each p orbital is assigned one of the following values −1, 0, +1.

The value for $m_s$ is either + 1/2 or − 1/2. If the arrows are pointing in the same direction, then the value of $m_s$ must be the same.

For example, an acceptable set of four quantum numbers for the first 1s electron is $n = 1$, $\ell = 0$, $m_1 = 0$, $m_s = +1/2$. Another acceptable set would be $n = 1$, $\ell = 0$, $m_\ell = 0$, $m_s = −1/2$.

# PERIODIC TRENDS IN ATOMIC PROPERTIES

(*Chemistry* 6th ed. pages 324–330 / 7th ed. pages 309–314)

## ATOMIC RADII

(*Chemistry* 6th ed. pages 329–330 / 7th ed. pages 313–314)

Atomic radii decrease in going from left to right across a period from the alkali metals to the halogens. This can be explained by an increasing effective nuclear charge, which causes the electrons to be drawn closer to the nucleus.

Atomic radius increases down a group because of the increases in the orbital sizes in successive principal quantum levels. Also, there is an increase in the shielding of the nucleus by the core electrons, which decreases the effective nuclear charge felt by the valence electrons.

**EXAMPLE:** Arrange the following atoms in order of increasing atomic radius.

C, Al, F, Si

F < C < Si < Al ; F and C are in the second period, which means that they have two shells or energy levels. Within a period, the radius decreases as the atomic number increases. This means F will be smaller than C. Al and Si are in the third period and have one more energy level, so they will have a larger radius than atoms in period two. Since Si has more protons than Al, it has a smaller radius.

## IONIZATION ENERGY

(*Chemistry* 6th ed. pages 324–327 / 7th ed. pages 309–312)

Ionization energy is the energy required to remove an electron from a gaseous atom or ion.

The first ionization energy is the energy required to remove the highest energy electron (the one bound least tightly) which is one of the valence electrons.

$$X(g) + energy \rightarrow X^{1+}(g) + e^-$$

For lithium, the first ionization energy is the energy required to remove an electron from the 2s orbital. The configuration for lithium is $1s^22s^1$ and the valence electron is the 2s electron.

In general, the first ionization energy increases across a period from left to right. This trend can be related to the trend in atomic radii. The smaller the atom, the greater its first ionization energy. In going across a period from left to right, atomic radii decrease as the charge on the nucleus increases. There are irregularities in this trend in ionization energy. The first ionization energy decreases from beryllium to boron because the electrons in the 2s orbital are much more effective at shielding the electrons in the 2p orbital than they are at shielding each other. In all atoms except for hydrogen, the 2p orbital is slightly higher in energy than the 2s orbital. The higher energy 2p electron of boron is easier to remove than one of the 2s electrons of beryllium.

The first ionization energy decreases in going from nitrogen to oxygen due to repulsion of paired electrons in the $p^4$ configuration of oxygen. (Nitrogen has a valence electron configuration $[He]2s^22p^3$ and oxygen has a valence electron configuration of $[He]2s^22p^4$) The first ionization energy decreases down a group with increasing atomic number because the valence electrons become easier to remove as the atoms become larger going down a group.

The second ionization energy is the energy required to remove the second outermost electron. This value is larger than the first ionization energy. The first ionization energy removes an electron from a neutral atom. The second ionization energy removes an electron from a positive ion.

$$X^{1+}(g) + energy \rightarrow X^{2+}(g) + e^-$$

The increase in positive charge binds the electrons more tightly, and the ionization energy increases as each successive electron is removed. Also, a slightly larger increase in ionization energy occurs when removing an electron from a higher orbital, such as a 3p orbital vs. a 3s orbital. Note a large increase occurs in ionization energy when a core electron from a full energy level is removed.

**EXAMPLE:** Explain why the difference between the third ionization energy and fourth ionization energy for aluminum is so large in comparison with the first and second and second and third ionization energies.

Aluminum has the valence electron configuration [Ne]$3s^2 3p^1$. The first electron removed is the 3p; the second electron removed is a 3s, the third electron removed is the remaining 3s electron. The fourth electron, a core electron (a 2p), requires much more energy for removal than the valence electrons, the 3s and the 3p electrons, because it is more tightly bound.

**EXAMPLE:** Arrange the following atoms in order of increasing first ionization energy. I, Rb, Na

Na will have the smaller radius. Rb < I < Na. Na has the largest $IE_1$ because the electron is removed from the 2s sublevel. Both Rb and I have valence electrons in the 5th energy level, but I has 53 protons and a smaller radius than Rb with only 37 protons which makes I smaller than Rb. The smaller the atom, the more energy required to remove the valence electron.

## ELECTRON AFFINITY

(*Chemistry* 6th ed. pages 327–329 / 7th ed. pages 312–313)

Electron affinity is the energy change associated with the addition of an electron to a gaseous atom.

$$X(g) + e^- \rightarrow X^-(g)$$

In general, electron affinity becomes increasingly negative or exothermic as we proceed across a period with increasing atomic number. The greater the charge on the nucleus of the atom, the greater the affinity for electrons. The addition to a noble gas, however, would require the added electron to reside in a new, higher energy level which is energetically unfavorable, so that stable, isolated X- ions do not exist. Electron affinities are not measured for noble gases.

Electron affinities become slightly more positive, less exothermic, going down a group. The electron–nucleus attraction decreases down a group due to increasing atomic size and shielding. But the added electron experiences less electron–electron repulsion since the orbital to which the electron is being added is increasingly spread out. One exception to this trend is fluorine. The value of electron affinity for F is more positive than that of Cl due to the small size of the F atom and the greater electron–electron repulsion in the smaller 2p sublevel.

**EXAMPLE:** Arrange the following elements from the least to the most exothermic electron affinity. Cl, Al, P.

From the least to the most exothermic is Al, P, Cl. In general, the greater the nuclear charge on the atom, the more exothermic the electron affinity. Atoms become smaller across the period as the atomic number increases.

## MULTIPLE-CHOICE QUESTIONS

No calculators are to be used in this section.

1. Which of the following sets of quantum numbers is unacceptable?

| | $n$ | $\ell$ | $m_l$ | $m_s$ |
|---|---|---|---|---|
| (A) | 4 | 3 | –2 | +1/2 |
| (B) | 3 | 0 | +1 | –1/2 |
| (C) | 3 | 0 | 0 | –1/2 |
| (D) | 3 | 1 | 1 | +1/2 |
| (E) | 2 | 0 | 0 | –1/2 |

ANSWER: B Quantum number assignments follow set patterns or rules. The angular momentum quantum number $\ell$ (also called the second quantum number by some chemists) is limited to a maximum number of whole number values, from 0 ... to ... $(n - 1)$. The magnetic quantum number $m_\ell$ is limited to integers from $-\ell$ to zero to $+\ell$. Finally, the spin quantum number can have only one of two values, +1/2 or –1/2 (*Chemistry* 6th ed. pages 307–308 / 7th ed. pages 293–294).

2. The differentiating electron in gallium, Ga, is the single $4p^1$ electron. A valid set of quantum numbers for this electron is

| | $n$ | $\ell$ | $m_l$ | $m_s$ |
|---|---|---|---|---|
| (A) | 3 | 2 | 1 | +1/2 |
| (B) | 3 | 1 | –1 | +1/2 |
| (C) | 4 | 0 | 0 | +1/2 |
| (D) | 4 | –1 | +1 | +1/2 |
| (E) | 4 | 1 | –1 | +1/2 |

ANSWER: E Since the last part of the electron configuration for Ga is $4p^1$, you can assign $n = 4$ from the 4p orbital; $\ell$ follows the pattern: $\ell = 0$ for all s orbitals, $\ell = 1$ for all p orbitals, $\ell = 2$ for all d orbitals, etc. If $\ell = 1$, as it does in this case, then $m$ can be only –1, 0, or +1 (representing the three orientations of a "p" orbital in space) (*Chemistry* 6th ed. pages 307–308 / 7th ed. pages 293–294).

3. In a given atom, no two electrons can have the same set of four quantum numbers. This statement is knows as the
(A) Pauli exclusion principle.
(B) Hund's rule.
(C) Einstein principle.
(D) Heisenberg uncertainty principle.
(E) Bohr law.

ANSWER: A This is also another way of saying that no more than two electrons are allowed in the same orbital in the same atom (*Chemistry* 6th ed. pages 310–312 / 7th ed. pages 296–298).

4. If the angular momentum (second) quantum number $\ell$ equals 3, the total number of allowed orbitals in just that subenergy level is
   (A) 1.
   (B) 3.
   (C) 5.
   (D) 7.
   (E) 9.

**Answer: D** If $\ell = 3$, then $m_\ell$ can have values of –3, –2, –1, 0, +1, +2, +3. That is seven possible values, one for each of the possible orbitals allowed (*Chemistry* 6th ed. pages 307–310 / 7th ed. pages 293–296).

5. The element with the ground state electron configuration (spectroscopic notation) of [Ar] $3d^7 4s^2$ is
   (A) Mg.
   (B) K.
   (C) Ar.
   (D) Co.
   (E) Ni.

**Answer: D** From the 4 in the $4s^2$ notation, you know that you are operating in the fourth period of the Periodic Table. Since the 3d subenergy level fills next and has seven electrons in it, you must be dealing with the seventh element in the transition metals level, hence cobalt, Co (*Chemistry* 6th ed. pages 317–321 / 7th ed. pages 302–306).

6. The electron configuration for the element antimony, Sb, #51, is
   (A) [Na] $3s^2 2d^{10} 3p^3$.
   (B) [Ar] $4s^2 3d^{10} 4p^5$.
   (C) [Ar] $4s^2 3d^{10} 4p^3$.
   (D) [Kr] $5s^2 4d^{10} 5p^3$.
   (E) [Kr] $5s^2 4d^{10} 5p^5$.

**Answer: D** First find Sb on the Periodic Table, and note that it is in the 5A family (also called the 15 family). This grouping from 3A to 8A is filling the "p" orbital subenergy level; hence Sb must end in $p^3$. But which "p" subenergy level? Now note that this element is in the fifth period; hence it ends in $5p^3$ (*Chemistry* 6th ed. pages 324–329, esp. Exercise 7.10 / 7th ed. pages 309–313, esp. Exercise 7.10).

7. Atomic radii decrease from left to right across a period because of
   (A) an increase in effective nuclear charge.
   (B) an increase in gross energy level ($n$).
   (C) an increase in subenergy level ($\ell$).
   (D) an increase in shielding.
   (E) a decrease in effective nuclear charge.

**Answer: A** While adding electrons as you proceed from left to right might seem to add to the atomic radii, remember that you are also adding protons to the nucleus while adding electrons to the same gross energy level; so the increase in nuclear charge is the determining factor in this trend (*Chemistry* 6th ed. pages 309–312 / 7th ed. pages 295–298).

8. The number of unpaired electrons (no counterspinning electron) in the chromium atom is
   (A) 1.
   (B) 2.
   (C) 4.
   (D) 5.
   (E) 6.

**ANSWER: E** Remember that chromium is an exception to the general rules of electron configuration and that it ends in $4s^1 3d^5$, so it has six unpaired electrons (*Chemistry* 6th ed. pages 317–324 / 7th ed. pages 302–309).

9. The correct ordering of atoms in progressively decreasing ionization energy is
   (A) F > O > C > Li > Na.
   (B) Na > Li > C > O > F.
   (C) F > O > C > Na > Li.
   (D) C > O > F > Li > Na.
   (E) O > F > C > Na > Li.

**ANSWER: A** The trends in decreasing ionization are similar to those in increasing in atomic radii. From left to right in one period, the atomic radius decreases and the ionization energy increases from top to bottom in a family; the ionization energy decreases as you move to greater energy levels (*Chemistry* 6th ed. pages 324–329, esp. Exercises 7.8 and 7.9 / 7th ed. pages 309–313, esp. Exercises 7.8 and 7.9).

10. Electron affinity is the
    (A) energy required to remove an electron from an atom in the gaseous state.
    (B) energy released when an electron is gained by an atom in its standard state.
    (C) maximum energy required to remove an electron from an atom in its standard state.
    (D) energy involved when an electron is gained by an atom in the gaseous state.
    (E) energy gained by an electron when it is absorbed by another electron.

**ANSWER: D** This is a basic definition you need to know (*Chemistry* 6th ed. pages 327–329 / 7th ed. pages 312–313).

## FREE-RESPONSE QUESTIONS

1. One of the emission lines that provides the green color in fireworks, which comes from barium salts, has a wavelength of 505 nm.
   (a) Determine the frequency of this green line.
   (b) Determine the energy associated with this green line.
   (c) Arrange the visible spectrum (from red to violet) in order of increasing:
       i. wavelength
       ii. frequency
       iii. energy

ANSWER:
   (a)

   Frequency = speed of light / wavelength

   $$= \frac{2.998 \times 10^8 \text{ m/s}}{505 \text{ nm} \times 10^{-9} \text{ m/nm}}$$
   $$= 5.94 \times 10^{14} \text{ s}^{-1}$$
   $$= 5.94 \times 10^{14} \text{ Hz.}$$

   (b) Energy = Planck's constant x frequency

   $$= 6.626 \times 10^{-34} \text{ J} \cdot \text{s} \times 5.90 \times 10^{14} \text{ s}^{-1}$$

   $$= 3.93 \times 10^{-19} \text{J}$$

   (c) From red → violet
       i. frequency increases
       ii. wavelength decreases
       iii. energy increases

(*Chemistry* 6th ed. pages 293–296 / 7th ed. pages 278–281).

2. You have studied the Bohr model of the atom and the quantum mechanical model.
   (a) What phenomena was Bohr able to explain using this model?
   (b) What was one of major shortcomings of Bohr's model?
   (c) Give two differences in the quantum mechanical model from the Bohr model.

ANSWER: The Bohr model of the atom involves a specific circular orbital for the electron, whereas the quantum mechanical model is an electron density plot or probability model of orbitals based on an understanding of standing waves. According to Bohr's model, as the electron revolves around the nucleus, it should emit light, lose energy, and be drawn into the nucleus. None of these things happen. (You might also note the differences in the terms "orbit" and "orbital." In an orbit, the electron is always found at the given distance from the nucleus; the orbital describes the most probable distance at which the electron is found (*Chemistry* 6th ed. pages 299–307 / 7th ed. pages 285–293).

3. The emission spectrum of hydrogen consists of several series of sharp emission lines. Why is the emission spectrum of hydrogen made of discrete wavelengths rather than a continuum of

wavelengths? Discuss this in terms of features of the electronic energies of the hydrogen atom.

**ANSWER:** The energy of the electron is restricted only to certain values; it is quantized. This means that only a certain few changes in the energy of the electron can occur and not all possible changes in energy. If this were so then the spectrum that you see would be a continuous spectrum made of all the colors in the visible region (*Chemistry* 6th ed. pages 297–299 / 7th ed. pages 284–285).

# 8

# BONDING, STRUCTURE, AND ORGANIC CHEMISTRY

All chemical reactions involve making and breaking chemical bonds. In this chapter, you will learn why different types of bonds form, the nature of those bonds, and you will study a model to predict the three-dimensional structure of molecules formed from covalent bonds.

Bond formation always releases energy and bond breaking always requires energy. The net enthalpy for a chemical reaction will depend on whether it takes more energy to break the old bonds in the reactants than is released when new bonds form in the products. The change in enthalpy for reactions will be calculated from bond energies.

Intermolecular forces and their effect on properties of molecules will also be reviewed.

An overview of organic chemistry will cover basic nomenclature, isomerism, and functional groups.

You should know:
- The characteristics of ionic and covalent bonding.
- The relative sizes of ions.
- The effect of lattice energy on melting points of ionic compounds.
- How to use electronegativity to predict the polarity of covalent bonds.
- How to draw Lewis symbols for atoms and Lewis structures for molecular compounds and polyatomic ions.
- How to draw resonance structures.
- How to assign molecular shapes using the VSEPR Theory for molecules, polyatomic ions, and multi-centered

molecules and polyatomic ions, through the octahedral shape.
- How to assign hybrid orbitals to central atoms of molecules and polyatomic ions.
- How to calculate the enthalpy of a reaction using bond energies.
- How to compare bond lengths in molecules.
- How to write formulas, names, and structures for organic compounds including alkanes, alkenes, alkynes, cyclic compounds and compounds with functional groups.
- How to draw and name isomers of alkanes and alkenes.
- How to identify the intermolecular forces in a substance.

## IONIC BONDING

(*Chemistry* 6th ed. page 349 / 7th ed. pages 330–331)

Ionic bonding is the result of electrostatic attraction of oppositely charged ions. Ionic bonds form when an atom which loses electrons easily reacts with an atom that has a high affinity for electrons. Most of the time, an ionic compound results when a metal reacts with a nonmetal.

## COULOMB'S LAW

(*Chemistry* 6th ed. page 349 / 7th ed. pages 330–331)

Coulomb's law expresses the energy of interaction between a pair of ions. This is directly proportional to the charge on each ion and inversely proportional to the distance between the ions. The greater the charge on the ions, the stronger the ionic bond, and if the charges are equal, then we expect a smaller ion to have a stronger ionic bond.

$$E = (2.31 \times 10^{-19} \text{ J} \cdot \text{nm})\left(\frac{Q_1 Q_2}{r}\right).$$

$E$ is the energy of interaction between a pair of ions, in joules.

$r$ is the distance in nanometers between ion centers.

$Q_1$ and $Q_2$ are the charges of the ions.

## IONS: ELECTRON CONFIGURATION AND SIZES

(*Chemistry* 6th ed. pages 357–362 / 7th ed. pages 338–342)

### PREDICTING FORMULAS OF IONIC COMPOUNDS; FORMATION OF IONIC COMPOUNDS

When a metal and a nonmetal react, ions form so that the valence electron configuration of the nonmetal achieves the electron configuration of the next noble gas atom by gaining electrons and the main group of metals lose all of their valence electrons. Usually both ions achieve noble gas configurations.

EXAMPLE: Explain how atoms of sodium and oxygen form an ionic compound.

Sodium and chlorine have the following valence electron configurations:

Na: $[Ne]3s^1$

O: $[He]2s^22p^4$

The electronegativity of oxygen is much greater than that of sodium, so electrons are transferred from sodium to oxygen. Oxygen needs two electrons to fill its 2p valence orbitals and to achieve the configuration of neon. Sodium loses one electron to achieve the electron configuration of neon. Two sodium atoms give one oxygen atom the two electrons necessary to fill its valence orbitals. The oxide ion has an electron configuration like that of neon.

$$2\,Na \rightarrow 2\,Na^+ + 2\,e^-$$
$$O + 2\,e^- \rightarrow O^{2-}$$

Using electron configurations:

$$Na \rightarrow Na^+ + e^- \qquad O + 2\,e^- \rightarrow O^{2-}$$
$$[Ne]3s^1 \rightarrow [Ne] + e^- \qquad [He]2s^22p^4 + 2\,e^- \rightarrow [He]2s^22p^6 = [Ne]$$

## SIZES OF IONS

(*Chemistry* 6th ed. pages 360–361 / 7th ed. pages 340–341)

Cations are smaller than their parent atoms and anions are larger than their parent atoms. The size of an ion depends on its nuclear charge, the number of electrons it contains, and the outer energy level of the atom. Positive ions are always smaller than neutral atoms because the nuclear charge is the same but the number of electrons has decreased increasing the effective nuclear charge felt by each remaining electron. When all the valence electrons are lost, then the change in volume is greater because as $n$ decreases, the radial probability is smaller so the radius is smaller.

Anions are larger than their parent atoms because the electrons gained cause the proton to electron ratio to decrease, and the effective nuclear charge felt by each electron decreases. Electron–electron repulsions also increase. The electrons spread out more to accommodate the additional electrons.

In an isoelectronic series, ions contain the same number of electrons.

EXAMPLE: Arrange the following ions in order of increasing size: $Ca^{2+}$, $S^{2-}$, $K^+$, $Cl^-$.

All of the ions mentioned above are isoelectronic, containing 18 electrons each. In order of increasing size $Ca^{2+} < K^+ < Cl^- < S^{2-}$. $Ca^{2+}$ is the smallest because it contains the greatest number of protons. The greater the positive charge on the nucleus, the stronger the attraction for electrons.

## LATTICE ENERGY

(*Chemistry* 6th ed. pages 364–366 / 7th ed. pages 344–346)

Lattice energy is the energy release that occurs when separated gaseous ions are packed together to form an ionic solid.

$$X^{x+}(g) + Y^{y-}(g) \rightarrow X_y Y_x (s) + energy$$

$$Lattice\ energy = k\left(\frac{Q_1 Q_2}{r}\right).$$

$k$ is a proportionality constant dependent on the structure of the solid and the electron configuration of the ions.

$r$ is the distance in nanometers between the ion centers.

$Q_1$ and $Q_2$ are the charges of the ions.

> EXAMPLE: Which has the most exothermic lattice energy, NaCl or KCl?
>
> NaCl has the most exothermic lattice energy since all of the ions in both compounds have either +1 or –1 charges (the distance between the charges needs to be considered). $Na^+$ is smaller than $K^+$. So the distance between the centers of the sodium and chloride ions is less than the distance between the potassium and chloride ions.

## COVALENT BONDING

(*Chemistry* 6th ed. page 350 / 7th ed. page 331)

Covalent bonding involves sharing electrons between nuclei with similar electronegativities. A covalent or molecular compound is usually formed by bonding two nonmetals. There are some metallic compounds that are covalent, such as beryllium chloride, $BeCl_2$.

## ELECTRONEGATIVITY

(*Chemistry* 6th ed. pages 352–354 / 7th ed. pages 333–335)

Electronegativity measures the ability of an atom in a molecule to attract shared electrons to itself. In general, the trend in electronegativity follows the trend in atomic size. The smaller the atom, the greater its electronegativity. Electronegativity increases across a period with increasing atomic number and decreases going down a group.

EXAMPLE: Place the following atoms in order of increasing electronegativity: Cs, Rb, S, Al, Sr, O.

In order of increasing electronegativity: Cs < Rb< Sr < Al< S < O.

Electronegativity increases from the lower left to the upper right of the periodic table.

# POLAR BOND

(*Chemistry* 6th ed. pages 351–357 / 7th ed. pages 332–338)

The type of bond formed between two atoms is related to the difference in electronegativity between the two atoms. For identical atoms, such as a molecule of hydrogen, the electronegativity difference is zero. There is an equal sharing of electrons between the atoms, and the bond formed is a nonpolar covalent bond. When the difference in electronegativities is very small, the bond is described as nonpolar. An important nonpolar covalent bond is the C–H bond in organic compounds.

When two atoms have a very large difference in electronegativity, an ionic bond forms. For example, atoms of lithium and chlorine have a large difference in electronegativity and form an ionic compound.

A polar covalent bond results when there is an unequal sharing of electrons. The difference in the electronegativies of atoms in the polar covalent bond is between that of the ionic and nonpolar bonds. An example of a polar covalent bond is HF. HF has the following charge distribution.

$$H\text{–}F$$
$$\delta + \delta -$$

EXAMPLE: Which of the following three bonds will be most polar? O–F, N–F, C–F.

The most polar bond will be the bond in which the difference in electronegativies is the greatest. The C–F bond has the greatest electronegativity difference and will be the most polar bond in the example. C is the least electronegative among C, N, and O.

## BOND POLARITY AND DIPOLE MOMENTS

Fluorine is the most electronegative element in the periodic table. When HF is placed in an electric field, the molecules tend to orient themselves with the fluoride end closest to the positive pole and the hydrogen end closest to the negative pole. HF is said to have a dipole moment, having two poles. An arrow points to the negative charge center and the tail of the arrow indicates the positive center of charge.

$$\delta + \qquad\qquad \delta -$$

A diatomic molecule with a polar bond has a dipole moment. Molecules with more than two atoms can also have dipolar behavior.

This is addressed in the section on molecular shapes where we look at the distribution of charge in the molecule.

## LOCALIZED ELECTRON BONDING MODEL

(*Chemistry* 6th ed. page 374 / 7th ed. page 353)

The localized electron model assumes that a molecule is composed of atoms that are bound together by sharing pairs of electrons using the atomic orbitals of the bound atoms. Lone pairs are pairs of electrons which are localized in an atom. Bonding pairs occupy the space between atoms.

## LEWIS STRUCTURES

(*Chemistry* 6th ed. pages 375–383 / 7th ed. pages 354–361)

The Lewis structure shows how the valence electrons are arranged among atoms in a molecule.

Steps for writing Lewis structures:
1. Sum the total valence electrons from each atom. For a polyatomic ion, add one electron for each negative charge on the anion, or take away one electron for each positive charge on the cation.

2. Place the least electronegative element in the middle. (Hydrogen can never be in the middle because it can form only one bond.) Draw the outer atoms around the central atom. Chemical formulas are often written in the order in which they are connected, such as HCN. The central atom is sometimes written first in a formula, when a central atom has a group of other atoms bonded to it, as in $SO_4^{2-}$.

3. Connect the central atom to the outer atoms using a line to indicate each pair of bonding electrons.

4. Count the valence electrons used in the bonds in step 3 and subtract these from the total valence electrons in step 1 to determine how many valence electrons remain.

   Valence electrons left =
   Total valence electrons (step 1) – Valence electrons used (step 3).

5. Arrange the remaining electrons around the atoms to follow the octet rule for families 4A–7A outer atoms to satisfy the duet rule for hydrogen (hydrogen needs only two electrons to have a filled orbital) and the octet rule for second-row elements and the periods following.

6. Determine the number of valence electrons remaining as in step 4. Place the remaining electrons on the central atom to satisfy the octet rule for elements in the second-row and after.

a.  If there are no electrons remaining to satisfy the central atom, then add a double bond for every pair of electrons you are "short" and readjust the number of electrons on each surrounding atom.

b.  Elements in periods higher than the second can have more than eight electrons. If $n > 2$, then d orbitals exist and the number of electrons that can surround the element can exceed eight.

7.  Check that the valence electrons drawn = the total valence electrons in step 1.

EXAMPLE: Draw the Lewis Structure for $NF_3$

Step 1:  Total valence electrons = $[1 \times N(5e^-) + 3 \times F(7e^-)]$ =

$5\ e^- + 21e^-\quad$ = 26 valence electrons.

Steps 2 and 3: Place nitrogen in the middle and connect the fluorine atoms to nitrogen using a line to represent a single bond which is a pair of electrons.

$$
\begin{array}{c}
\text{F} \\
| \\
\text{F} \!-\!\! \text{N} \!-\!\! \text{F}
\end{array}
$$

Step 4: Count the valence electrons used. Count two electrons for each single bond ($2 \times 3$ single bonds = 6 electrons used). Count the electrons left. $26 - 6 = 20$ electrons left.

Step 5: Satisfy the octet rule for each of the outer fluorine, F, atoms.  Each F atom needs six more to complete the octet.

$$
\begin{array}{c}
:\!\ddot{\text{F}}\!: \\
| \\
:\!\ddot{\text{F}} \!-\!\! \text{N} \!-\!\! \ddot{\text{F}}\!:
\end{array}
$$

Step 6: Count electrons remaining after satisfying the outer atoms. $20 - 6\ (3)$ used = 2 electrons left. Place remaining electrons on central atom, nitrogen, and its octet is completed.

$$
\begin{array}{c}
:\!\ddot{\text{F}}\!: \\
| \\
:\!\ddot{\text{F}} \!-\!\! \ddot{\text{N}} \!-\!\! \ddot{\text{F}}\!:
\end{array}
$$

Step 7: Check that electrons drawn, 26, equals the total valence electrons in step 1.

## RESONANCE

(*Chemistry* 6th ed. pages 383–385 / 7th ed. pages 362–363)

Resonance occurs when more than one Lewis structure can be drawn for a molecule or ion.

**EXAMPLE:** Draw the Lewis structure for the nitrate ion, $NO_3^-$.

Step 1: Total valence electrons = 5 + 3(6) + 1 = 24 valence electrons (5 from N and 6 from each O and 1 from the negative charge on the ion).

Step 2: Place nitrogen in the middle and connect the oxygen atoms to nitrogen.

Step 4: Count valence electrons used, 2 for each single bond (2 × 3 single bonds = 6 electrons used). Count the electrons left. 24 – 6 = 18 electrons left.

Step 5: Satisfy the octet rule for each of the outer atoms of oxygen. Each O atom needs six more to complete the octet.

Step 6: Count electrons remaining after satisfying the outer atoms. 18 – 6(3) used = 0 electrons left. The central atom, N, still needs two more electrons to satisfy the octet rule. You can add a double bond for every pair of electrons you are "short."

One double bond forms between the N and the O. A double bond is made of two pairs of electrons. However, the nitrate ion does not have one double bond and two single bonds. All of the three bonds are the same length, somewhere intermediate between a single bond and a double bond. The nitrate ion exists as an average of three resonance structures as pictured below.

Step 7: Check that electrons drawn, 26, equals the total valence electrons in step 1.

## EXCEPTIONS TO THE OCTET RULE

Boron tends to form compounds in which boron can have less than eight electrons, such as boron trifluoride, $BF_3$. Boron can have six electrons around it.

Some elements in period 3 of the periodic table and beyond can exceed the octet rule. For example, $ClF_3$, has 10 electrons around the central atom.

Elements in the third period have 3s, 3p, and 3d valence orbitals. The valence electrons in chlorine occupy the 3s and 3p orbitals. The 3d orbital is empty and can hold extra electrons. Elements in the second period have only 2s and 2p orbitals. These elements cannot have more than eight electrons in their valence orbitals because they do not have d orbitals available.

## MOLECULAR STRUCTURE: THE VSEPR MODEL

(*Chemistry* 6th ed. pages 389–400 / 7th ed. pages 367–379)

The valence shell electron-pair repulsion (VSEPR) model predicts the geometries of molecules and polyatomic ions. The structure around the central atom is determined by minimizing electron-pair repulsions. Bonding and nonbonding pairs of electron are positioned as far apart as possible. Nonbonding or lone pairs of electrons require more room than bonding pairs and tend to compress the angle between bonding pairs.

Steps to apply the VSEPR model:

1. Draw the Lewis structure of the molecule.

2. Count the bonding and nonbonding electron pairs around the central atom. (For structures with multiple bonds, count the multiple bond as one effective pair of electrons.)

3. Count the number of atoms attached to the central atom. Look in the chart on the next page to find the arrangement that minimizes that repulsion, placing the electrons as far apart as possible. For example, $BeH_2$, another exception to the octet rule, has two electron pairs and two atoms attached to the central atom. If the two bonded pairs of electrons are placed $180^0$ apart, the resulting linear geometry has minimal repulsion.

## Molecular Shapes and Bond Angles

| No. of Electron Pairs | Electron Pair Geometry (Bond Angle) | No. of Bonding Electron Pairs | No. of Lone Electron Pairs | Molecular Geometry | Formula | 2D Structure | Hybrid Orbital |
|---|---|---|---|---|---|---|---|
| 2 | Linear (180°) | 2 | 0 | Linear | $BeH_2$ | | sp |
| 3 | Trigonal planar (120°) | 3 | 0 | Trigonal planar | $CO_3^{2-}$ | | sp² |
| 3 | Trigonal planar (120°) | 3 | 1 | Bent | $NO_2^-$ | | sp² |
| 4 | Tetrahedral (109.5°) | 4 | 0 | Tetrahedral | $CH_4$ | | sp³ |

| No. of Electron Pairs | Electron Pair Geometry (Bond Angle) | No. of Bonding Electron Pairs | No. of Lone Electron Pairs | Molecular Geometry | Formula | 2D Structure | Hybrid Orbital |
|---|---|---|---|---|---|---|---|
| 4 | Tetrahedral (>109.5°) | 3 | 1 | Trigonal pyramidal | $NH_3$ | | $sp^3$ |
| 4 | Tetrahedral (>>109.5°) | 2 | 2 | Bent | $H_2O$ | | $sp^3$ |
| 5 | Trigonal bipyramidal (90°, 120°) | 5 | 0 | Trigonal bipyramidal | $PCl_5$ | | $sp^3d$ |
| 5 | Tetrahedral (90° and >120°) | 4 | 1 | Seesaw unsymmetrical tetrahedron | $SF_4$ | | $sp^3d$ |

| No. of Electron Pairs | Electron Pair Geometry (Bond Angle) | No. of Bonding Electron Pairs | No. of Lone Electron Pairs | Molecular Geometry | Formula | 2D Structure | Hybrid Orbital |
|---|---|---|---|---|---|---|---|
| 5 | Tetrahedral (90° and 180°) | 3 | 2 | T-shaped | $BrF_3$ | | $sp^3d$ |
| 5 | Tetrahedral (180°) | 2 | 3 | Linear | $ICl_2^-$ | | $sp^3d$ |
| 6 | Octahedral (90°) | 6 | 6 | Octahedral | $SF_6$ | | $sp^3d^2$ |
| 6 | Octahedral (90°) | 5 | 1 | Square pyramidal | $BrF_5$ | | $sp^3d^2$ |

| No. of Electron Pairs | Electron Pair Geometry (Bond Angle) | No. of Bonding Electron Pairs | No. of Lone Electron Pairs | Molecular Geometry | Formula | 2D Structure | |
|---|---|---|---|---|---|---|---|
| 6 | Octahedral (90°) | 4 | 2 | Square planar | $ICl_4^-$ | | $sp^3d^2$ |

**EXAMPLE:** Predict the molecular structure of the carbon dioxide molecule. Is this molecule expected to have a dipole moment?

First we must draw the Lewis structure for the $CO_2$ molecule.

$$\ddot{O}=C=\ddot{O}$$

$$\longleftrightarrow$$

In this structure for $CO_2$, there are two effective pairs around the central atom (each double bond is counted as one effective pair). There are two atoms attached to the central atom.

According to the table, a linear arrangement is required.

Each C–O bond is polar, but, since the molecule is linear, the dipoles cancel out and the molecule is nonpolar.

**EXAMPLE:** Draw the Lewis structure for $PF_5$ and identify the molecular geometry.

$PF_5$ has 40 valence electrons. Connecting 5 F atoms to P, the central atom, uses 10 electrons in five single bonds. The 30 remaining electrons can be used to satisfy the octet rule for each F atom. There are 5 electron pairs around P and 5 atoms attached to P, resulting in a trigonal bipyramidal shape.

**EXAMPLE:** Draw the Lewis structure for $CH_3OH$.

This molecule has more than one central atom, the carbon and the oxygen. Hydrogen cannot be in the middle since it can form only one bond. Draw the correct Lewis structure and then using the rules for VSEPR, assign the molecular geometry around each central atom. The carbon atom has four electron pairs around it and four atoms attached resulting in tetrahedral geometry. Oxygen also has four electron pairs around it, but only has two atoms attached. This leads to bent geometry around oxygen.

EXAMPLE: Arrange the following molecules in order of increasing bond angles: $H_2S$, $CCl_4$, $NF_3$.

First, draw the correct Lewis structure for each molecule and then assign an approximate bond angle for each.

Each of the four molecules has four electron pairs around the central atom, resulting in a tetrahedral arrangement of the electron pairs about the central atom and approximately a 109.5° bond angle. $H_2S$ has two lone pairs on the central atom; $NF_3$ has one lone pair on the central atom. These lone pairs repel more than the bonded electron pairs and will cause the bond angle to be smaller than the expected $109.5^0$. The more lone pairs, the smaller the bond angle.

In order of increasing bond angles, the answer is $H_2S$ <$NF_3$ < $CCl_4$.

## HYBRIDIZATION

Hybridization describes the mixing of atomic orbitals to form special orbitals that share electrons during bond formation. For example, in methane ($CH_4$), the 2s orbital and three 2p orbitals of carbon mix together to form four $sp^3$ hybrid orbitals. Each $sp^3$ orbital of carbon overlaps the 1s orbital of hydrogen forming a sigma bond. A sigma bond, σ, is formed from an overlap of orbitals. Single bonds are sigma bonds.

In general, the sum of the superscripts on the hybrid orbitals equals the number of electron groups (multiple bonds count as one group) around the central atom. The carbon atom in $CH_4$ has 4 electron groups around the central atom. The sum of the superscripts in $sp^3$ equals 4.

The carbon atom of the double bond in *cis*-2-pentene on page 158 of this guide is $sp^2$ hybridized. There are 3 electron groups around the C atom of the double bond. The sum of the superscripts in $sp^2$ is 3. The 2s orbital and two of the 2p orbitals of carbon mix to form $sp^2$, leaving one 2p orbital unused. Each carbon atom of the double bond has an unhybridized p orbital perpendicular to the $sp^2$ orbital. These parallel p orbitals can share a pair of electrons above and below the carbon-carbon bond, forming a pi (π) bond.

A double bond, such as the one in *cis*-2-pentene, consists of one sigma bond because of the head-on overlap of $sp^2$ orbitals of each carbon atom and one pi bond due to the parallel p orbitals.

The hybrid orbitals are oriented in space according to the VSEPR Theory. For example, the four $sp^3$ orbitals have a tetrahedral arrangement (there are 4 electron groups around the central atom).

## BOND ENERGIES AND ENTHALPY

(*Chemistry* 6th ed. pages 371–374 / 7th ed. pages 350–353)

Estimate the change in enthalpy, $\Delta H$, for the following reaction using the table of bond energies on page 372 of the text:

$$N_2 + 3\,H_2 \rightarrow 2NH_3$$

First, draw the Lewis structures for the reactants and products.

In the reactants, one triple bond between the atoms of nitrogen and one single bond between atoms of hydrogen are broken. Bond breaking is an endothermic process that has a positive value for enthalpy. Energy must be added to break the bonds.

Three single bonds between hydrogen and nitrogen are formed in the products. Bond formation is exothermic having a negative value for enthalpy.

$\Delta H = \Sigma$ energies to break bonds + $\Sigma$ energies released to form new bonds.

### REACTANT BONDS BROKEN

| | | |
|---|---|---|
| $N_2$: 1 mol N≡N | 1 mol × 941 kJ/mol | = 941 kJ |
| $H_2$: 3 mol H – H | 3 mol × 432 kJ/mol | = 1296 kJ |
| | Total energy required | = 2237 kJ |

### PRODUCT BONDS FORMED

| | | |
|---|---|---|
| $NH_3$: 3 N – H | 6 mol × –391 kJ/mol | = –2346 kJ |

$\Delta H = 2237$ kJ + –2346 kJ = –109 kJ

## BOND LENGTH

(*Chemistry* 6th ed. page 372 / 7th ed. page 352)

As the number of bonds between two atoms increases, the bond grows shorter and stronger.

EXAMPLE: Arrange the following molecules in order of decreasing C–C bond length: $C_2H_4$, $C_2H_2$, $C_2H_6$.

In order of decreasing C–C bond length: $C_2H_6$, $C_2H_4$, $C_2H_2$.

First, draw the Lewis structure for each molecule. The C–C triple bond is the shortest, followed by the double bond, and the single bond.

## ORGANIC CHEMISTRY

In this section, the structure of carbon-containing compounds and their properties will be reviewed. You should know how to name the compounds, draw their structures inclusive of isomers, and identify trends in physical properties of the compounds based on their intermolecular forces.

## Hydrocarbons
(*Chemistry* 6th ed. page 1044 / 7th ed. pages 997–998)

Hydrocarbons contain hydrogen and carbon. Saturated hydrocarbons have all carbon–carbon single bonds. Unsaturated hydrocarbons contain carbon–carbon multiple bonds.

## Alkanes
(*Chemistry* 6th ed. pages 1044–1047 / 7th ed. pages 997–1001)

Each alkane has the general formula $C_nH_{2n+2}$; n is the number of carbon atoms. This is a saturated hydrocarbon. In straight-chain hydrocarbons, also called unbranched or normal, all of the carbon atoms are connected in a long chain. The name of each alkane ends in –ane. The names of the first ten members of the alkane series appear in the table below.

### Alkanes: Names and Formulas

| Name | Formula | Name | Formula |
| --- | --- | --- | --- |
| Methane | $CH_4$ | Hexane | $C_6H_{14}$ |
| Ethane | $C_2H_6$ | Heptane | $C_7H_{16}$ |
| Propane | $C_3H_8$ | Octane | $C_8H_{18}$ |
| Butane | $C_4H_{10}$ | Nonane | $C_9H_{20}$ |
| Pentane | $C_5H_{12}$ | Decane | $C_{10}H_{22}$ |

### Isomerism in Alkanes
(*Chemistry* 6th ed. pages 1048–1050 / 7th ed. pages 1001–1003)

Structural isomers occur when two molecules have the same chemical formulas but different structural arrangements of atoms. These structural differences result in differences in melting point, viscosity, and other properties that we examine later in this chapter.

**Example:** Draw all the isomers of hexane and name each one.
Step 1: The first isomer is the straight chain alkane, $C_6H_{14}$.

CH₃-CH₂-CH₂-CH₂-CH₂-CH₃

This is the condensed structural formula. Each C has 4 bonds.

Step 2: Take one C atom off of the end of the chain and place it anywhere along the chain except on the end carbon which will result in the same straight chain isomer.

CH₃-CH-CH₂-CH₂-CH₃
|
CH₃

Step 3: To name this isomer, circle the longest continuous chain that determines the root or parent name. In this case the longest chain is pentane. This is the last part of the alkane name,

Next, name the alkyl groups which are substituted on the parent chain. Count the number of carbons in the alkyl group. Using the root of name for the number of carbon atoms (meth, eth, and so on) drop the -ane from the hydrocarbon and change it to –yl. In the example here the branch or substituent group is called the methyl group.

Step 4: Specify the location of the substituent groups by numbering the longest chain of carbon atoms sequentially, starting at the end closest to the branching. Note, numbering left to right allows the methyl group to be at position number 2. If the longest chain is numbered from right to left, the methyl group would be on position number 4. The substituent groups should always have the lowest number possible, so the chain must be numbered from left to right.

The name of this branched alkane is 2-methylpentane.
A hyphen separates the number from the name.

Step 5: Continue to place the substituted group, methyl in this case, in as many places as you can along the chain such that you arrive at a different name. If the name of the branched alkane is the same as the ones you've already found, then a new isomer has not been found. The methyl group in this example can also be placed on carbon #3 of the parent chain. The name of the resulting alkane is 3-methyl pentane.

Note that placing the methyl group on carbon #4 yields the same name as 2-methylpentane which has already been found. Remember, the parent chain must be numbered to give the lowest number for the substituents.

Step 6: When you have identified all possibilities in step 5, remove two C atoms off of the straight chain hydrocarbon in step 1. Place these C atoms along the remaining parent chain, naming each one. A new correct name, means a new isomer.

$$CH_3\text{-}CH\text{-}CH\text{-}CH_3 \quad 2,3-\text{dimethylbutane.}$$
$$\quad\quad | \quad\ |$$
$$\quad\quad CH_3 \ CH_3$$

The longest chain is 4 C atoms. Numbers are separated by commas. Prefixes di-, tri-, and so on, are used to indicate multiple identical substituents.

Another isomer

$$CH_3 - \underset{\underset{CH_3}{|}}{\overset{\overset{CH_3}{|}}{C}} - CH_2 - CH_3 \quad 2,2-\text{ dimethylbutane.}$$

All isomers based on the parent chain butane have been identified. If you try to remove an additional atom, for a total of three, from the straight chain hexane and place the remaining three on the parent chain propane, you end up duplicating isomers which have already been identified. For example, if you draw 2-ethyl-2-methylpropane and then name it, circling the longest parent chain, the correct name is 2,2–dimethylbutane.

To summarize the naming of branched alkanes: The location and number of each substituent are followed by the root alkane name, the substituents are listed in alphabetical order, and the prefixes di-, tri-, and so on, are used to indicate multiple, identical substituents.

## ALKENES

(*Chemistry* 6th ed. pages 1052–1053 / 7th ed. pages 1005–1006)

Alkenes are hydrocarbons that contain at least one double bond. Alkenes have the general formula, $C_nH_{2n}$.

NOMENCLATURE OF ALKENES The root ending for alkenes is –ene instead of –ane.

For alkenes which have more than three carbon atoms, the lowest numbered carbon atom is the location of the double bond.

Isomers of 2-pentene appear on the next page. The longest continuous chain of carbon atoms is five. So the parent name of the alkene will be pentene. There are no substituted groups on the pentene chain. Numbering the chain from left to right will yield a parent name of 2-pentene. The double bond must have the lowest number of carbon atoms involved in the bond.

Alkenes exhibit *cis–trans* isomerism because the rotation of the double bond in alkenes is restricted. Look at each C atom of the double bond. The heavier of the two groups (or atoms) attached to each carbon is circled in the structure below. So on C atom #2, $CH_3$ is heavier than H. In the *cis* isomer, the heavier groups are on the same side of the double bond. In the *trans* isomer, the heavier groups are on opposite sides of the double bond.

*cis*-2-pentene                                    *trans*-2-pentene

The $CH_3$ and $CH_2CH_3$ groups which are in the longest chain containing the C=C lie on the same side of the C=C.

The $CH_3$ and $CH_2CH_3$ groups which are in the longest chain containing the C=C lie across the C=C.

## ALKYNES

*(Chemistry 6th ed. page 1053 / 7th ed. page 1006)*

Alkynes are hydrocarbons that contain at least one triple bond. An example of an alkyne is ethyne, $C_2H_2$, more commonly called acetylene. Alkynes have the general formula $C_nH_{2n-2}$ and are named similarly to the alkenes; the –yne ending is substituted for the –ane on the longest alkane chain.

## CYCLIC HYDROCARBONS

*(Chemistry 6th ed. pages 1051–1052, 1054 / 7th ed. pages 1004–1005, 1006–1007)*

Carbon atoms also form ring-shaped structures.
Cyclic alkanes and alkenes are represented below.
A carbon atom is located at the vertex of each polygon.
For each cyclic alkane, only two hydrogen atoms can attach to each carbon atom.

Cyclopropane    Cyclobutane    Cyclopentane    Cyclohexene
$C_3H_6$         $C_4H_8$         $C_5H_{10}$         $C_6H_{10}$

Substitution can also occur in cyclic hydrocarbons.
            methylcyclobutane

     $CH_3$

## AROMATIC HYDROCARBONS

*(Chemistry 6th ed. page 1055 / 7th ed. pages 1007–1008)*

Aromatic compounds are a special class of cyclic unsaturated hydrocarbons. The simplest structure is benzene, $C_6H_6$, which appears below. Benzene has alternating single and double bonds. Benzene exhibits resonance. More complex aromatic hydrocarbons can be thought of as consisting of a number of "fused" benzene rings.

## FUNCTIONAL GROUPS

*(Chemistry 6th ed. pages 1058–1063 / 7th ed. pages 1010–1015)*

Many organic molecules contain elements in addition to hydrogen and carbon. Functional groups are atoms or groups of atoms attached to the hydrocarbon which have characteristic chemistry.

| Functional Group | Name of Group | Example of IUPAC name, Common Name |
|---|---|---|
| –X | Halide | $CH_3CH_2Br$ (bromoethane) |
| –OH | Alcohol | $CH_3OH$ (methanol) |
| –O– | Ether | $CH_3OCH_3$ (dimethyl ether) |
| $\overset{O}{\underset{\parallel}{-C-H}}$ | Aldehyde | $CH_3CHO$ (ethanal, acetaldehyde) |
| $\overset{O}{\underset{\parallel}{-C-}}$ | Ketone | $CH_3COCH_3$ (propanone, acetone) |
| $\overset{O}{\underset{\parallel}{-C-OH}}$ | Carboxylic acid | $CH_3CO_2H$ (ethanoic acid, acetic acid) |
| $\overset{O}{\underset{\parallel}{-C-O-}}$ | Ester | $CH_3CO_2CH_3$ (methyl ethanoate, methyl acetate) |
| $\overset{O}{\underset{\parallel}{-C-NH_2}}$ | Amide | $CH_3CONH_2$ (acetamide) |
| $-\underset{\vert}{N}-$ | Amine | $CH_3NH_2$ (methylamine) Note: H or an alkyl group is attached to N. |

## AP tip

You should know how to name compounds containing functional groups, draw their structures, identify bond angles, determine changes in enthalpy from bond energy values and predict trends in properties based on intermolecular forces. On the AP exam organic chemistry is often included in other types of problems such as thermochemistry and stoichiometry of combustion reactions.

# INTERMOLECULAR FORCES

*(Chemistry 6th ed. pages 450–453 / 7th ed. pages 426–429)*

Intermolecular forces are attractions between molecules.

## DIPOLE–DIPOLE ATTRACTIONS

Polar molecules attract each other, lining up so that their positive and negative poles are close to each other. Polar molecules generally have higher boiling points than nonpolar molecules of similar molar mass because they have dipole–dipole attractions in addition to London forces. An example of a polar molecule is $SO_2$.

## HYDROGEN BONDNG

Hydrogen bonding is an unusually strong dipole–dipole force among molecules in which hydrogen is bound to a highly electronegative atom such as nitrogen, oxygen, or fluorine. The hydrogen bond occurs between the hydrogen on one atom and the O, F, or N on another molecule. This is the strongest type of intermolecular force. Examples of molecules which exhibit H-bonding include $H_2O$, $NH_3$, and HF.

## LONDON DISPERSION FORCES

London dispersion forces exist between all molecules and account for the boiling points of the noble gases. If there were no attractions between molecules, then they would never liquefy. We believe that the electron cloud can experience temporary shifts that result in one side of the molecule becoming more negative than the other, causing a temporary dipole. This instantaneous dipole can induce a similar dipole on a neighboring atom.

We'd expect that the greater the charge on the nucleus and the larger the number of electrons in the molecule, the greater the induced dipole. As the atomic number increases down a group, atoms become more polarizable, that is, the electron cloud can become polarized due to the instantaneous dipole. For example, the halogens all experience London dispersion forces, but the force becomes stronger toward the bottom of the group. $F_2$ and $Cl_2$ are found as gases in nature, $Br_2$ is a liquid, and $I_2$ is a solid due to increasing London forces as the atomic number and size of the electron cloud increases in the group.

> **EXAMPLE:** Arrange $C_2H_6$, $CH_4$, $C_4H_{10}$ in order of increasing boiling point.
>
> In order of increasing boiling point $CH_4$, $< C_2H_6$, $< C_4H_{10}$. All of these molecules exhibit London dispersion forces. The higher the molar mass, the more polarizable the molecule becomes. $C_4H_{10}$ has the greatest molar mass, followed by $C_2H_6$, and has the strongest London dispersion forces.

## MULTIPLE-CHOICE QUESTIONS

No calculators are to be used in this section.

1. The compound most likely to be ionic is
   (A) KF.
   (B) $CCl_4$.
   (C) $CO_2$.
   (D) ICl.
   (E) $CS_2$.

**ANSWER: A** Bonds formed from elements with greater differences in electronegativity are most likely ionic in nature. (Electronegativity increases for elements as you move up and to the right on the Periodic Table.) (*Chemistry* 6th ed. pages 358–359 / 7th ed. pages 339–340)

2. Ranking the ions $S^{2-}$, $Ca^{2+}$, $K^+$, $Cl^-$, from smallest to largest gives the order as
   (A) $S^{2-}$, $Cl^-$, $K^+$, $Ca^{2+}$.
   (B) $Ca^{2+}$, $K^+$, $Cl^-$, $S^{2-}$.
   (C) $K^+$, $Ca^{2+}$, $Cl^-$, $S^{2-}$.
   (D) $Cl^-$, $S^{2-}$, $K^+$, $Ca^{2+}$.
   (E) $Ca^{2+}$, $K^+$, $Cl^-$, $S^{2-}$.

ANSWER: **A** Note that all ions in this isoelectronic series have an Ar electron configuration; therefore the nuclear charge determines the size (*Chemistry* 6th ed. pages 360–362 / 7th ed. pages 340–342).

3. The type of bonding within a water molecule is
   (A) ionic bonding.
   (B) polar covalent bonding.
   (C) nonpolar covalent bonding.
   (D) metallic bonding.
   (E) hydrogen 'bonding.'

ANSWER: **B** This question is asking you to consider the nature of the bonding between hydrogen and oxygen in water (note the term *within* in the question!). With an electronegativity difference (H = 2.1 and O = 3.5) of 1.4, this suggests a bond of a very polar covalent nature due to the uneven sharing of the pair of electrons between hydrogen and oxygen. (Note: The force <u>between</u> one water molecule and another water molecule is known as a hydrogen 'bond.') (*Chemistry* 6th ed. pages 354–358 / 7th ed. pages 335–339).

4. Dinitrogen oxide, $N_2O$, has two double bonds. The general structure is N=N=O. The formal charge on the oxygen atom in this molecule is
   (A) zero.
   (B) positive one (+1).
   (C) positive two (+2).
   (D) negative one (-1).
   (E) negative two (-2).

ANSWER: **A** To determine the formal charge on an atom you need to determine the difference in the number of electrons assigned to the atom in the molecule and the number of outermost (valence) electrons on the free atom. In this case, the double bond between nitrogen and oxygen allows oxygen to "own" two (one-half of the four) electrons in the bond, plus totally own the other four electrons. (If this is not clear to you, draw the Lewis structure for this molecule.) Since the number of electrons assigned to oxygen in the molecule and as a free atom is the same, the formal charge is zero (*Chemistry* 6th ed. pages 385–388 / 7th ed. pages 363–367).

5. The Lewis structure of which molecule requires resonance structures?
(A) $MgCl_2$
(B) $PCl_5$
(C) $SiO_2$
(D) $SO_2$
(E) $OCl_2$

ANSWER: **D** Watch for *resonance* when it is possible to draw more than one valid Lewis structure. This tends to occur when the same kind of atoms are bonded once as a single bond and again with a multiple bond holding the same kind of atoms together. In this case there is a single bond between sulfur and one of the oxygens, and a double bond between sulfur and the other oxygen (*Chemistry* 6th ed. pages 383–385 / 7th ed. pages 362–363).

6. The predicted geometry (shape) of $PH_3$, according to the VESPR theory, is
(A) linear.
(B) bent or angular.
(C) trigonal pyramidal.
(D) tetrahedral.
(E) trigonal planar.

ANSWER: **C** It is the position of the nuclei upon which we base the shape of a molecule. From the Lewis structure, you can see an unshared pair of electrons on the P; it is this unshared pair which repels the bonded pairs (between P and H) down toward the corners of a tetrahedron. However, the top pair does not contribute to the shape; hence the geometry is a trigonal pyramid (*Chemistry* 6th ed. pages 389–400 / 7th ed. pages 367–379).

7. Rank the following from lowest to highest boiling temperature:
   I $C_2H_6$  II $C_2H_5OH$  III $C_2H_5Cl$
(A) I<II<III
(B) II<III<I
(C) III<II<I
(D) III<I<II
(E) I<III<II

ANSWER: **E** Consider the IMF (intermolecular forces) between molecules. $C_2H_6$, with the lowest boiling temperature is nonpolar and hence has a low IMF. $C_2H_5OH$ has hydrogen 'bonding' between molecules. Watch for these high IMF when hydrogen is bonded to O, N, or F. $C_2H_5Cl$ is polar covalent and has intermediate strength IMF between molecules. (Measured boiling temperatures are as follows: for $C_2H_6$ it is -88.3°C, for $C_2H_5OH$ it is +78.5°C, and for $C_2H_5Cl$ it is +12.3°C.) (*Chemistry* 6th ed. pages 450–453 / 7th ed. pages 429–429).

8. The total number of lone pairs in $PCl_3$ is
   (A) 1.
   (B) 8.
   (C) 10.
   (D) 12.
   (E) 14.

ANSWER: **C** A sketch of the Lewis structure will show three lone pairs on each of the chlorine atoms and one more lone pair on the phosphorus atom, for a total of 10 lone pairs (*Chemistry* 6th ed. pages 357–359 / 7th ed. pages 338–340).

9. The Lewis structure for carbon dioxide is
   (A) :C=O-O:
   (B) :C̈=O-O:
   (C) :C̈-Ö-Ö
   (D) :Ö=C=Ö:
   (E) :Ö=C=Ö:

ANSWER: **D** First count the number of total outermost (valence) electrons. Each oxygen has six and the carbon has four, for a total of sixteen. Only response D, with two double bonds, has sixteen valence electrons (*Chemistry* 6th ed. pages 374–382 / 7th ed. pages 353–361).

10. Which of the following molecules has a net dipole (dipole moment)?
    I. $CO_2$   II. $SO_3$   III. $CCl_4$   IV. $PF_3$.

    (A) I and II only
    (B) II and III only
    (C) III only
    (D) IV only
    (E) II, III, and IV

ANSWER: **D** In order to have a dipole moment there must be polar covalent bonds present and an assymetrical distribution of charge. In order to determine this, you must consider both the nature of the bonding (extent of polarity) and the arrangement of these bonds (the molecular shape). If all the vectors, resulting from polar bonds, cancel each other, then there is zero dipole moment. In the case of $PF_3$, the shape is that of a trigonal pyramid, the vectors do not cancel each other, and the molecule is polar (*Chemistry* 6th ed. pages 354–357 / 7th ed. pages 335–338).

## FREE-RESPONSE QUESTIONS

1. Is it possible to have
   (a) a nonpolar molecule that is made of polar bonds?
   (b) a polar molecule that is made of nonpolar bonds?

   Answer both parts (a) and (b). Support your answers with discussion.

   (c) Sodium iodide boils at 1300°C. Sulfur hexafluoride sublimes at −64°C. Yet the difference in electronegativity between sodium and iodine (1.6) is almost exactly the same as that between sulfur and fluorine (1.5). Explain why there are such differences in physical properties between NaI and $SF_6$.

ANSWERS:

(a) Yes, it is possible to have a nonpolar molecule that contains polar bonds. This is possible if the polar bonds are arranged so that their charge distributions cancel each other, thereby having no separation between the centers of positive and negative charge,

(b) No, it is not possible to have a polar molecule formed from only nonpolar bonded atoms. Without the polarity of some unequal sharing, there is no way for the molecule to have any charge separation.

(c) NaI is ionic; in crystal form, each positive Na ion is surrounded by six negative I ions, and each $I^-$ is surrounded by six $Na^+$. Powerful ionic bonds lead to a high boiling temperature for NaI. $SF_6$ is a covalent compound. In this compound, six fluorine atoms surround one sulfur atom to form a neutral molecule. The symmetry of $SF_6$ (an octahedral molecule) and the high electron density of fluorine lead to little attraction, hence low IMF, and a lower boiling temperature (*Chemistry* 6th ed. pages 351–357, 362–368, 450–453, 478–483, and 489 / 7th ed. pages 332–338, 342–348, 426–429, 454–459, and 462).

2. The combinations of nitrogen and oxygen forming different molecules are many and interesting. Their properties differ greatly. For this question consider nitrogen dioxide.
   (a) Draw the Lewis structure for $NO_2$, which is an odd electron compound.
   (b) What kind of hybridization does the nitrogen display in $NO_2$? Explain how you arrived at this answer. Compare this with the hybridization in the linear $NO_2^+$ ion.
   (c) Estimate the O–N–O bond angle in $NO_2$. What is this geometric shape called? Explain how you arrived at this answer.
   (d) Discuss the reactivity of the $NO_2$ molecule based on its structure. Explain how you arrived at this answer.

ANSWERS:

(a) :Ö=Ṅ-Ö:

(b) Nitrogen displays $sp^2$ hybridization (leaving a 'p' orbital unhybridized for the pi bonding). The hybridization of nitrogen in the ion (:O=N⁺=O:) is sp, allowing for formation of the double bonds between nitrogen and each of the oxygens.

(c) This would usually suggest a bond angle of 120°. However, because the lower repulsion between the lone nonbonded electron and the bonded pairs is lower than the repulsion between the bonded pairs on the oxygen, the O-N-O angle should be greater than 120° (Linus Pauling, in his book *Nature of the Chemical Bond* gives 140°). The shape is 'bent' (angular).

(d) The properties of this molecule would certainly be related to that lone electron on the nitrogen. For example, this material is very reactive and dimerizes (reacts with itself to form $N_2O_4$). (*Chemistry* 6th ed. pages 375–383, 385–388, 391–398, and 421 / 7th ed. pages 354–362, 363–367, 369–376, and 397–398).

3. (a) Show the complete equation for the combustion of the flammable gas butene, $C_4H_8$.
   (b) Given the following table of bond energies, estimate the enthalpy change, $\Delta H$, for the reaction noted in 3(a).
   (c) Often the heat of reaction (enthalpy) calculated from bond energies differs by 10–20% or more from the laboratory determined values. Suggest why this is so.
   (d) Show two of the several isomers of butene.
   (e) Draw a Lewis structure diagram of butene. Indicate the kinds of hybridization of carbon in this molecule. Indicate all bond angles.

ANSWERS:

(a)    $C_4H_8(g) + 6O_2(g) \rightarrow 4CO_2(g) + 4H_2O(g)$

     (*Chemistry* 6th ed. pages 102–106)

(b)    Average Bond energies (kJ/mol)

     C-H    413

     C-C    347 (single)

     C=C    614 (double)

     C≡C   839 (triple)

     C-O    358

     C=O    799

     H-O    467

     H-H    432

     O=O    495

From $CH_2CHCH_2CH_3$ + 6O-O → 4O=C=O + 4H-O
$$\backslash$$
$$H$$

Bonds broken (reactants):

1 C=C  1 mol × 614 kJ/mol  = 614 kJ

2 C-C  2      × 347        = 694

8 C-H  8      × 413        = 3304

6 O-O  6      × 495        = 2970

Total energy required to break bonds = +7582 kJ.

Bonds formed (products):

8 C=O   8 mol × 799 kJ/mol =  6392 kJ

8 H–O   8      × 467        = 3736

Total energy released as bonds form = –10128 kJ.

Note that it is a negative value, denoting an exothermic reaction.

$\Delta H$ = –2546 kJ/mol $C_4H_8$ reacting.

(*Chemistry* 6th ed. pages 372–374 / 7th ed. pages 351–353)

(c)  The surrounding bonds often affect the strength of a given bond. For example, the average C=O bond energy is 745 kJ/mol, but the C=O bond energy in $CO_2$ is 799 kJ/mol (the value given for your use in the table).  Examining the structure of butane, you might guess that the C-H bond energies of bonds near a C=C double bond would be somewhat different from those with a C-C single bond attached to the same carbon.  The environment of the bond does make a difference (*Chemistry* 6th ed. page 371 / 7th ed. pages 350–351).

(d)  1-butene is $CH_3$-$CH_2$-$CH=CH_2$

2-butene is $CH_3$-$CH=CH-CH_3$

(this form might also be shown as both *cis*- and *trans*-isomers)

2=methyl propene   $\begin{array}{c} CH_2=C-CH_3 \\ | \\ CH_3 \end{array}$

(*Chemistry* 6th ed. pages 1003–1009, 1053 / 7th ed. pages 960–965, 1006)

(e)

```
   H      H H H
    \     / | |
     C = C - C - C - H
    /       | |
   H        H H
```

The first and second carbons display sp$^2$ hybridization with bond angles of 120°.

The other two carbons are sp$^3$ hybridized with bond angles of 109° (*Chemistry* 6th ed. pages 1052–1054 / 7th ed. pages 1005–1006).

# 9

# LIQUIDS AND SOLIDS

The forces that hold together solids and liquids, the condensed states of matter, are similar. The effect of these forces on properties such as surface tension and vapor pressure will be reviewed. The bonding models, structure, and properties of liquids and solids will be discussed. Changes in state from solid to liquid to gas will also be considered.

You should be able to

- Explain how intermolecular forces such as surface tension, capillary action, viscosity, vapor pressure, and boiling point affect the properties of liquids.
- Rank substances in order of decreasing or increasing properties such as boiling points.
- Understand the different types of solids and know examples of each.
- Know characteristic points of heating curves and phase diagrams.

## AP tip

Although the topics in this chapter can appear in the multiple-choice section of the exam, it is often the case that many of these topics such as the effect of intermolecular forces on properties of liquids and the structures and properties of solids appear in the essay section.

## INTERMOLECULAR FORCES

(*Chemistry* 6th ed. pages 450–453 / 7th ed. pages 426–429)

Intermolecular forces were introduced in the last chapter. In this section, you will see the effect of intermolecular forces on properties of liquids such as boiling point, surface tension, and vapor pressure.

When molecular compounds such as water undergo changes in state, a disruption of forces between molecules is involved. The covalent bonds in the molecules are not broken. When ionic compounds such as sodium chloride change state, ionic bonds are broken.

## PROPERTIES OF LIQUIDS

### SURFACE TENSION

(*Chemistry* 6th ed. page 454 / 7th ed. page 429)

Molecules at the surface of a liquid are pulled inward by the molecules beneath them. The surface tension of a liquid is the resistance of a liquid to increasing its surface area. Liquids with strong intermolecular forces tend to have high surface tensions.

EXAMPLE: Equal-sized drops of Hg, $Br_2$, or $H_2O$ are placed onto the same glass surface. Which of the three liquids will have the most spherical shape?

ANSWER: Hg will have the most spherical shape due to the metallic bonding which occurs between mercury atoms. Metallic bonding is stronger than both the hydrogen bonding in water and the dispersion forces of bromine.

### CAPILLARY ACTION

(*Chemistry* 6th ed. page 454 / 7th ed. page 429)

The rise of liquids up very narrow tubes is called capillary action. Water and dissolved nutrients move upward through plants via capillary action.

Adhesive forces, the attractions between a liquid and its container, cause the liquid to creep up the walls of the container. Adhesive forces are opposed by cohesive forces, the intermolecular forces in the liquid that try to decrease the liquid's surface area.

EXAMPLE: Two different liquids are placed in separate glass tubes. Liquid A forms a concave meniscus. Liquid B forms a convex meniscus. Why do the meniscis of liquids A and B have two different shapes? Suggest a possible identity for each liquid.

ANSWER: Liquid A's adhesive forces are stronger than its cohesive forces, as in water. Liquid B's cohesive forces are stronger than its adhesive forces, as in mercury.

## VISCOSITY

(*Chemistry* 6th ed. page 455 / 7th ed. pages 429–430)

Viscosity is a measure of a liquid's resistance to flow. The greater its resistance, the more slowly it will flow. Molasses, motor oil, and glycerol are viscous liquids. Highly viscous liquids have large intermolecular forces or are very complex molecules with large molar masses such as polymers. For example, grease is viscous because it contains carbon chains 20 to 25 carbons long which can get tangled. Gasoline consists of shorter carbon chains three to eight carbons long and is therefore a nonviscous liquid. Viscosity decreases with temperature because more molecules will have greater average kinetic energy to overcome the intermolecular forces of the liquid.

## VAPOR PRESSURE

(*Chemistry* 6th ed. pages 483–487 / 7th ed. pages 459–463)

In the process of vaporization (evaporation), liquid molecules leave the surface and form a vapor. The heat of vaporization (enthalpy of vaporization, $\Delta H_{vap}$) is the energy required to vaporize 1 mole of a liquid at its normal boiling point (1 atm). This quantity is endothermic because energy is required to overcome the intermolecular forces. In the reverse process, condensation, energy is released as the IMFs form between the particles when the gas molecules become closer together in the liquid.

In a closed container, a liquid is at equilibrium when the rate of evaporation equals the rate of condensation. The equilibrium vapor pressure is the pressure above the liquid from the molecules of the liquid in the gas phase.

Volatile liquids are liquids with high vapor pressures. The vapor pressure at a given temperature is determined by the intermolecular forces of the liquid. The stronger the forces, the lower the vapor pressure. The weaker the forces, the higher the vapor pressure because liquid molecules are not held as strongly by intermolecular forces within a liquid and can readily escape from the surface.

The vapor pressure of a liquid increases with increasing temperature. At higher temperatures, more molecules will have sufficient kinetic energy to overcome intermolecular forces.

The dependence of vapor pressure on temperature is represented in the figure on the next page.

> **EXAMPLE:** Explain why diethyl ether, $CH_3CH_2OCH_2CH_3$, is more volatile than ethanol, $CH_3CH_2OH$, at 25°C.

> **ANSWER:** The dipole forces in diethyl ether are weaker than the hydrogen bonding in ethanol. The weaker the intermolecular forces, the higher the liquid's vapor pressure and volatility.

## BOILING

(*Chemistry* 6th ed. pages 491–492 / 7th ed. page 466)

The boiling of a liquid occurs in a container open to the atmosphere when the vapor pressure of the liquid equals atmospheric pressure.

Bubbles containing the vapor form within the liquid and rise to the surface due to their reduced density. During boiling, the temperature remains constant until all of the liquid boils away. The normal boiling point of a liquid is the temperature at which the vapor pressure of a liquid is equal to standard atmospheric pressure (1 atm = 760 mm Hg). To determine the normal boiling point of a liquid, interpolate the vapor pressure curve for the liquid at 1 atmosphere.

> **EXAMPLE:** What are the normal boiling points of diethyl ether and ethanol?
>
> **ANSWER:** The normal boiling points for diethyl ether and ethanol are approximately 34.6°C and 78.4°C, respectively. The normal boiling point occurs at 1 atm and can be found by interpolating the vapor pressure curves above.
>
> **EXAMPLE:** At a location 19,340 ft above sea level, the atmospheric pressure is 350 torr. At what temperature will water boil? Will it take more or less time to cook an egg at this location than at sea level or 1 atm?
>
> **ANSWER:** Using the vapor pressure curve above, at 350 torr, water will boil at about 80°C. It will take more time to hard-boil an egg at this elevation because water boils at a temperature lower than that at sea level.
>
> **EXAMPLE:** Pressure cookers heat a small amount of water under high constant pressure. What effect does using a pressure cooker have on cooking time?
>
> **ANSWER:** Food in a pressure cooker will take less time to prepare because it will cook at a higher temperature. When the pressure is increased, the boiling point will increase until the vapor pressure of the liquid equals the pressure above the liquid. Once the boiling point is reached, the temperature will not change as long as there is liquid present. Vapor pressure is directly proportional to temperature.

## USING VAPOR PRESSURE DATA

(*Chemistry* 6th ed. pages 486–488 / 7th ed. pages 461–463)

The Clausius-Clapeyron equation can be used to determine the vapor pressure of a liquid at a different temperature or to find the heat of vaporization of a substance.

$$\ln\left(\frac{P_{\text{vap},T_1}}{P_{\text{vap},T_2}}\right) = \frac{\Delta H_{\text{vap}}}{R}\left(\frac{1}{T_2} - \frac{1}{T_1}\right).$$

In an experiment, plotting ln$P$ vs. $1/T$ ($K^{-1}$) will result in a line whose slope is equal to $-\Delta H_{vap}/R$, where $R = 8.3145$ J mol$^{-1}$K$^{-1}$.

> **EXAMPLE:** Determine the boiling point of water in Breckenridge, Colorado, where the atmospheric pressure is 520 torr. For water, $\Delta H_{vap} = 40.7$ kJ/mol.
>
> **ANSWER:** 362 K or 89°C.
>
> At 100°C, 373 K, the vapor pressure of water is 1.00 atm.
>
> $$\ln\left(\frac{520 \text{ torr}}{760 \text{ torr}}\right) = \left[\frac{(4.07 \times 10^4 \text{ J/mol})}{(8.3145 \text{ J/K} \times \text{mol})}\right] \times \left(\frac{1}{373 \text{ K}} - \frac{1}{T_2}\right)$$
>
> $$T_2 = 362 \text{ K or } 89°C.$$

# STRUCTURE AND TYPES OF SOLIDS

(*Chemistry* 6th ed. pages 456, 459–461/ 7th ed. pages 430–432, 435–436)

## CRYSTALLINE SOLIDS

Crystalline solids are classified according to the type of particle that occupies the lattice points: atoms, molecules, or ions. The four types of crystalline solids are molecular, ionic, metallic, and network covalent. The table below describes each of these solids.

| Type of Solid | Molecular | Ionic | Atomic | Metallic | Network Covalent |
|---|---|---|---|---|---|
| Examples | $C_6H_{12}O_6$ | NaCl | Ar | Cu | C (diamond) |
| Particles occupying lattice points | Molecule | Ions | Atoms | Atoms | Atoms |
| Bonds or forces between lattice points | Dispersion forces | Ionic | Dispersion forces | Delocalized nondirectional metallic | Covalent |
| Melting points | Low | High | Low | High (1083°C) | Very high (3500°C) |
| Electrical conductivity as solid | None | None | None | Yes | None |
| Electrical conductivity when melted | None | Yes | None | Yes | None |
| Other properties | | brittle | | Malleable and ductile | Insulator very hard |

A lattice is a three-dimensional system of points designating the points of the components (atoms, molecules, or ions) that make up the substance. A unit cell is the smallest repeating unit of a lattice.

## AMORPHOUS SOLIDS

Amorphous solids, such as rubber and glass, have particles with no orderly structure.

## STRUCTURE AND BONDING IN METALS

It is said that the bonding is nondirectional; the metallic atoms can easily slide over each other because the environment will be the same. Metals are malleable, can be formed into thin sheets, are ductile, and can be pulled into wires.

## ALLOYS

Alloys are mixtures of metallic elements and have metallic properties. Substitutional alloys, like brass and sterling silver, have host metal atoms replaced by atoms of similar size. Brass is made of copper and zinc. Interstitial alloys have smaller atoms occupying the holes in the metal structure. The presence of the smaller atoms changes the properties of the host making it harder and stronger. Steel contains iron and carbon.

# CHANGES IN STATE

## HEATING CURVES

(*Chemistry* 6th ed. pages 488–492 / 7th ed. pages 463–466)

The chapter on thermochemistry discussed the quantitative aspects of the heating curve. Energy is added at a constant rate to a solid and a plot of temperature vs. time is drawn. The first plateau on the heating curve represents the melting point of the substance. The second plateau on the heating curve represents the boiling point of the substance. The temperature does not change during melting or boiling because the energy added is used to overcome intermolecular forces.

Usually, the plateau for the boiling point is longer in duration than the melting plateau because it takes more energy to separate the molecules from a liquid into a gaseous state than from a solid into a liquid state. This greater energy is related to the greater forces that must be overcome.

## PHASE DIAGRAMS

(*Chemistry* 6th ed. pages 492–497 / 7th ed. pages 467–471)

Phase diagrams graphically summarize the conditions under which equilibria exist between different states of matter of a substance. The state of matter, which is stable at a given set of temperature and pressure conditions, can be determined from the phase diagram.

You should be able to interpret and sketch the phase diagram of a substance.

EXAMPLE: Consider the phase diagram for water. At 1.0 atmosphere, describe the phases of matter present as the temperature is increased from very low to high temperatures in experiment 1.

ANSWER: At low temperatures, solid water (ice) is present. As the temperature is increased to 0°C, water reaches its normal melting point at 1 atm. Water in its solid and liquid state is present along the solid/liquid line from point A to point B. The line from point A through point B represents the change in melting point with pressure.

Line AB, the solid–liquid equilibrium line for water, has a negative slope. Ice is less dense than liquid water. Carbon dioxide, on the contrary, has a solid–liquid line with a positive slope; solid $CO_2$ is more dense than liquid $CO_2$. The triple point of water is lower than that of $CO_2$. This means that water exists as a solid, liquid, and a gas at about 1 atm. Because the triple point of solid $CO_2$ is at 5.11 atm, it does not melt at pressures less than 5.11 atm; it sublimes and becomes a gas.

Line AC represents the vapor pressure curve of the liquid. Point C is the critical point of the liquid. It corresponds to the critical temperature and the critical pressure. The critical temperature is the highest temperature at which a liquid can exist. Beyond the critical temperature, no amount of pressure will liquefy a gas. The critical pressure is the pressure required to liquefy a vapor at the critical temperature. The liquid vapor line on the phase diagram ends at this point.

At point A, the triple point, all three states of matter are present. Under these conditions, the three states of the substance will coexist in a closed system.

## MULTIPLE-CHOICE QUESTIONS

No calculators are to be used in this section.

1.  Water has a higher capillary action than mercury due to
    (A) higher dipole-dipole forces between the water molecules.
    (B) strong cohesive forces within water.
    (C) very significant induced intermolecular attractions.
    (D) weak adhesive forces in water.
    (E) strong cohesive forces in water which work with strong adhesive forces.

ANSWER: (E) The strong adhesive forces lead to a creeping effect as water moves up the narrow tubing and the strong cohesive forces attempt to minimize the surface area (*Chemistry* 6th ed. pages 454–455 / 7th ed. pages 429–430).

2.  The Clausius-Clapeyron equation indicates that
    I.    the greater the vapor pressure, the higher the entropy of
          vaporization.
    II.   the higher the heat of vaporization, the steeper the slope of $\ln P$
          vs. $1/T$.
    III.  the enthalpy of vaporization is equal to the slope of the vapor
          pressure vs. temperature.

    (A) I only
    (B) II only
    (C) III only
    (D) I and II only
    (E) I, II, and III

ANSWER: (B)  The Clausius-Clapeyron equation relates the ln vapor pressure to $-\Delta H_{vapor}$. It is most often used to determine the value of $\Delta H_{vapor}$ by measuring changes in vapor pressure with changes in temperature (see figure at right). You should be able to use this important equation (*Chemistry* 6th ed. pages 484–488 / 7th ed. pages 459–463).

3.  Small drops of water tend to bead up because of
    (A) high capillary action.
    (B) the shape of the meniscus.
    (C) the resistance to increased surface area.
    (D) low London dispersion forces.
    (E) weak covalent bonds.

ANSWER: (C)  This is a description of surface tension, which is a result of high dipole–dipole forces between water molecules. These intermolecular forces are also called hydrogen "bonds" (*Chemistry* 6th ed. pages 454–455 / 7th ed. pages 429–430).

4.  The vapor pressure increases in a predictable order as shown as
    (A) $CH_4 < C_2H_5OH < C_2H_5-O-C_2H_5 < Ne$.
    (B) $Ne < CH_4 < C_2H_5-O-C_2H_5 < C_2H_5-OH$.
    (C) $C_2H_5-OH < C_2H_5-O-C_2H_5 < CH_4 < Ne$.
    (D) $C_2H_5-O-C_2H_5 < C_2H_5-OH < CH_4 < Ne$.
    (E) $C_2H_5-O-C_2H_5 < C_2H_5-OH < Ne < CH_4$.

ANSWER:  (C)  Examining the IMF will suggest that only the alcohol ($C_2H_5-OH$) has an exposed –OH, suggesting strong hydrogen "bonding." The other three are essentially controlled by weaker London dispersion forces, which are greater in the larger, more massive compound. Realize also that the higher the IMF, the lower the vapor pressure (*Chemistry* 6th ed. pages 484–485 / 7th ed. pages 459–461).

5. Several liquids are compared by adding them to a series of 50-mL graduated cylinders, then dropping a steel ball of uniform size and mass into each. The time required for the ball to reach the bottom of the cylinder is noted. This is a method used to compare the differences in a property of liquids known as
(A) surface tension.
(B) buoyancy.
(C) capillary action.
(D) viscosity.
(E) surface contraction.

ANSWER: (D) The resistance to flow of any fluid is called viscosity. As you would predict, liquids with high viscosity (e.g., maple syrup) have large intermolecular forces (*Chemistry* 6th ed. page 455 / 7th ed. pages 429–430).

6. The properties of solids vary with their bonding. An example of this is shown by
(A) ionic solids with strong electrostatic attractions called ionic bonds, which have high melting temperatures.
(B) molecular solids with high intermolecular forces which have high melting temperatures.
(C) ionic solids with highly mobile ions which have high conductance.
(D) amorphous solids with strong London dispersion forces and high vapor pressure.
(E) network solids with low melting temperatures due to strong covalent bonds.

ANSWER: (A) Ionic bonds are unusually strong, requiring high temperatures to melt ionic substances. Table salt, for example, which is $Na^+Cl^-$, melts at 804°C (*Chemistry* 6th ed. pages 478–483, esp. Table 10.7 / 7th ed. pages 454–459, esp. Table 10.7).

7. As you go down the noble gas family on the periodic table, the boiling temperature increases. This trend is due mainly to
(A) an increase in hydrogen "bonding."
(B) a decrease in dipole–dipole forces.
(C) the lower atomic masses as you go down the family.
(D) an increase in London dispersion forces.
(E) an increase in lattice energy.

ANSWER: (D) These very symmetrical atoms are nonpolar; except for the induced dispersion forces, they would all boil at zero Kelvin. As it is, He boils at 4K and Ne at 25K (*Chemistry* 6th ed. pages 450–453 / 7th ed. pages 426–429).

8. Carbon dioxide and silicon dioxide have properties which differ dramatically. This is due to
   (A) the hardness of silica compounds.
   (B) the size difference of carbon and silicon, resulting in less effective overlap with the smaller orbits of oxygen.
   (C) shared $CO_2$ and $CO_4$ tetrahedra.
   (D) the formation of a glass when silicates are heated.
   (E) the repeating sigma bonds formed in carbon dioxide.

ANSWER: **B** Carbon dioxide acts as individual molecules and has a very low boiling temperature; $SiO_2$ is a network solid with chains of $SiO_4$ tetrahedra sharing oxygen atoms; it has both a very high melting and boiling temperature (*Chemistry* 6th ed. pages 470–472, esp. Figure 10.26 / 7th ed. pages 446–448, esp. Figure 10.26).

9. This question refers to the phase diagram shown. Select the true statement.
   (A) Point E represents both critical temperature and critical pressure.
   (B) Moving from point A to point F is sublimation.
   (C) Point C represents both solid and vapor states.
   (D) Point B represents both solid and vapor states.
   (E) Point D represents a point where, no matter how much pressure is exerted, a liquid cannot form.

ANSWER: **B** Be sure you know the details of common pressure vs. temperature phase diagrams, which sections represent each physical state, and what process the movement from one section to another represents. In this case, moving in a line from point A to point F represents going from the solid to vapor (gaseous) state (*Chemistry* 6th ed. pages 492–497 / 7th ed. pages 467–471).

10. The densities of certain compounds are greater as liquids than as solids. This means that increasing the pressure will result in
    (A) a solid becoming a liquid.
    (B) a liquid becoming a solid.
    (C) lowering of the critical temperature.
    (D) elevation of the freezing temperature.
    (E) lowering of the triple point.

ANSWER: **A** Since the density of the solid is less than that of the liquid, the slope of the solid/liquid line must be negative (like water, unlike carbon dioxide and most other substances). If the pressure is increased, then you move from the solid section of the phase diagram into the liquid section (*Chemistry* 6th ed. pages 492–497 / 7th ed. pages 467–471).

## FREE-RESPONSE QUESTION

Calculators and Equation Tables may be used.

1.
   a. Compare the structural models used to describe solids, liquids, and gases. Include (1) distance between particles, (2) relative strength of intermolecular forces, and (3) the extent of entropy.
   b. (i) Is vaporization an endothermic or exothermic process? Support your answer with an explanation.
      (ii) In what ways is the high heat of vaporization of water important to each of us personally?

ANSWER: 1a

|        | Distance | Forces    | Entropy |
|--------|----------|-----------|---------|
| Solid  | Shortest | Strongest | Lowest  |
| Liquid | Short    | Strong    | Low     |
| Gas    | Greatest | Weakest   | Highest |

ANSWER: 1b

Because vaporization is a bond breaking process (overcoming intermolecular forces) and thereby forms a higher potential energy state, it must be endothermic ($\Delta H$ = +). The high value of $\Delta H_{vapor}$ for water becomes important in many ways, from removing heat from our bodies (perspiration) to moderating temperature changes in areas near large bodies of water (*Chemistry* 6th ed. pages 455–456, 483–484 / 7th ed. pages 430, 459–460).

2. The vapor pressure of water at 25°C is 23.8 torr; $\Delta H_{vapor}$ = 43.9 kJ / mol.
   a. Determine the vapor pressure of water at 70°C.
   b. Determine the percentage error if your experimental results for vapor pressure at 70°C give a result of 225 torr.
   c. Give structural reasons that support the fact that the boiling temperature of $C_2H_5OH$ is 78°C and water is 100°C, (both at 760 torr).

ANSWER: 2a

$$\ln\left(\frac{P_{v1}}{P_{v2}}\right) = \frac{\Delta H_{\text{vapor}}}{R}\left(\frac{1}{T_2} - \frac{1}{T_1}\right)$$

$$\ln\left(\frac{23.8}{x}\right) = \frac{43900 \text{ J/mol}}{8.314 \text{ J/mol} \cdot \text{K}}\left(\frac{1}{343} - \frac{1}{298}\right) = -2.32$$

$$\frac{23.8}{x} = 0.0983$$

$$x = 242 \text{ torr.}$$

(Note: Be sure to change the kJ/mol for $\Delta H_{\text{vapor}}$ to J.)

ANSWER: 2b

$$\% \text{ error} = \left(\frac{\text{your experimental value} - \text{accepted value}}{\text{accepted value}}\right) \times 100\%$$

$$= \left(\frac{225 - 242}{242}\right) \times 100\% = -7.0\%.$$

ANSWER: 2c

The molecules of water are smaller and each has two –OH portions (twice as many as the alcohol), giving the opportunity for stronger hydrogen "bonding" (*Chemistry* 6th ed. pages 485–488, 451–452 / 7th ed. pages 460–463, 426–428).

# 10

# SOLUTIONS

In this section, the properties of liquid solutions will be reviewed. Solution composition, factors affecting solubility, and colligative properties such as freezing point depression, boiling point elevation, and osmotic pressure will be covered.

You should be able to

- Perform calculations with different solution concentrations such as molarity, mass percent, molality, and mole fraction.
- Discuss the effects of temperature, pressure, and structure on solubility.
- Perform calculations with Raoult's law.
- Understand colligative properties such as boiling point elevation, freezing point depression, and osmotic pressure.
- Use colligative properties to determine the molar mass of a solute.

## AP tips

Questions from this section may appear on both the free-response and multiple-choice sections of the AP Exam. Questions on the solubility or miscibility of substances in one another may appear in the essay section. Questions involving the determination of molar mass by freezing point depression could also appear in the lab essay.

## SOLUTIONS AND THEIR COMPOSITIONS

(*Chemistry* 6th ed. pages 511–514 / 7th ed. pages 485–487)

### TYPES OF SOLUTIONS

(*Chemistry* 6th ed. page 511 / 7th ed. page 485)

A solution is a homogeneous mixture. The table below summarizes the different types of solutions that can exist. In this chapter, the focus will be on liquid solutions.

**Various Types of Solutions**

| Example | State of Solution | State of Solute | State of Solvent |
|---|---|---|---|
| Air, natural gas | Gas | Gas | Gas |
| Vodka in water, antifreeze | Liquid | Liquid | Liquid |
| Brass | Solid | Solid | Solid |
| Carbonated water (soda) | Liquid | Gas | Liquid |
| Seawater, sugar solution | Liquid | Solid | Liquid |
| Hydrogen in platinum | Solid | Gas | Solid |

A solute, the substance being dissolved, is added to a solvent, which is present in the largest amount when referring to a liquid–liquid or gas–gas solution.

### COMPOSITION

(*Chemistry* 6th ed. pages 511–513 / 7th ed. pages 485–487)

MOLARITY  The molarity, M, of a solution is the number of moles of solute per liter of solution.

MASS PERCENT  Mass percent, also called weight percent, is the percent by mass of solute in a solution.

MOLE FRACTION  Mole Fraction, $\chi$, is the ratio of moles of a given component to the total number of moles of solution. For a two-component solution, where $n_A$ and $n_B$ are the moles of the two components:

$$\text{Mole fraction of component A} = \chi_A = \frac{n_A}{n_A + n_B}$$

**MOLALITY** Molality, $m$, is the number of moles of solute per kilogram of solvent.

$$\text{Molality} = \frac{\text{moles solute}}{\text{kilogram of solvent}}$$

**EXAMPLE:** A solution is prepared by mixing 30.0 mL of butane ($C_4H_{10}$, $d = 0.600$ g/mL) with 65.0 mL of octane ($C_8H_{18}$, $d = 0.700$ g/mL). Assuming that the volumes add in mixing, calculate the following for butane:

(a) molarity

(b) mass percent

(c) mole fraction

(d) molality

**SOLUTION:**

(a) The molarity of butane is 3.26 M.

$$30.0 \text{ mL} \times \frac{0.600 \text{ g}}{1 \text{ mL}} \times \frac{1 \text{ mol}}{58.1 \text{ g}} = \frac{0.310 \text{ mol } C_4H_{10}}{0.0950 \text{ L}} = 3.26 \text{ M}$$

(b) The mass percent is 28.3% butane.

Mass solution = mass solute + mass of solvent

$$= \left(30.0 \text{ mL} \times \frac{0.600 \text{ g}}{1 \text{ mL}}\right) + \left(65.0 \text{ mL} \times \frac{0.700 \text{ g}}{1 \text{ mL}}\right) = 63.5 \text{ g}$$

$$\text{Mass solute} = 30.0 \text{ mL} \times \frac{0.600 \text{ g}}{1 \text{ mL}} = 18.0 \text{ g}$$

$$\text{Mass percent} = \frac{18.0 \text{ g}}{63.5 \text{ g}} \times 100\% = 28.3\% \text{ butane}$$

(c) The mole fraction of butane is 0.437.

$$\text{Moles of octane} = 65.0 \text{ mL} \times \frac{0.700 \text{ g}}{1 \text{ mL}} \times \frac{1 \text{ mol}}{114 \text{ g}} = 0.399 \text{ mol octane}$$

$$\text{Mole fraction of butane} = \frac{0.310 \text{ mol butane}}{(0.310 \text{ mol butane} + 0.399 \text{ mol octane})} = 0.437$$

(d) 0.310 moles butane/0.0455 kg = 6.81 $m$

## FACTORS AFFECTING SOLUBILITY

(*Chemistry* 6th ed. pages 519–524 / 7th ed. pages 492–496)

The formation of a liquid solution begins by separating the solute into its individual components. Next, the solvent's intermolecular forces must be overcome to make room for the solute. The solute and solvent then interact to form the solution.

### STRUCTURAL EFFECTS

The phrase "like dissolves like" means that solubility is favored if the solute and solvent have similar polarities, as determined by their structure. The table below summarizes the solubilities of different types of solutes in different types of solvents.

| Type of Solute | Type of Solvent | Solubility | Example |
|---|---|---|---|
| Ionic | Polar | Usually soluble | LiCl in $H_2O$ |
| Polar | Polar | Soluble (miscible) | $CH_3OH$ in $H_2O$ |
| Nonpolar | Polar | Immiscible | $C_6H_{14}$ in $H_2O$ |
| Nonpolar | Nonpolar | Miscible | $C_6H_{14}$ in $CCl_4$ |

**EXAMPLE:** Discuss the solubility of each of the following solutes in carbon tetrachloride: ammonium nitrate, 1-pentanol, and pentane. Explain why each solute will or will not dissolve.

**SOLUTION:** Carbon tetrachloride is nonpolar and nonpolar solutes will dissolve in it. Ammonium nitrate is ionic and will not dissolve in a nonpolar solvent. It will dissolve in a polar solvent like water. 1-Propanol is polar and will not be soluble in carbon tetrachloride. Pentane is nonpolar and will be miscible (will form a homogeneous mixture) with carbon tetrachloride.

### PRESSURE EFFECTS

Pressure has little effect on the solubilities of liquids and solids. Gases become more soluble in liquids when the pressure of the gas above the liquid increases.

### TEMPERATURE EFFECTS

The solubility of most solids increases with temperature. However, some solids decrease solubility with temperature. The energy needed to separate the solute–solute particles is usually greater than the energy released from the solute–solvent attractions. It is difficult to predict the temperature dependence of the solubilities of solids. Gases, on the other hand, always decrease in solubility with increasing temperature.

# THE VAPOR PRESSURES OF SOLUTIONS

(*Chemistry* 6th ed. pages 524–531 / 7th ed. pages 497–504)

A nonvolatile solute lowers the vapor pressure of the solvent. The nonvolatile solute decreases the escaping tendency of the solvent molecules.

## RAOULT'S LAW

Raoult's law states that the vapor pressure of a solution is directly proportional to the mole fraction of the solvent present.

$$P_{soln} = \chi_{solvent}P^{o}_{solvent}$$

$P_{soln}$ is the observed vapor pressure of the solution.

$\chi_{solvent}$ is the mole fraction of the solvent.

$P^{o}_{solvent}$ is the vapor pressure of the pure solvent.

For an ideal solution, a plot of $P_{soln}$ vs. $\chi_{solvent}$ at constant temperature gives a straight line with a slope equal to $P^{o}_{solvent}$.

EXAMPLE: The vapor pressure of a solution containing 53.6 g of glycerin, $C_3H_8O_3$, in 133.7 g of ethanol, $C_2H_5OH$ is 113 torr at 40°C. Calculate the vapor pressure of the pure ethanol at 40°C assuming that glycerin is a nonvolatile, nonelectrolyte solute in ethanol.

SOLUTION: The vapor pressure of pure ethanol at 40°C is 136 torr.

$$P_{ethanol} = \chi_{ethanol}P^{o}_{ethanol}$$

$$\chi_{ethanol} = \frac{\text{mol ethanol}}{\text{total mol in solution}}$$

$$\frac{133.7 \text{ g } C_2H_5OH \times 1 \text{ mol}}{46.07 \text{ g}} = 2.90 \text{ mol } C_2H_5OH$$

$$\frac{53.6 \text{ g } C_3H_8O_3 \times 1 \text{ mol}}{92.09 \text{ g}} = 0.582 \text{ mol } C_3H_8O_3$$

$$\text{Total mol} = 0.582 \text{ mol} + 2.90 \text{ mol} = 3.48 \text{ mol}$$

$$P^{o}_{ethanol} = \frac{P_{ethanol}}{\chi_{ethanol}} = \frac{113 \text{ torr}}{(2.90 \text{mol} / 3.48 \text{ mol})} = 136 \text{ torr}$$

REAL VS. IDEAL SOLUTIONS Ideal solutions obey Raoult's law. Real solutions best approximate the behavior of ideal solutions when the solute concentration is low and the solute and solvent have similar types of intermolecular forces and molecular sizes.

Deviations from Raoult's law occur when the interactions between the solute and solvent are extremely strong, as in hydrogen bonding, and the vapor pressure of the solution is lower than predicted. Also, the vapor pressure of a real solution is greater than predicted when the intermolecular forces between the solute and solvent are weaker than the intermolecular forces between the solute–solute or solvent–solvent.

## COLLIGATIVE PROPERTIES

(*Chemistry* 6th ed. pages 531–542 / 7th ed. pages 504–513)

Colligative properties depend on the number, not the identity, of the solute particles in an ideal solution.

### BOILING POINT ELEVATION

A nonvolatile solute elevates the boiling point of a solvent. Because a nonvolatile solute lowers the vapor pressure of a solution, it must be heated to a temperature higher than the boiling point of the pure solvent. The normal boiling point is the point where the vapor pressure of liquid equals 1 atmosphere. The change in boiling point due to the presence of the nonvolatile solute is represented by

$$\Delta T_b = K_b m.$$

$\Delta T$ is the difference between the boiling point of the solution and that of the pure solvent.

$K_b$ is the molal boiling point elevation constant of the solvent.

$m$ is the molality of the solute in solution.

> **EXAMPLE:** 2.00g of a large biomolecule was dissolved in 15.0 g of carbon tetrachloride. The boiling point of this solution was 77.85°C. Calculate the molar mass of the biomolecule. For carbon tetrachloride, the boiling point constant is 5.03°C kg/mol, and the boiling point of pure carbon tetrachloride is 76.50°C.
>
> **SOLUTION:** The molar mass is 498 g/mol.

$$\Delta T_b = 77.85°C - 76.50°C = 1.35°C.$$

$$m = \frac{1.35°C}{(5.03°C \text{ kg/mol})} = 0.268 \text{ mol/ kg.}$$

Mol of biomolecule = 0.0150 kg solvent × 0.268 mol hydrocarbon/kg solvent

$$= 4.02 \times 10^{-3} \text{ mol.}$$

$$\text{Molar mass} = \frac{2.00 \text{ g}}{4.02 \times 10^{-3} \text{ mol}} = 498 \text{ g/mol.}$$

## FREEZING POINT DEPRESSION

A nonvolatile solute decreases the freezing point of a solvent.

$$\Delta T_f = K_f m.$$

$\Delta T$ is the difference between the boiling point of the solution and that of the pure solvent.

$K_f$ is the molal freezing point depression constant of the solvent.

> EXAMPLE: Which of the following solutions would have the largest freezing point depression? Explain your answer.
>
> (a) 0.010 $m$ NaCl
>
> (b) 0.010 $m$ Na$_3$PO$_4$
>
> (c) 0.010 $m$ Al$_2$(SO$_4$)$_3$
>
> (d) 0.010 $m$ C$_6$H$_{12}$O$_6$

SOLUTION: The solution with the lowest freezing point depression is 0.010 $m$ Al$_2$(SO$_4$)$_3$, the solution which yields the highest molality (or the most) particles when dissolved. 0.010 $m$ Al$_2$(SO$_4$)$_3$ yields five particles (ions) when dissolved or the total concentration of solute particles is 0.050 m. Two particles (or 0.020 $m$) are present in a solution of 0.010 $m$ NaCl. Four particles (or 0.040 $m$) are present in a solution of 0.010 $m$ Na$_3$PO$_4$. One particle (or 0.010 $m$) is present in a solution of 0.010 $m$ C$_6$H$_{12}$O$_6$.

## OSMOTIC PRESSURE

Osmosis is the flow of a pure solvent into a solution through a semipermeable membrane. Osmostic pressure is the pressure that must be applied to a solution to stop osmosis.

$$\pi = MRT.$$

$\pi$ is the osmotic pressure in atmospheres.

$M$ is the molarity of the solution.

$R$ is the gas law constant, 0.08206 L atm/mol K.

$T$ is temperature in Kelvin.

> EXAMPLE: An aqueous solution of 10.00 g of catalase, an enzyme found in the liver, has a volume of 1.00 L at 27°C. The solution's osmotic pressure at 27°C is 0.745 torr. Calculate the molar mass of the catalase.

SOLUTION: The molar mass of catalase is $2.51 \times 10^5$ g/mol.

$$M = \frac{\pi}{RT} = \frac{0.745 \text{ torr} \times \left(\frac{1 \text{ atm}}{760 \text{ torr}}\right)}{\left[0.08206 \text{L} \times \frac{\text{atm}}{(\text{mol} \times \text{K})}\right]300\text{K}} = 3.98 \times 10^{-5} \text{ M}$$

$$1.00 \text{ L} \times 3.98 \times 10^{-5} \text{ mol/L} = 3.98 \times 10^{-5} \text{ mol}$$

$$\text{Molar mass} = \frac{10.00 \text{ g}}{3.98 \times 10^{-5} \text{ mol}} = 2.51 \times 10^5 \text{ g/mol}$$

## MULTIPLE-CHOICE QUESTIONS

No calculators may be used on this part of the exam.

1. As temperature decreases
   (A) solutes always become less soluble in solvents.
   (B) polar solutes always become less soluble in polar solvents.
   (C) solid solutes always become less soluble in liquid solvents.
   (D) gaseous solutes always become more soluble in liquid solvents.
   (E) there is little effect on solubility.

ANSWER: D Only in the case of a gas dissolved in a liquid can we always be certain that as the temperature decreases, the gas will become more soluble (*Chemistry* 6th ed. pages 522–523 / 7th ed. pages 495–496).

2. The aqueous solution with the lowest freezing temperature (greatest freezing temperature depression) would be
   (A) 0.010 M NaCl.
   (B) 0.020 M NaCl.
   (C) 0.030 M $C_6H_{12}O_6$.
   (D) 0.030 M $NaNO_3$.
   (E) 0.030 M $HC_2H_3O_2$.

ANSWER: D Freezing temperature depression is a colligative property. The greater the concentration of particles in solution, the greater the freezing temperature lowering. With response (A) 0.010 M NaCl, you expect 0.0100 M $Na^+$ ions and 0.010 M $Cl^-$ ions, for a total concentration of 0.020 M ions. In response (D), we would have the greatest expected concentration of ions, 0.030 M $Na^+$ and 0.030 M $NO_3^-$ (*Chemistry* 6th ed. pages 533–534 / 7th ed. pages 506–507).

3. The expected value for $i$, the van't Hoff factor, is often only a first approximation due to
   (A) ion pairing.
   (B) moles of undissolved solute.
   (C) an exothermic heat of solution.
   (D) a variation in the freezing or boiling temperature constant.
   (E) complete dissociation into component ions.

ANSWER: **A** Ion pairing suggests that a few of the ionic solute ions stay together, acting as one unit rather than two (or more) particles. This is an important effect in more concentrated solutions (*Chemistry* 6th ed. pages 540–542 / 7th ed. pages 512–513).

4. The number of moles of $Al(NO_3)_3$ which must be added to water to form 2.00 L of 0.30 M $NO_3^-$ ions is
   (A) 0.30 mol.
   (B) 0.60 mol.
   (C) 0.20 mol.
   (D) 2.4 mol.
   (E) 8.0 mol.

ANSWER: **C**

Since every $Al(NO_3)_3$ unit contains 3 units of $NO_3^-$ ions, you need 1/3 as many $Al(NO_3)_3$ units as $NO_3^-$ units. Needed are 0.30 mol/L of $NO_3^-$ × 2.00 L = 0.60 mol $NO_3^-$. Hence 1/3 × 0.60 mol = 0.20 mol $Al(NO_3)_3$ required (*Chemistry* 6th ed. pages 541–542 / 7th ed. pages 512–513).

5. Refer to the solubility graph above. When the temperature of a saturated solution of potassium nitrate is lowered from 80°C to 40°C, the grams of $KNO_3$ lost from 2.00 g of water would be about
   (A) 2 g.
   (B) 7 g.
   (C) 17 g.
   (D) 100 g.
   (E) 200 g.

ANSWER: **A** The solubility at 80°C is about 170 g $KNO_3$/100 g $H_2O$, dropping to 70 g $KNO_3$/100 g $H_2O$ at 40°C, for a loss of 100 g $KNO_3$. If only 2/100 as much water were used, then only 2/100 of the 100 g of $KNO_3$ would come out of solution (*Chemistry* 6th ed. pages 521–523 / 7th ed. pages 493–496).

6. Not all liquid–liquid solutions are ideal.  Those that are usually have
    (A) solute–solvent interactions that are weaker than the interactions in pure liquid.
    (B) solute–solvent interactions that are stronger than the interactions in pure liquid.
    (C) similar liquids with an enthalpy of solution near zero.
    (D) an observed vapor pressure lower than that predicted by Raoult's law.
    (E) high dipole–dipole interactions between solute and solvent molecules.

ANSWER: C  Liquid–liquid solutions that obey Raoult's law are said to be ideal solutions. This is most likely when the solute–solute, solvent–solvent, and solute–solvent interactions are of about the same magnitude. This means that the tendency of solvent molecules to escape, thereby producing a vapor pressure, is about what is expected (*Chemistry* 6th ed. pages 529-531 / 7th ed. pages 501–504).

7. Mole fraction is a method of expressing concentration that is needed in several kinds of problems, such as those dealing with vapor pressure of a solution. To calculate the mole fraction, it is necessary to
    (A) divide the total number of moles of solution by 2.
    (B) divide the moles of solute by moles of solvent.
    (C) divide the moles of one component by the sum of the moles of all components.
    (D) divide the mass of the solute by the mass of the solution.
    (E) divide the mass of the solute by 1000 g of solvent.

ANSWER: C  Mole fraction is the ratio of the number of moles of a given component to the total number of moles of solute plus solvent; it is most often applied to two-component solutions of liquids in liquids at this level of the study of chemistry (*Chemistry* 6th ed. pages 512–513 / 7th ed. pages 485–487).

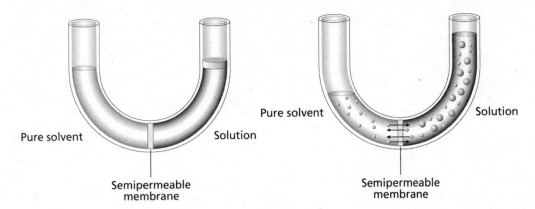

Pure solvent

Solution

Semipermeable membrane

Pure solvent

Solution

Semipermeable membrane

8.  Refer to the two diagrams above. Osmotic pressure may be defined as
    I.   the excess pressure shown by different liquid levels on the two sides.
    II.  the pressure which must be applied to stop osmosis.
    III. flow through a semipermeable membrane.
    (A) I only
    (B) II only
    (C) III only
    (D) I and II only
    (E) I, II, and III

ANSWER: D  Selection III defines osmosis, not osmotic pressure. Both of the other two definitions are valid (*Chemistry* 6th ed. pages 535–536 / 7th ed. page 508–509).

9.  In the following chart, which compares a KCl aqueous solution with pure water, the row which correctly identifies the salt solution's characteristics is

| Row | Boiling Temp. | Freezing Temp. | Vapor Pressure |
|-----|---------------|----------------|----------------|
| (A) | Lower         | Lower          | Lower          |
| (B) | Lower         | Lower          | Higher         |
| (C) | Lower         | Higher         | Lower          |
| (D) | Higher        | Lower          | Higher         |
| (E) | Higher        | Lower          | Lower          |

ANSWER: E  Boiling temperature is always elevated, freezing temperature is always depressed, and vapor pressure is lower for an ionic aqueous solution than for the pure solvent (*Chemistry* 6th ed. pages 529–535 / 7th ed. pages 501–507).

10. Toluene has half the molecular mass (g/mol) of glucose. If 100. g of each of these two substances are mixed and all of the glucose dissolves in the toluene, the mole fraction of toluene in this solution would be
    (A) 1/3.
    (B) 1/2.
    (C) 2/3.
    (D) 3/2.
    (E) 5/6.

ANSWER: C   Assume that you have a mass (grams) of one mol of glucose. That same mass (in grams) would be two moles of toluene. Of the total 3 moles, 2 moles would be toluene, hence $2/(1+2) = 2/3$ for the mol fraction (*Chemistry* 6th ed. page 512 / 7th ed. pages 485–486).

## FREE-RESPONSE QUESTIONS

1. (a) Molarity is a temperature dependent quantity, whereas molality is not. Explain why this is so.
   (b) When the concentration of very dilute aqueous solutions are calculated, the values of molarity and molality are the same. Explain.
   (c) The heat of solution of salts may be either endothermic (which most are) or exothermic. Explain what factors determine whether $\Delta H_{sol'n}$ is endothermic or exothermic.
   (d) As a solution freezes, the freezing temperature continues to decrease. Why is this so?

ANSWER: 1(a)   Molarity depends upon the volume of the solution and this property changes (generally rather slightly with liquid solvents) with temperature. Molality depends on mass which is not temperature dependent (*Chemistry* 6th ed. pages 511–512 / 7th ed. pages 485–486).

ANSWER: 1(b)   For very dilute aqueous solutions, the solvent (water) mass is 1000 g, which is a volume of 1 liter; because the solution is "very dilute," the solution is almost totally solvent (*Chemistry* 6th ed. pages 512–513 / 7th ed. pages 485–486).

ANSWER: 1(c)   The heat of solution, $\Delta H_{sol'n}$ may be considered as the sum of the energy required to separate the ions in the salt (plus, perhaps, the energy to separate the particles of the liquid) and the energy given off when the ions become "bonded" to liquid particles. In this expression, $\Delta H_{sol'n} = \Delta H_{c.e.} + \Delta H_{hyd}$, the crystal energy (or lattice energy) is always positive (the energy to separate charged particles must be added, as in "bond breaking"), and energy is always given off as charged particles move nearer to each other ("bond forming"). In summary, $\Delta H_{c.e.}$ is always (+) and $\Delta H_{hyd}$ is always (–); whichever of these dominates determines the sign of $\Delta H_{sol'n}$ (*Chemistry* 6th ed. pages 515-518 / 7th ed. pages 488–492).

ANSWER: 1(d)   As a solution freezes, the solvent molecules are removed from the solution which causes an increase in the concentration. This causes further freezing temperature depression (*Chemistry* 6th ed. pages 531-532 / 7th ed. pages 504–505).

2. You are to determine the formula and the molecular mass for an organic compound from the following information: 5.00 g of the organic solid was dissolved in 100.0 g of benzene. The boiling temperature of this solution was 82.3°C. The organic compound is 46.7% nitrogen, 6.67% hydrogen, 26.7% oxygen, and the remainder is carbon. The boiling temperature of pure benzene is 80.2°C; $K_b$ = 2.53 °C kg / mol.
   (a) Determine the molecular mass of the organic solid.
   (b) Determine the molecular formula of the solid.
   (c) Determine the mole fraction of the organic solid in the solution.
   (d) If the density of this solution is 0.8989 g / mL, calculate the molarity of the solution.

ANSWER: 2(a)    The molecular mass of the organic solid:

$$\Delta T_b = 82.3°C - 80.2°C = 2.1°C.$$

$$\text{Mass of solvent} = \frac{100.0\,g \times 1\,kg}{1000\,g} = 0.1000\,kg \text{ of solvent.}$$

$$m = \frac{\Delta T}{k} = \frac{2.1°C}{2.53°C \times kg/mol} = 0.83\,m = 0.83\,mol/kg.$$

$$\frac{5.00\,g}{0.1000\,kg} = 50.0\,g/kg = 0.83\,mol/kg$$

Hence, $\dfrac{50.0\,g}{0.83\,mol} = 60.\,g/mol.$

(*Chemistry* 6th ed. pages 531–533 / 7th ed. pages 504–505).

ANSWER: 2(b)    The formula:

N $\dfrac{46.7\,g}{14.0\,g/mol} = 3.34$ mol N.

H $\dfrac{6.67\,g}{1.00\,g/mol} = 6.67$ mol H.

O $\dfrac{26.7\,g}{16.0\,g/mol} = 1.67$ mol O.

C $\dfrac{19.9\,g}{12.0\,g/mol} = 1.66$ mol C.

$$\frac{3.34}{1.67} = 2 \text{ mol N.}$$

$$\frac{6.67}{1.67} = 4 \text{ mol H.}$$

$$\frac{1.67}{1.67} = 1 \text{ mol O and 1 mol C.}$$

This gives a formula of $CON_2H_4$, which has a molecular mass [see part (a)] of 60.g/mol.

This means that this is both the empirical and the molecular formula (*Chemistry* 6th ed. pages 98–101 / 7th ed. pages 93–95).

**Answer: 2(c)**    Mole fraction:

$$0.83 \; m = \frac{0.83 \text{ mol of organic solid}}{1000 \text{ g of benzene}}; \quad \frac{1000 \text{ g}}{78.0 \text{ g/mol}} = 12.8 \text{ mol benzene.}$$

$$\frac{0.83 \text{ mol}}{(0.83 + 12.8 \text{ mol})} = \frac{0.83}{13.6} = 0.061.$$

(*Chemistry* 6th ed. page 515 / 7th ed. pages 487–488).

**Answer: 2(d)**    Molarity:

Mass of solution = 5.00 g solute + 100.0 g benzene = 105.0 g of solution.

$$105.0 \text{ g} \times \frac{1 \text{ mL}}{0.8989 \text{ g}} = 116.8 \text{ mL of solution.}$$

$$\frac{0.083 \text{ mol}}{\left(116.8 \text{ mL} \times \dfrac{1 \text{ L}}{1000 \text{ mL}}\right)} = \frac{0.083 \text{ mol}}{0.1168 \text{ L}} = 0.71 \text{ M.}$$

(*Chemistry* 6th ed. pages 512–514 / 7th ed. pages 485–487).

# 11

# KINETICS

Chemical kinetics is the study of the rates of reactions and the mechanisms, the series of steps by which each reaction occurs. You will review the factors which affect the rate of a reaction and the determination of the rate law and mechanism for a reaction.

In this chapter, you should be able to:

- Identify factors which affect reaction rates.
- Calculate the rate of production of a product or consumption of a reactant using mole ratios and the given rate.
- Determine the rate law for a reaction from given data, overall order, and value of the rate constant, inclusive of units.
- Determine the instantaneous rate of a reaction.
- Use integrated rate laws to determine concentrations at a certain time, $t$, and create graphs to determine the order of a reaction. Also, determine the half-life of a reaction.
- Write the rate law from a given mechanism given the speeds of each elementary step.
- Write the overall reaction for a mechanism and identify catalysts and intermediates present.
- Determine the activation energy for the reaction using the Arrhenius equation.
- Graphically determine the activation energy using the Arrhenius equation.

## FACTORS AFFECTING REACTION RATES

Increasing the concentrations of reactants increases the chances for more molecular collisions which make products. Usually the increase in concentrations of the reactants will increase the rate of a reaction.

Increasing the surface area of the reactants will also increase the number of collisions. For example, granular zinc will react more quickly with hydrochloric acid than a strip of zinc.

Increasing the temperature of a reaction will also speed it up. Increasing the temperature increases the rate of both forward and reverse reactions, increasing the number of collisions over a certain period of time.

In the next section, the method by which a catalyst speeds up a reaction will be discussed.

## COLLISON THEORY

(*Chemistry* 6th ed. pages 582–583 / 7th ed. pages 552–553)

The collision theory model accounts for the observed characteristics of reaction rates. For a reaction to occur, the molecules must collide in the correct orientation with sufficient energy. The theory assumes that all but the simplest reactions take place in a series of two-particle collisions. This sequence of two-particle collisions is called the reaction mechanism.

The potential energy diagram for a reaction, the reaction progress, is shown in the graph above. At the top of the "hill," or barrier, is the activated complex, or transition state. Once the energy in the transition state, the activation energy, is overcome, the reactants can become products. Activation energy is the amount of energy required to form an activated complex.

Because products have a higher potential energy than reactants, the reaction represented in the above diagram is endothermic. Recall that $\Delta H_{rxn} = \Delta H_{products} - \Delta H_{reactants}$. The difference between the potential energies of the products and the reactants gives the enthalpy of the reaction.

If asked to draw the reaction progress for an exothermic reaction, the energy of the products would be lower than that of the reactants.

The diagram shows the pathway for both catalyzed and uncatalyzed reactions. A catalyst speeds up a reaction by lowering the activation energy of the reaction without being consumed itself.

# REACTION RATES

(*Chemistry* 6th ed. pages 557–561 / 7th ed. pages 527–532)

The reaction rate is the change in the concentration of a reactant or product per unit time.

Consider the reaction, A ⟶ B.

$$Rate = -\frac{\Delta[A]}{\Delta t}$$

The concentration of A in moles per liter is represented by [A]. The change in time is represented by $\Delta t$. The quantity, $-\Delta[A]/\Delta t$, is negative because the reactants are disappearing and forming products.

### EXAMPLE:

Consider the reaction $4PH_3(g) \longrightarrow P_4(g) + 6 H_2(g)$.

If 0.0048 mol of $PH_3$ is consumed in a 2.0-L container during each second of the reaction, what are the rates of production for $P_4$ and $H_2$?

The rate at which $PH_3$ is being consumed is $-\dfrac{\Delta[PH_3]}{\Delta t}$.

0.0048 mol $PH_3$ / (2.0 L × s) = 0.0024 mol $L^{-1}$ $s^{-1}$ $PH_3$

The rate at which $P_4$ and $H_2$ are being produced can be determined by using mole ratios:

$$\frac{0.0024 \text{ mol } PH_3}{L \times s} \times \frac{1 \text{ mol } P_4}{4 \text{ mol } PH_3} = 0.0060 \text{ mol } L^{-1} \text{ } s^{-1} \text{ } P_4$$

$$\frac{PH_3}{L \times s} \times \frac{6 \text{ mol } H_2}{4 \text{ mol } PH_3} = 0.0036 \text{ mol } L^{-1} \text{ } s^{-1} \text{ } H_2$$

## RATE LAWS

*(Chemistry* 6th ed. pages 561–578 / 7th ed. pages 532–547)

Rate laws show how the rate of a reaction depends on the concentrations of reactants. Except for an elementary reaction (one in which the balanced equation represents the mechanism), the rate law cannot be determined from the balanced equation.

For the reaction

$$2NO_2(g) \longrightarrow 2NO(g) + O_2(g),$$

the rate law can be written as

$$\text{Rate} = k[NO_2]^n.$$

The proportionality constant, $k$, is called the rate constant and is determined by experiment. For a given reaction at a given temperature, this value is constant. Its units depend on the order of the reactants.

The order, $n$, of the reactant must also be determined by experiment. It is the power to which the reactant concentration much be raised in the rate law. For example, if the reaction A → B is first order, then the rate law is Rate = $k[A]$; doubling the concentration of the reactant doubles the rate of the reaction. If the reaction is second order, then Rate= $k[A]^2$; doubling the concentration of the reactant will result in the rate quadrupling. If the reactant concentration is changed and the rate is not affected, the order of the reactant is zero. The rate law would be Rate = $k[A]^0$.

## INSTANTANEOUS RATES

*(Chemistry* 6th ed. pages 558–559 / 7th ed. pages 529–530)

One way to determine the rate of a reaction at a particular time, the instantaneous rate, is to plot the reactant concentration versus time and take the slope of the tangent to the curve at time $t$.

If the slopes of tangents to the curve at two different concentrations are calculated, the rate law of a reaction can be determined by comparing the changes in rate to the changes in concentration.

In the graph to the right, when the concentration of the reactant, $N_2O_5$, is halved, the rate is also halved. The reaction is first order. The rate law for the reaction is Rate = $k[N_2O_5]$.

## METHOD OF INITIAL RATES

*(Chemistry* 6th ed. pages 564–567 / 7th ed. pages 535–538)

The initial rate of a reaction is the instantaneous rate just after the reaction begins (just after $t = 0$ and before the initial concentrations of the reactants have changed.)

The rate law of a reaction can be determined by performing a few trials with different reaction concentrations and measuring the initial rate for each trial. To find the order of one reactant, change its concentration while holding the concentration of the other reactants constant.

EXAMPLE: Determine the rate law and the value of the rate constant for the reaction at –10°C:

$$2NO(g) + Cl_2(g) \longrightarrow 2NOCl(g)$$

| Trial | $[NO]_o$ (mol/L) | $[Cl_2]_o$ (mol/L) | Initial rate ( mol/L • s) |
|-------|------------------|--------------------|---------------------------|
| 1 | 0.10 | 0.10 | 0.18 |
| 2 | 0.10 | 0.20 | 0.36 |
| 3 | 0.20 | 0.20 | 1.45 |

First, write the general form of the rate law: Rate = $k[NO]^x[Cl_2]^y$

Your goal is to determine $x$ and $y$.

To find $x$, the order with respect to NO, pick two trials in which NO changes and $Cl_2$ remains the same.

Compare Trial 3 to Trial 2.

Write a ratio of the rate law of trial 3 to the rate of the rate law for trial 2. Substitute the values for the rates and known concentrations and solve for the order, $x$.

$$\frac{\text{Rate (Trial 3)}}{\text{Rate (Trial 2)}} = \frac{k[NO]^x \cancel{[Cl_2]^y}}{k[NO]^x \cancel{[Cl_2]^y}}; \quad \frac{1.45}{0.36} = \frac{(0.20)^x}{(0.10)^x}; \quad 4.0 = 2^x; x = 2.0$$

The order, $x$, with respect to NO is 2. This means that when the concentration of NO is doubled, the rate quadruples. Note that the rate constant and the concentration of $Cl_2$ cancel out because they are the same in Trials 2 and 3.

To find $y$, the order with respect to $Cl_2$, pick two trials in which $Cl_2$ changes and NO remains the same.

Compare Trial 2 to Trial 1.

$$\frac{\text{Rate (Trial 2)}}{\text{Rate (Trial 1)}} = \frac{\cancel{k[NO]^x}[Cl_2]^y}{\cancel{k[NO]^x}[Cl_2]^y}; \quad \frac{0.36}{0.18} = \frac{(0.20)^x}{(0.10)^x}; \quad 2.0 = 2^y; y = 1$$

The order, $y$, with respect to $Cl_2$, is 1. This means that when the concentration of $Cl_2$ is doubled, the rate also doubles. Note that the rate constant and the concentration of NO cancel out because they are the same in Trials 1 and 2.

The rate law for the reaction is Rate = $k[NO]^2[Cl_2]$.

The overall order of a reaction is the sum of the reaction orders. For this example, the overall order is $2 + 1 = 3$.

Note: The order with respect to each reactant will not always be the same as the coefficients in the balanced equation.

To determine the value of the rate constant, including its units, use the rate law and experimental data from any given trial.

$k$ = Rate / $([NO]^2[Cl_2])$

Using values from Trial 1:

= $(0.18$ M s$^{-1}) / (0.10M)^2(0.10M)$ = 180 L$^2$/(mol$^2$ s).

It helps to work out the units separately.

$$\left(\frac{mol}{L \cdot s}\right)\left(\frac{L^2}{mol^2}\right)\left(\frac{L}{mol}\right) = \frac{L^2}{mol^2 \cdot s}$$

If the rate constant is known and the order with respect to one of the reactants has been determined, the order with respect to the other can be calculated even if its concentration is not held constant between any of two trials.

## INTEGRATED RATE LAWS

(*Chemistry* 6th ed. pages 568–577 / 7th ed. pages 538–547)

An integrated rate law, derived from the differential rate law, expresses the reactant concentration as a function of time. The table below summarizes the integrated rate laws for the reaction

A $\longrightarrow$ Products.

| | Order | | |
|---|---|---|---|
| | *Zero* | *First* | *Second* |
| Rate Law: | Rate = $k$ | Rate = $k[A]$ | Rate = $k[A]^2$ |
| Integrated Rate Law: | $[A] = -kt + [A]_0$ | $\ln[A] = -kt + \ln[A]_0$ | $\frac{1}{[A]} = kt + \frac{1}{[A]_0}$ |
| Plot Needed to Give a Straight Line | $[A]$ versus $t$ | $\ln[A]$ versus $t$ | $\frac{1}{[A]}$ versus $t$ |
| Relationship of Rate Constant to the Slope of Straight Line: | Slope = $-k$ | Slope = $-k$ | Slope = $k$ |
| Half-life: | $t_{1/2} = \frac{[A]_0}{2k}$ | $t_{1/2} = \frac{0.693}{k}$ | $t_{1/2} = \frac{1}{k[A]_0}$ |

You can determine the order of the reactant graphically if you know the concentration of A at various times, $t$, during the reaction. For example, if you plot $\ln[A]$ vs. $t$ and obtain a straight line, the reaction is first order in A. If the graph is not linear, then the reaction is not first order. The table above summarizes what is graphed to test for the order of the reactant.

The half-life of a reactant, $t_{1/2}$, is the time required for a reactant to reach half of its original concentration. The general equations for the half-lives for each reactant order appear in the table above.

EXAMPLE: The rate of the reaction

$$NO_2(g) + CO(g) \longrightarrow NO(g) + CO_2(g)$$

depends only on the concentration of nitrogen dioxide below 225°C. At a temperature below 225°C, the following data were collected.

| Time (s) | $[NO_2]$ (M) | $\ln [NO_2]$ | $1/[NO_2]$ ($M^{-1}$) |
|---|---|---|---|
| 0 | 0.500 | –0.693 | 2.00 |
| $1.20 \times 10^3$ | 0.444 | –0.812 | 2.25 |
| $3.00 \times 10^3$ | 0.381 | –0.965 | 2.62 |
| $4.50 \times 10^3$ | 0.340 | –1.079 | 2.94 |
| $9.00 \times 10^3$ | 0.250 | –1.386 | 4.00 |
| $1.80 \times 10^4$ | 0.174 | –1.749 | 5.75 |

Assume that the data are first order and see if the plot of $\ln[NO_2]$ vs. time is linear. If this isn't linear, try the second-order plot of $1/[NO_2]$ vs. time. The data and plots follow.

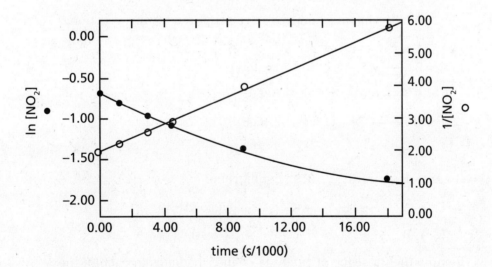

The plot of $1/[NO_2]$ vs. time is linear, so the slope of this line gives the value of $k$.

$$\text{Slope} = k = \frac{\Delta y}{\Delta x} = \frac{(5.75 - 2.00)\ M^{-1}}{(1.8 \times 10^4 - 0)\ s} = 2.08 \times 10^{-4}\ \text{L/mol s}$$

To determine $[NO_2]$ at $2.70 \times 10^4$ s, use the integrated rate law where $1/[NO_2]_o = 1/0.500\ M = 2.00\ M^{-1}$.

$$1/[NO_2] = kt + 1/[NO_2]_o$$

$$\frac{1}{[NO_2]} = \frac{2.08 \times 10^{-4} \text{ L}}{\text{mol s}} \times 2.70 \times 10^4 \text{ s} + 2.00 \text{ M}^{-1}$$

$$\frac{1}{[NO_2]} = 7.62, \quad [NO_2] = 0.131 \text{ M}$$

## MECHANISMS

(*Chemistry* 6th ed. pages 579–582 / 7th ed. pages 549–552)

A mechanism is a series of steps by which a reaction occurs. The elementary steps, each step in the mechanism, must add up to give the overall balanced equation for the mechanism. The slowest step in the mechanism must agree with the experimentally determined rate law.

Once the rate law has been determined experimentally, then chemists propose several mechanisms consistent with that rate law and design experiments to determine which matches the data. You will have to know how to identify catalysts and intermediates in a mechanism and write the overall equation for a mechanism. You will also need to be able to write a rate law from a given mechanism.

EXAMPLE: Consider the balanced equation for the hypothetical reaction

$2A + 2B \longrightarrow 2C + D_2$.

The experimentally determined rate law is Rate = $k[A]^2[B]$.

The two following mechanisms were proposed for this reaction:

I. $A + B \longrightarrow E + C$        (slow)

   $E + A \longrightarrow E_2$            (fast)

   $E_2 + B \longrightarrow D_2 + C$    (fast)


II. $A + B \rightleftharpoons E$    (fast, equilibrium)

    $E + A \longrightarrow E_2 + C$      (slow)

    $E_2 + B \longrightarrow D_2 + C$     (fast)

Identify the presence of catalysts or intermediates in each of the mechanisms.

A catalyst is present in the reactants and then appears in the products. Recall that the catalyst is not consumed in the reaction. The catalyst does not appear in the overall reaction. In the two mechanisms above, there are no catalysts. An example of a mechanism involving a catalyst involves the decomposition of ozone by atoms of chlorine.

$$Cl(g) + O_3(g) \longrightarrow ClO(g) + O_2(g)$$
$$\underline{O(g) + ClO(g) \longrightarrow Cl(g) + O_2(g)}$$
$$O(g) + O_3(g) \longrightarrow 2O_2(g)$$

An intermediate is produced in one elementary step and then consumed in the next. It also does not appear in the overall reaction for the mechanism because it cancels out. In both mechanisms, E is an intermediate. It is produced in the first step and consumed in the second.

Which of the following two mechanisms is consistent with the rate law? For the proposed mechanism to be correct, the overall reaction and the rate law for the mechanism must agree with the given reaction and the experimentally determined rate law.

Write the overall reaction for each mechanism. Add up the elementary steps. Cancel out the catalysts and intermediates that appear.

I. A + B $\longrightarrow$ E + C          (slow)

~~E~~ + A $\longrightarrow$ ~~E₂~~          (fast)

~~E₂~~ + B $\longrightarrow$ D₂ + C          (fast)

2A + 2B $\longrightarrow$ 2C + D₂ is the overall reaction.

II. A + B $\rightleftharpoons$ ~~E~~          (fast, with equal rates)

~~E~~ + A $\longrightarrow$ ~~E₂~~ + C          (slow)

~~E₂~~ + B $\longrightarrow$ D₂ + C          (fast)

2A + 2B $\longrightarrow$ 2C + D₂ is the overall reaction.

The overall reaction for each mechanism matches the given reaction.

### DETERMINE THE RATE LAW FROM THE MECHANISM

Begin by writing the rate law for the slow step of each mechanism.

The rate law for the slow step of mechanism I is Rate = $k[A][B]$ which is not consistent with the rate law determined by experiment. Mechanism I is not possible for this reaction.

The rate law for the slow step of mechanism II is Rate = $k[E][A]$.

[E] is an intermediate and may not be included in the rate law. Because the first step in the mechanism is reversible and fast, the rate of the forward reaction equals the rate of the reverse reaction.

Rate (forward) = $k_f[A][B]$

Rate (reverse) = $k_r[E]$

Rate (forward) = Rate(reverse), therefore $k_f[A][B] = k_r[E]$

[E] = ($k_f$/$k_r$ ) [A][B] = $k'$ [A][B]

Substitute for [E] in the rate law for the slow step in mechanism II to get Rate = $kk'$[A]²[B] or $k''$[A]²[B] where $k''$ is the rate constant of the experimentally determined rate law. Mechanism II is possible because its rate law agrees with the experimentally determined rate law.

## DETERMINATION OF ACTIVATION ENERGY

(*Chemistry* 6th ed. pages 585–588 / 7th ed. pages 554–557)

Activation energy can be determined using the Arrhenius equation

$$\ln(k) = -\frac{E_a}{R}\left(\frac{1}{T}\right) + \ln(A).$$

The rate constant is represented by $k$.

$E_a$ is the activation energy.

$R$ is the gas constant, 8.3145 J/(molK).

$T$ is the Kelvin temperature.

One way to determine the activation energy, $E_a$, is to measure the rate constant, $k$, at several different temperatures, and then graph $\ln(k)$ vs. $1/T$ which gives a straight line with the slope equal to $-E_a/R$.

The activation energy can also be calculated from the values of $k$ at only two temperatures using the equation:

$$\ln\frac{k_2}{k_1} = \frac{E_a}{R}\left(\frac{1}{T_1} - \frac{1}{T_2}\right)$$

EXAMPLE: A first order reaction has constants of $4.6 \times 10^{-2}$ s$^{-1}$ and $8.1 \times 10^{-2}$ s$^{-1}$ at 0°C and 20.°C, respectively. Calculate the value of the activation energy.

$$\ln\left(\frac{8.1\times10^{-2}\,s^{-1}}{4.6\times10^{-2}\,s^{-1}}\right) = \frac{E_a}{8.3145\ \text{J/mol} \bullet \text{K}}\left(\frac{1}{273\ \text{K}} - \frac{1}{293\ \text{K}}\right)$$

$0.57 = E_a/8.3145\ (2.5 \times 10^{-4})$

$E_a = 1.9 \times 10^4$ J/mol = 19 kJ/mol

## MULTIPLE-CHOICE QUESTIONS

No calculators are to be used in this section.

Each of the following questions or incomplete statements has five suggested responses. Select the one which best answers the question or incomplete statement.

1. The reaction $H_2(g) + Cl_2(g) \longrightarrow 2HCl(g)$
   (A) has a high positive $\Delta G$ value and is therefore a very fast reaction.
   (B) has a high negative $\Delta G$ value and is therefore a very fast reaction.
   (C) has $\Delta G$ = zero indicating the rate of reaction is not great.
   (D) has a $\Delta G$ value which does not relate to the reaction rate.
   (E) has a positive $\Delta G$ value indicating a spontaneous reaction.

ANSWER: D  While it may seem counterintuitive, spontaneity [$\Delta G$ = (–)] refers to the tendency to react and not to the rate of reaction (*Chemistry* 6th ed. pages 555 / 7th ed. page 527).

2. The rate constant
   (A) always shows an exponential increase with the Kelvin or absolute temperature.
   (B) increases with increasing concentration.
   (C) usually increases with increased pressure for gases.
   (D) never changes (it is a constant).
   (E) is the same for a given reaction at the same Kelvin or absolute temperature.

ANSWER:  E  In this question be sure to note that it is the rate constant that is involved.  Refer to the Arrhenius equation for the specific relationship (*Chemistry* 6th ed. pages 582–586 / 7th ed. pages 552–555).

3. A reaction mechanism
   (A) is the sum of all steps in a reaction except the rate determining step.
   (B) has a $\Delta H$ equal to the $\Delta H$ for the most energy demanding step.
   (C) always has a rate determining step.
   (D) may be absolutely proven from the rate law.
   (E) is determined from the balanced expression only.

ANSWER: C  In all series of steps (the mechanism for the overall reaction), there must always be a slowest step. The rate determining step is the slowest step (*Chemistry* 6th ed. page 581 / 7th ed. page 550–551).

4. If both the $\Delta H$ and $E_a$ for this forward reaction are known, the reverse reaction would have an $E_a$
   (A) of $(-\Delta H \longrightarrow ) + E_{a\rightarrow}$ .
   (B) of $\Delta H \longrightarrow + E_{a\rightarrow}$ .
   (C) equal to $E_a$ for the forward reaction.
   (D) equal to $(-E_a)$ for the forward reaction.
   (E) but none of the above describe the value of $E_a$.

ANSWER: A  This activation energy for the reverse reaction is difference between the potential energy of the products and the (top of the curve) energy of the activated complex, hence the sum of the reverse of the enthalpy change of the forward reaction plus the activation energy of the forward reaction. This can be understood by starting with the energy of the products and tracing the curve and resulting energy change to the top of the curve (*Chemistry* 6th ed. page 584 / 7th ed. page 553).

5. The rate of a chemical reaction
   (A) is always dependent on the concentration of all reactants.
   (B) is always increased with increasing temperatures.
   (C) is directly proportional to the value of $\Delta E$.
   (D) is greater with higher activation energies.
   (E) is independent of surface area.

ANSWER: B  The collision model indicates that molecular collision is necessary for reaction. Because an increase in temperature raises molecular velocities and the percentage of effective collisions, the reaction rate increases (*Chemistry* 6th ed. pages 582–585 / 7th ed. pages 552–555).

6. The values for the change in enthalpy, $\Delta H$, and the activation energy, $E_a$, for a given reaction are known. The value of $E_a$ for the reverse reaction equals
   (A) $E_a$ for the forward reaction.
   (B) $-(E_a)$ for the forward reaction.
   (C) the sum of $-\Delta H$ and $E_a$.
   (D) the sum of $\Delta H$ and $E_a$.
   (E) the difference of $\Delta H$ and $E_a$.

ANSWER: C Examine the diagram that accompanies Question 4 in this chapter for a clear understanding (*Chemistry* 6th ed. pages 582–586 / 7th ed. pages 552–555).

7. For two first-order reactions of different substances A and X

   $$A \longrightarrow B \quad t_{1/2} = 30.0 \text{ min}$$

   $$X \longrightarrow Y \quad t_{1/2} = 60.0 \text{ min}$$

   This means that
   (A) doubling the concentration of A will have 1/2 the effect on half-life that doubling the concentration of B will have on its half-life.
   (B) a certain number of grams of A will react twice as fast as the same number of grams of X.
   (C) a certain number of grams of X will react twice as fast as the same number of grams of A.
   (D) the rate constant for $A \longrightarrow B$ is lower than the rate constant of $X \longrightarrow Y$.
   (E) 3 moles of A will react more rapidly than 3 moles of X.

ANSWER: E It takes half as much time for A to form B as for X to form Y, as seen by the smaller half-life. Note that option "B" would be incorrect as the grams of A and the grams of X are not the same number of moles (*Chemistry* 6th ed. pages 571–573 / 7th ed. pages 541–543).

8. A reaction is first order with respect to [X] and second order with respect to [Y]. When [X] is 0.20 M and [Y] = 0.20 M the rate is $8.00 \times 10^{-3}$ M/min. The value of the rate constant, including correct units, is
   (A) $1.00 \text{ M min}^{-1}$.
   (B) $1.00 \text{ M}^{-2} \text{ min}^{-1}$.
   (C) $2.00 \text{ M}^{-1} \text{ min}^{-1}$.
   (D) $2.0 \text{ M}^{-2} \text{ min}^{-1}$.
   (E) $8.00 \times 10^{-3} \text{ min}^{-3}$.

ANSWER: B From these data, it follows that the rate law is Rate = $k[X][Y]^2$. Solving for the rate constant and substituting data for this reaction:

   k = Rate / $[X][Y]^2$
   = $8.00 \times 10^{-3}$M/min/$(0.200M)(0.200M)^2$
   = 0.008 M/min/0.008 M$^3$
   = $1.00 \text{ M}^{-2}\text{min}^{-1}$.

(*Chemistry* 6th ed. pages 564–570, 578 / 7th ed. pages 535–541, 548)

9. The activation energy for this reaction, X + 2Y ⟶ 3Z, shown in the potential energy diagram, could be
(A) increased by increasing [X].
(B) increased by increasing [X] and [Y].
(C) increased by increasing the temperature.
(D) decreased by removing Z from the system as it forms.
(E) decreased by adding a suitable catalyst.

ANSWER: **E**   Adding a catalyst suitable for this reaction will lower the energy barrier (activation energy) by forming a different activated complex which has a lower potential energy (*Chemistry* 6th ed. pages 582–590 / 7th ed. pages 552–559).

10. For all zero-order reactions
(A) a plot of time vs. concentration squared is linear.
(B) $E_a$ is very low.
(C) the concentration of reactants is constant.
(D) the rate is independent of time.
(E) the rate constant is zero.

ANSWER: **D**   For zero-order reactions, Rate = $k[X]^0$.  Because anything raised to the zero power is equal to one, Rate = $k$.  This is another way of saying that the rates of zero-order reactions do not change; they do not speed up and they do not slow down, they either take place or they do not (*Chemistry* 6th ed. page 576 / 7th ed. page 546).

## FREE-RESPONSE QUESTIONS

Calculators may be used for this section.

1. The data below are for the reaction 2X + 2Y ⟶ $Q_2$ + 2Z. The rate has been determined in terms of [$Q_2$] formed for various concentrations of the two reactants. Temperature is held constant.

| Exp. | Initial [X] | [Y] | Rate $\Delta[Q_2]/\Delta t$ |
|---|---|---|---|
| 1 | 0.20 M | 0.10 M | 0.048 M s$^{-1}$ |
| 2 | 0.40 M | 0.10 M | 0.19 M s$^{-1}$ |
| 3 | 0.40 M | 0.20 M | 0.38 M s$^{-1}$ |
| 4 | 0.60 M | 0.60 M | ? |

(a) Determine the rate law for X + 2Y ⟶ $Q_2$ + 2Z based on the above information.
(b) Indicate the order of this reaction.
(c) Determine the value of the rate constant.
(d) Calculate the rate for Exp. 4.
(e) Determine the rate for Exp. 3 in terms of $\Delta[X]/\Delta t$.

ANSWER: **1(a)**   To determine how each of the concentrations of the reactants is related to the reaction rate, only the one reactant concentration may change.  For example, as we go from Exp. 1 to Exp.

2, only [X] is changing. As it doubles, the reaction rate becomes four times greater, suggesting that rate is directly related to the square of the concentration of X. In like manner, compare the concentration changes from Exp. 2 to 3: [X] is held constant, [Y] is doubled, and the rate doubles. This suggests that rate is proportional to [Y]. Therefore, the rate law is Rate = $k[X]^2[Y]$.

ANSWER: **1(b)** From the rate law, we can say that this reaction is second order with respect to X, first order with respect to Y, and third order (2+1) overall.

ANSWER: **1(c)** Solving the rate law for $k$ yields:

$$k = \text{Rate} / ([X]^2[Y])$$

$$= 0.048 \text{ M s}^{-1} / (0.20M)^2(0.10M)$$

$$= 12 \text{ M}^{-2} \text{ s}^{-1}.$$

ANSWER: **1(d)** Rate = 12 $\text{M}^{-2}\text{s}^{-1}$ (0.60 M)$^2$ (0.60M) = 2.6 M s$^{-1}$.

ANSWER: **1(e)** Note that every time a mole of Q is formed, two moles of X disappear, so
$$0.38 \text{ M s}^{-1} \times 2 = 0.76 \text{ M s}^{-1.}$$
(*Chemistry* 6th ed. pages 566–576 / 7th ed. pages 537–546)

2.  Three major methods used to increase the rate of a reaction are
    (a) adding a catalyst,
    (b) increasing the temperature, and
    (c) increasing the concentration of a reactant.

    From the perspective of collision theory, explain how each of these methods increases the reaction rate.

    (d) For the same reaction run at two different temperatures, the rate constant is shown to change as:

| Temp. | $k$ |
|---|---|
| 25°C | $4.2 \times 10^{-5}$ |
| 37°C | $2.6 \times 10^{-4}$ |

    Determine the activation energy for this reaction from these data.

ANSWER: **2(a)** Adding a suitable catalyst lowers the energy demands of the reaction (the activation energy) by providing a less demanding pathway. It does this by forming a different activated complex with a lower activation energy.

ANSWER: **2(b)** Increasing the temperature increases the average kinetic energy of the molecules. However, what is most important here is that the fraction of molecules that have the energy necessary to react (the activation energy) is increased.

ANSWER: **2(c)** Increasing the concentration of a reactant increases the probability of collisions, and therefore increases the possibility of reaction, for the particles must collide to react. Note that not all collisions result in successful reactions.

ANSWER: **2(d)** The activation from $k$ values taken at different temperatures:

$$\ln\frac{k_2}{k_1} = \frac{E_a}{R}\left[\frac{(T_2-T)_1}{(T_2T_1)}\right];$$

the form of the equation given on the formula sheet for the AP exam is

$$\ln k = -\frac{E_a}{R}\left(\frac{1}{T}\right) + \ln A$$

$$\ln\left(\frac{2.6\times10^{-4}}{4.2\times10^{-5}}\right) = \frac{E_a}{8.314\text{ J/molK}}\left(\frac{310-298}{310\times298}\right)$$

$$E_a = \frac{8.314\times1.82}{1.30\times10^{-4}} = 120,000\text{ J} = 120\text{ kJ}$$

Note: Because the two-point form of this equation is not given on the equation sheet, it is more likely that you will be given a graph and asked what the slope represents or asked to label the axes. (*Chemistry* 6th ed. pages 582–588 / 7th ed. pages 552–557)

# 12

# EQUILIBRIUM

In this chapter, you will review the characteristics of equilibrium and calculations involving the concentrations of reactants and products for a given system at equilibrium. Your knowledge of the basic equilibrium problems in this chapter will enable you to solve for the pH of acids, bases, salts, and buffers as well as the solubility of ionic compounds.

## AP tip

The first section of the AP Exam, worth 25% of the free response score, is always an equilibrium problem. It can be one of the types of problems found in any of the three equilibrium chapters in this guide.

You should be able to
- Write equilibrium expressions for a given reaction.
- Calculate $Q$ and compare it to $K$ to determine if a reaction is at equilibrium.
- Manipulate $K$ if a reaction is reversed or multiplied by a coefficient.
- Calculate $K$ from given equilibrium concentrations, or if given $K$ and all except one equilibrium concentration, solve for the missing value.
- Calculate equilibrium concentrations (or one of the missing variables) if given any two of the following values : $K$, the initial concentrations, one equilibrium concentration.
- Do calculations involving gaseous equilibria, partial pressures, and $K_p$.
- Use LeChâtlier's principle to determine in what direction the position of equilibrium will shift when a change is imposed.
- Calculate the value of $K$ from thermodynamic values such as $\Delta G°$.

## EQUILIBRIUM CONDITION

Many reactions are reversible. When equilibrium is reached, the rate of the forward reaction equals the rate of the reverse reaction. The ratio of the product concentrations to the reactant concentrations is constant at equilibrium.

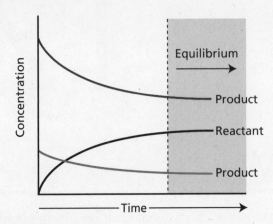

### THE EQUILIBRIUM EXPRESSION

(*Chemistry* 6th ed. pages 613–614 / 7th ed. pages 582–583)

The equilibrium constant, $K$, measures the ratio of product concentrations to reactant concentrations.

For a hypothetical reaction $aA + bB \rightleftharpoons cC + dD$

$$K = \frac{[A]^a[B]^b}{[C]^c[D]^d}.$$

$K$ is the equilibrium constant, usually given without units. [ ] represents the concentration of the reactants and products at equilibrium in moles/liter. The concentrations of products and reactants are raised to the powers of their respective coefficients in the balanced chemical equation.

You will be expected to write the equilibrium expression for a given reaction. Remember, the concentrations of solids and liquids are not included.

EXAMPLE: Write the equilibrium expression for the reaction

$$P_4(s) + 5O_2(g) \rightleftharpoons P_4O_{10}(s).$$

SOLUTION: The equilibrium expression for the reaction is

$$K = \frac{1}{[O_2]^5}.$$

The concentrations of the solids are not included in the expression because the concentration of a solid or liquid is a constant, so it is included in the value of $K$.

The value of the equilibrium constant measures the extent to which a reaction occurs.

$K > 1$: The concentrations of the products are greater than the concentrations of the reactants.

$K < 1$: The concentrations of the reactants are greater than the concentrations of the products. The reaction hardly proceeds toward completion.

# REACTION QUOTIENT

(*Chemistry* 6th ed. pages 624–625 / 7th ed. page 593)

When reactants and products are mixed together, they may not be at equilibrium. The reaction quotient, $Q$, compared to the equilibrium constant, $K$, will determine which way a system will shift to reach equilibrium. If a system is not at equilibrium, it will move in a direction to reach equilibrium.

If a reactant or product in a reaction is not present and is at zero concentration, the reaction will move in the direction that produces the missing component.

If all reactants and products are present and have an initial concentration, you must determine the value of $Q$, the reaction quotient.

To determine if the reaction is at equilibrium or the direction it will shift to attain equilibrium, plug all of the initial concentrations into the reaction quotient, which is the same as the equilibrium expression (also called the law of mass action), and compare the value of $Q$ to $K$.

## Comparison of $Q$ to $K$

If $Q = K$, the reaction is at equilibrium.

If $Q > K$, the reaction will shift to the left. A shift toward the reactants will consume products.

If $Q < K$, the reaction will shift to the right. A shift to produce more products will consume reactants.

EXAMPLE: For the reaction $2NO(g) \rightleftharpoons N_2(g) + O_2(g)$ $K = 2.4 \times 10^3$ at a particular temperature.

(a) If the initial concentrations are 0.024 M NO, 2.0 M $N_2$, and 2.6 mol $O_2$, is the system at equilibrium?

(b) If it is not at equilibrium, in which direction will the reaction shift?

SOLUTION: The reaction is not at equilibrium, $Q > K$, so it will shift to the left.

$$Q = \frac{[N_2][O_2]}{[NO]^2} = \frac{(2.0)(2.6)}{(0.024)^2} = 9.0 \times 10^3$$

## TYPES OF EQUILIBRIUM PROBLEMS

(*Chemistry* 6th ed. pages 614, 628–636 / 7th ed. pages 583, 596–604)

The types of equilibrium problems you can expect to solve in this chapter and the two chapters which follow include:

1. Manipulation of the equilibrium constant, $K$
   a. If the reaction is reversed, the equilibrium expression is the reciprocal of the expression for the forward reaction.
   EXAMPLE:

$$A \rightleftharpoons B: K = \frac{[B]}{[A]}$$

$$B \rightleftharpoons A: K' = \frac{1}{K} = \frac{[A]}{[B]}$$

   b. If the coefficients in a balanced equation are multiplied by a number, $n$, the equilibrium constant is raised to the power $n$.
   EXAMPLE:

$$A \rightleftharpoons B: K = \frac{[B]}{[A]}$$

$$2A \rightleftharpoons 2B: K'' = K^2 = \frac{[B]^2}{[A]^2}$$

2. If given all equilibrium concentrations, calculate the value of the constant, $K$.
3. If given the value of the equilibrium constant, $K$, and all but one of the equilibrium concentrations, solve for the missing concentration.
4. If given the value of the initial concentrations of the reactants and one of the equilibrium concentrations of either the reactants or products, solve for all equilibrium concentrations and the value of $K$.
5. If given the initial concentrations and the value of $K$, solve by approximation for the equilibrium concentrations.

The first four types of problems listed above can be solved using basic algebra. The fifth type of problem can be solved using the method of approximation.

EXAMPLE: (Type 1) For the reaction

$N_2(g) + 3H_2(g) \longrightarrow 2NH_3(g)$, $K = 1.3 \times 10^{-2}$ at a certain temperature.

Calculate the value of $K$, called $K'$, for the reaction

$NH_3(g) \longrightarrow \frac{1}{2} N_2(g) + \frac{3}{2} H_2(g)$.

SOLUTION: The reaction is the reverse and one half of the one which is given.

$$K' = [N_2]^{1/2}[H_2]^{3/2}] / [NH_3] = (1/K)^{1/2}; \left(\frac{1}{1.3 \times 10^{-2}}\right)^{\frac{1}{2}} = 8.8.$$

EXAMPLE: (Type 4) At a particular temperature, $K = 1.00 \times 10^2$ for the reaction

$$H_2(g) + I_2(g) \rightleftharpoons 2HI(g).$$

In an experiment, 1.00 mol $H_2$, 1.00 mol $I_2$, and 1.00 mol HI are introduced into a 1.00-L container. Calculate the equilibrium concentrations of all reactions and products.

SOLUTION: To begin, write the balanced equation for the reaction and the equilibrium expression, omitting pure solids and liquids.

|  | $H_2(g) +$ | $I_2(g)$ | $\rightleftharpoons$ | $2HI(g)$ | $K = \dfrac{[HI]^2}{[H_2][I_2]}$ |
|---|---|---|---|---|---|
| I: | 1.00 | 1.00 |  | 1.00 |  |
| C: | $-x$ | $-x$ |  | $+2x$ |  |
| E: | $1.00 - x$ | $1.00 - x$ |  | $1.00 + 2x$ |  |

Make an ICE chart under the balanced chemical equation.

**I = initial concentration** in mol/L (note units are omitted from chart)

**C = the change to reach equilibrium** represented by + or $-x$. A minus sign indicates a decrease in concentration, a plus sign indicates an increase. The coefficient in front of the reactant in the balanced equation is placed in front of the $x$ in the change line.

In this example, $Q = 1$ which is less than $K$, so the reaction will shift to the right.

**E = Equilibrium concentrations** which are obtained by adding the I and C lines together.

Plug the equilibrium values into the expression and solve for $x$:

$$K = 100. = \frac{(1.00 + 2x)^2}{(1.00 - x)^2}.$$

Taking the square root of both sides:

$$10.0 = \frac{1.00 + 2x}{1.00 - x}, \ 10.0 - 10.0\,x = 1.00 + 2x, \ 12x = 9.0, \ x = 0.75 \text{ M}.$$

Use this value of $x$ to solve for the equilibrium concentrations of all reactants and products.

$[H_2] = [I_2] = 1.00 - 0.75 = 0.25$ M; $[HI] = 1.00 + 2\,(0.75) = 2.50$ M

Last, check your equilibrium concentrations by making sure that they equal the correct value of $K$.

**EXAMPLE:** (Type 5) For the reaction $N_2O_4(g) \rightleftharpoons 2NO_2(g)$,

$K = 4.0 \times 10^{-7}$ at a specific temperature.

In an experiment, 1.0 mol of $N_2O_4$ is placed in a 10.0-L vessel.

Calculate the equilibrium concentrations of $NO_2$ and $N_2O_4$.

**SOLUTION:** At equilibrium, $[N_2O_4] = 0.10$ M and $[NO_2] = 2.0 \times 10^{-4}$ M. To solve this problem, proceed as in the previous example.

| | $N_2O_4(g)$ | $\rightleftharpoons$ | $2NO_2(g)$ |
|---|---|---|---|
| I | 0.10 | | 0 |
| C | $-x$ | | $+2x$ |
| E | $0.10 - x$ | | $2x$ |

$$K = \frac{[NO_2]^2}{[N_2O_4]} = \frac{(2x)^2}{(0.10-x)} = 4.0 \times 10^{-7}$$

Since the value of $K$ is much smaller than 1, you can assume that the change from the initial concentration, $x$ in $0.10 - x$, is so small that it is negligible, that is, $0.10 - x$ is about equal to 0.10.

This greatly simplifies the math to

$$K = \frac{(2x)^2}{(0.10)} = 4.0 \times 10^{-7}$$

$4x^2 = 4.0 \times 10^{-8}$, $x = 1.0 \times 10^{-4}$ M

It is okay to make this assumption if the change from the initial concentration, in this case, $x$ is less than 5% of the initial concentration.

$$\frac{x}{0.10} \times 100\% = \frac{1.0 \times 10^{-4}}{0.10} \times 100\% = 0.10\%$$

This is less than 5%, so, it is okay to make the assumption.

The equilibrium concentrations are as follows:

$[N_2O_4] = 0.10 - x = 0.10 - 1.0 \times 10^{-4} = 0.10$ M

$[NO_2] = 2x = 2(1.0 \times 10^{-4}) = 2.0 \times 10^{-4}$ M

## GASEOUS EQUILIBRIUM

(*Chemistry* 6th ed. pages 617–620, 626–628 / 7th ed. pages 586–588, 594–596)

For equilibrium in the gas phase, the equilibrium expression can be written in terms of the partial pressures of the gases.

For the reaction $AsH_3(g) \rightleftharpoons 2As(s) + 3H_2(g)$

$$K_p = \frac{\left(P_{H_2}\right)^3}{P_{AsH_3}}$$

$K_p$ is the equilibrium constant in terms of the partial pressures of the gases.

$P$ represents the partial pressure of the gases raised to their coefficients in the balanced chemical equation.

EXAMPLE: Given the following reaction:

$2NO(g) + Br_2(g) \rightleftharpoons 2NOBr(g)$ $K_p = 109$ at 25°C.

The equilibrium partial pressure of $Br_2 = 0.0159$ atm and

NOBr = 0.0768 atm. Calculate the equilibrium partial pressure of NO.

$$Kp = \frac{\left(P_{NOBr}\right)^2}{\left(P_{NO}\right)^2 \left(P_{Br_2}\right)}$$

$$109\text{ atm}^{-1} = \frac{(.0768\text{ atm})^2}{\left(P_{NO}\right)^2 (0.0159\text{ atm})}$$

$$P_{NO} = 0.0583\text{ atm}$$

The relationship between $K_p$ and $K$ is given by $K_p = K(RT)^{\Delta n}$.

$K_p$ is the equilibrium gas constant.

$\Delta n$ is the sum of the coefficients of the gaseous products minus the sum of the coefficients of the gaseous reactants.

EXAMPLE: $2NO(g) + Cl_2(g) \rightleftharpoons 2NOCl(g)$ $K_p = 1.9 \times 10^3$ at 25°C. Calculate the value for $K_c$ at 25°C.

$\Delta n = 2 - (2+1) = -1$

$K_p = K(RT)^{-1} = K/RT$

$K = K_pRT = (1.9 \times 10^3)(0.08206)(298) = 4.6 \times 10^4$.

## LeCHÂTELIER'S PRINCIPLE

(*Chemistry* 6th ed. pages 636–642 / 7th ed. pages 604–610)

The value of $K$ is a constant at a particular temperature. The only factor that changes the value of $K$ is temperature. Pressure, a catalyst, and changes in concentration will not affect the value of $K$.

LeChâtelier's principle states that if a change is imposed in a system at equilibrium, the position of the equilibrium will shift in a direction that will counteract the change.

**Example:** Consider the following changes on the system below at equilibrium.

$$2\,SO_3\,(g) \;\rightleftharpoons\; 2\,SO_2(g) + O_2(g) \quad \Delta H = 197\ kJ.$$

(a) Addition of $O_2$

The equilibrium will shift to the left to form more reactants. The addition of oxygen increases the rate of the reverse reaction so more reactants will form, until the rate of the forward reaction again equals the rate of the reverse reaction. Another way to explain the shift is that $Q > K$ so that must decrease the products and increase reactants for $Q = K$ again.

(b) Removal of $SO_2$

Removal of an equilibrium component such as $SO_2$ at constant pressure and temperature will cause the equilibrium to shift toward the removed component to increase its concentration. The reaction in this example will shift to the right.

(c) Increase in temperature

To consider the effect of changes in temperature on the equilibrium constant, treat the heat of the reaction as a reactant (in an endothermic process) or as a product (in an exothermic process). The direction of the shift can be predicted in the same way as the addition or removal of a reactant or product.

In the reaction for the decomposition of $SO_3$, the reaction is endothermic. Treating heat as a reactant, an increase in temperature will cause the reaction to shift right, producing more products, increasing the value of $K$. A decrease in temperature will cause a shift to the left, increasing the reactant concentrations, and lowering the value of $K$.

(d) An increase in pressure.

If the pressure is increased in an equilibrium system, the reaction will shift toward the side with fewer moles of gas. Decreasing the volume has the same effect because the only way the pressure can be increased without changing the temperature or number of moles is to decrease the volume. In this example, the reaction will shift to the left, toward $SO_3$, when the pressure is increased or the volume is decreased. If the moles of gas are the same on both sides of the reaction, no shift will occur.

e) Addition of a solid or inert gas, such as Ne.

If a solid or an inert gas (with no change in volume) is added to the reaction, there will be no shift in equilibrium. Neither the solid nor the inert gas are part of the equilibrium expression. The concentrations of all the components of the equilibrium expression remain unchanged.

f) An inert gas, such as Ne, is added at constant pressure.

There will be a shift in the equilibrium to the side of the equation with more moles of gas. To add Ne at constant pressure, the volume of the container must increase so the concentrations or partial pressures of all gases have decreased. If the moles of gas are greater in the reactants, then $Q < K$, so equilibrium can be reestablished only by increasing the products.

## RELATIONSHIP OF $K$ TO FREE ENERGY, $\Delta G^O$

(*Chemistry* 6th ed. pages 810–814 / 7th ed. pages 774–778)

The equilibrium constant at standard temperature can be determined from the thermodynamic values of a reaction using the equation:

$\Delta G^0 = -RT \ln K$

$R = 8.314$ J/(mol•K)

EXAMPLE: Calculate the value of $K$ at 25.0°C for the reaction $2NO_2(g) \rightleftharpoons N_2O_4(g)$

The values of $\Delta H^\circ$ and $\Delta S^\circ$ are –58.03 kJ/mol and –176.6 J/K•mol, respectively.

SOLUTION:

At 25°C, $\Delta G^\circ = \Delta H^\circ - T \Delta S^\circ$

$\qquad = -58.03 \times 10^3$ J/mol – (298 K)(–176.6 J/(K•mol))

$\qquad = -5.40 \times 10^3$ J/mol.

$\Delta G^0 = -RT \ln K$; $\ln K = -\Delta G/RT = 2.18$ ; $K = 8.8$.

## MULTIPLE-CHOICE QUESTIONS

No calculators are to be used in this section.

1. In the reaction $3W + X \rightleftharpoons 2Y + Z$, all substances are gases. The reaction is initiated by adding an equal number of moles of W and of X. When equilibrium is reached,
   (A) [Y] = [Z].
   (B) [X] = [Y].
   (C) [W] =[X].
   (D) [X] > [W].
   (E) [W] + [X] = [Y] + [Z].

ANSWER: D Since every time a mole of X reacts, 3 moles of W must react, so the amount of W remaining must be less than the amount of X remaining (recall that you started with an equal number of moles of W and of X) (*Chemistry* 6th ed. pages 610–613 / 7th ed. pages 579–582).

2. Consider Rx.I   X $\rightleftharpoons$ Y $K = 1 \times 10^8$

   Rx.II  Z $\rightleftharpoons$ Y $K = 1 \times 10^5$

   both at the same temperature:
   (A) I is 3 times faster than II.
   (B) I is 1000 times faster that II.
   (C) II is 3 times faster than I.
   (D) II is 1000 times faster than I.
   (E) The size of $K$ and the time required to reach equilibrium are not directly related.

ANSWER: **E**  There is not a direct relationship between $K$ and reaction rate (*Chemistry* 6th ed. pages 624–625 / 7th ed. page 593).

3. The reaction $3H_2(g) + N_2(g) \rightleftharpoons 2NH_3(g)$ has an enthalpy of change of –92 kJ. Increasing the temperature of this equilibrium system causes
   (A) an increase in $[NH_3]$.
   (B) an increase in $[N_2]$.
   (C) a decrease in $[H_2]$.
   (D) an increase in $K$.
   (E) a decrease in pressure at constant volume.

ANSWER: **B**  Increasing the temperature causes the equilibrium to shift to the left. For an exothermic reaction, increasing the temperature increases the rate of both the forward and reverse reactions, but proportionally makes a greater increase in the reverse reaction since it has the higher activation energy, favoring the formation of more hydrogen gas and more nitrogen gas, and lowering the concentration of the ammonia. This forms more gaseous particles; therefore, the pressure increases at constant volume (*Chemistry* 6th ed. pages 641–642 / 7th ed. pages 609–610).

4. Consider $N_2(g) + O_2(g) \rightleftharpoons 2NO(g)$. The reaction was initiated by adding 15.0 moles of NO to a 1.0-L flask. At equilibrium, 3.0 moles of oxygen are present in the 1.0-L flask. The value of $K$ must be
   (A) 0.33.
   (B) 3.0.
   (C) 5.0.
   (D) 9.0.
   (E) 81.

ANSWER: **D**  If 3.0 moles of oxygen are formed, 6.0 moles of NO must have reacted, leaving 9.0 moles of NO at equilibrium (15.0 – 6.0 = 9.0 mol/L for [NO]). Each time 3.0 moles of oxygen form, the same number of moles of nitrogen are produced. Since the reaction (take care here) is written showing NO as a product, the equilibrium constant expression is $K = \dfrac{[NO]^2}{[N_2][O_2]} = \dfrac{(9.0)^2}{(3.0)(3.0)} = \dfrac{81}{9.0} = 9.0$.

(*Chemistry* 6th ed. pages 628–633 / 7th ed. pages 596–601)

5. At a certain temperature, the synthesis of ammonia gas from nitrogen and hydrogen gases, shown as $N_2 + 3 H_2 \rightleftharpoons 2 NH_3$, has a value for $K$ of $3.0 \times 10^{-2}$. If $[H_2] = [N_2] = 0.10$ M and $[NH_3] = 0.20$ M,
   (A) the reaction would shift toward the ammonia.
   (B) the reaction would shift toward the $N_2$ and the $H_2$.
   (C) the system is at equilibrium, therefore no shifting will occur.
   (D) the reaction will shift toward a new equilibrium position, but the direction cannot be determined from these data.
   (E) the equilibrium may shift but it is not possible to calculate $Q$ without knowing the temperature.

ANSWER: B

$$Q = \frac{\left[NH_3\right]^2}{\left[H_2\right]^3 [N_2]} = \frac{(0.20)^2}{(0.10)^3 (0.10)} = 400$$

Since $Q$ is greater than $K$, the reaction will shift toward $N_2$ and $H_2$. Note that this setup for $Q$ is easily handled if seen as $\frac{\left(2 \times 10^{-1}\right)^2}{\left(1 \times 10^{-1}\right)^3 \left(10^{-1}\right)} = \frac{4 \times 10^{-2}}{1 \times 10^{-4}} = 4 \times 10^2$.

(*Chemistry* 6th ed. pages 625–626 / 7th ed. pages 593–594)

6. The equilibrium constant, $K$, may be used to determine the $K$ of other reactions.

   I. When the equation is reversed the reciprocal of the original $K$ becomes the value of $K$ for the new equation.

   II. When the equation is doubled the square root of the original $K$ becomes the value for the new equation.

   III. When the equation for a reaction is tripled the equilibrium expression for the new equation is triple the original $K$.

   Of the above three statements, those which are always true are
   (A) I only.
   (B) II only.
   (C) III only.
   (D) I and III.
   (E) I, II, and III.

ANSWER: A Statement II is not valid; when the equation is doubled the K value becomes the square of the original K, $(K' = K^2)$. When the equation is tripled the value for K for the new equation is the cube of the original K $(K' = K^3)$. (*Chemistry* 6th ed. page 615 / 7th ed. page 584).

7. At a certain temperature, it has been determined that $K = 8.0$ for $H_2O(g) + CO(g) \rightleftharpoons CO_2(g) + H_2(g)$. If we also determine that the equilibrium mixture contains 0.80 mol of $H_2O(g)$, 0.080 mole of $CO_2(g)$, and 0.080 mole of $CO(g)$, in an 8.0-L flask, what must be the number of moles of $H_2(g)$ at equilibrium?
(A) 0.010 mole.
(B) 0.80 mole.
(C) 6.4 moles.
(D) 8.0 moles.
(E) 64 moles.

ANSWER: C

$$K = \frac{[CO_2][H_2]}{[H_2][CO]} = 8.0 = \frac{\left(\dfrac{0.080}{8.0}\right)(X)}{\left(\dfrac{0.80}{8.0}\right)\left(\dfrac{.080}{8.0}\right)}$$

$$\frac{X}{0.10} = 8.0$$

$$X = 0.80M$$

$$0.80\frac{mol}{L} \times 8.0L = 6.4 \text{ mole of } H_2$$

(*Chemistry* 6th ed. pages 628–631 / 7th ed. pages 596–599).

8. Which of the following systems at equilibrium are not affected by a change in pressure caused by changing the volume at constant temperature?
(A) $H_2(g) + Cl_2(g) \rightleftharpoons 2HCl(g)$

(B) $H_2(g) + I_2(s) \rightleftharpoons 2HI(g)$

(C) $N_2(g) + 3H_2(g) \rightleftharpoons 2NH_3(g)$

(D) $2NH_3(g) \rightleftharpoons N_2(g) + 3H_2(g)$

(E) $3O_2(g) \rightleftharpoons 2O_3(g)$

ANSWER: A Because you are seeking an equilibrium system which has not changed with a change in pressure due to a volume change, look for a system with an equal number of moles of both gaseous reactants and products (*Chemistry* 6th ed. pages 638–640 / 7th ed. pages 606–608).

9. The equilibrium $P_4(g) + 6Cl_2(g) \rightleftharpoons 4PCl_3(l)$ is established at $-10°C$. The equilibrium constant expression is

(A) $K = \dfrac{[PCl_3]}{[P_4][Cl_2]}$.

(B) $K = \dfrac{[PCl_3]^4}{[P_4][Cl_2]^6}$.

(C) $K = \dfrac{[P_4][Cl_2]^6}{[PCl_3]^4}$.

(D) $K = \dfrac{1}{[PCl_3]^4}$.

(E) $K = \dfrac{1}{[P_4][Cl_2]^6}$.

ANSWER: E  The equilibrium constant is a ratio of the concentration of products divided by the concentration of reactants, each taken to a power represented by their coefficients. Pure liquids and solids are not shown in the equilibrium constant expression (*Chemistry* 6th ed. pages 613–614, 620–622 / 7th ed. pages 582–583, 588–590).

10. Ammonium hydrogen sulfide will decompose into ammonia gas and hydrogen sulfide gas when heated. Consider the equilibrium system

$$NH_4HS(s) \rightleftharpoons NH_3(g) + H_2S(g),$$

which is developed from 1.000 mole of $NH_4HS$ in a 100-L cylinder. At equilibrium the total pressure is found to be 0.400 atm. $K_p$ will be equal to

(A) $2.00 \times 10^{-1}$.
(B) $1.00 \times 10^{-2}$.
(C) $4.00 \times 10^{-2}$.
(D) $4.00$.
(E) a value impossible to calculate from only these data.

ANSWER: C  The two gases are formed in equal molar amounts (1:1); therefore half of the pressure is due to each gas $\left(\dfrac{0.400 \text{ atm}}{2} = 0.200 \text{ atm}\right)$. $K_p = \left(P_{NH_3}\right)\left(P_{H_2S}\right) = (0.200) \times (0.200) = 0.0400$ $= 4.00 \times 10^{-2}$. (*Chemistry* 6th ed. pages 618–622 / 7th ed. pages 587–590)

## FREE-RESPONSE QUESTIONS

1. Consider the reaction $2HI(g) \rightleftharpoons H_2(g) + I_2(s)$ which is in equilibrium.
   (a). How will adding more hydrogen gas at constant volume affect the equilibrium?
   (b) Some of the iodine is removed. How will this affect the equilibrium?
   (c) The pressure is increased by pumping in pure neon gas. How will this affect the equilibrium? Explain.
   (d) The pressure is increased by decreasing the volume. How will this affect the equilibrium? Explain.
   (e) A suitable catalyst is added. How will this affect the equilibrium?

ANSWER 1(a): Adding hydrogen causes the equilibrium to shift toward the reactant forming more HI; this is due to the increase in $[H_2]$, which means more collisions between $H_2$ and $I_2$ therefore more reactions forming HI. $Q > K$ will cause more reactants to form.

ANSWER 1(b): There is no effect on equilibrium since the concentration of iodine, a pure solid, is constant.

ANSWER 1(c): There is no effect. The neon gas will not react with any product or reactant in this system. Since the volume does not change, there is no change in the concentrations of the gases.

ANSWER 1(d): Note that there are two moles of gas forming one mole of gas. Increasing the pressure by decreasing the volume ($Q < K$) causes the reaction to shift toward fewer moles of gas, (toward the products in this case), thereby opposing the pressure change by lowering this pressure somewhat.

ANSWER 1(e): There is no effect from adding a catalyst once the system is in equilibrium. (The addition of a catalyst will may cause the system to reach equilibrium more rapidly, but the equilibrium concentrations will remain the same) (*Chemistry* 6th ed. pages 636–642 / 7th ed. pages 604–610).

2. Refer to the system $PCl_3(g) + Cl_2(g) \rightleftharpoons PCl_5(g)$. To an empty 15.0-L cylinder, 0.500 moles of gaseous $PCl_5$ are added and allowed to reach equilibrium. The concentration of $PCl_3$ is found to be 0.0220M. Assume a temperature of 375 K.
   (a) How many moles of $PCl_5$ remain at equilibrium?
   (b) Write the equilibrium constant expression for the above reaction.
   (c) Determine the value of $K$.
   (d) Determine the value of $K_p$ for this same system at the same temperature.
   (e) How would the value of $K_p$ be effected by increasing the temperature of the system at equilibrium for this exothermic reaction?

ANSWER 2(a): 0.0220 mol/L $PCl_3$ × 15.0 L = 0.330 mole $PCl_3$ = 0.330 mole $PCl_5$ that reacted.

0.500 – 0.330 = 0.170 mole $PCl_5$ remains.

ANSWER 2(b): $K_c = \dfrac{[PCl_5]}{[PCl_3][Cl_2]}$

ANSWER 2(c): $\dfrac{\left(\dfrac{0.170}{15.0\ L}\right)}{(0.0220)(0.0220)} = 23.4$

ANSWER 2(d): $K_p = K_c\,(RT)^{\Delta n}$

In this case, 1 + 1 mol of gas $\longrightarrow$ 1 mole of gas, so $\Delta n = -1$ moles

$K_p = 23.4\,(0.08206 × 375)^{-1} = 0.760$

ANSWER 2(e): Raising the temperature of an exothermic reaction opposes the forward reaction. Further, from $K_p = K_c\,(RT)^{\Delta n}$, if $T$ increases, the value of $K_p$ decreases (*Chemistry* 6th ed. pages 626–628, 618–619 / 7th ed. pages 594–596, 587–588).

# 13

# ACIDS, BASES, AND SALTS

Concepts from the last chapter will be applied in this section covering the equilibria of weak acids, weak bases, and salts. In addition, acid base theories and properties of acids will be reviewed. Calculations involved in acid-base titrations, preparation of buffers, and salt hydrolysis will be demonstrated.

You should be able to

- Understand the acid-base theories of Arrhenius, Brønsted-Lowry, and Lewis.
- Identify strong acids and bases and calculate their pH's.
- Calculate the pH of a weak acid or base.
- Calculate the concentration of a strong or weak acid or base from its pH.
- Calculate the pH and ion concentrations in a polyprotic acid.
- Predict the pH of a salt from its formula and then calculate the pH of the salt.
- Identify the components of a buffer and perform calculations involving the preparation of a buffer and the addition of strong acid or strong base to a buffer.
- Perform calculations involving strong acid-strong base titrations as well as weak acid-strong base and weak base-strong acid calculations.
- Be familiar with titration curves and selection of an acid-base indicator.

> ## AP tip
>
> Be systematic. Acid-base equilibrium problems require a step-by-step process. Although the problems in this chapter have many similarities, you must make note of the differences.

## ACID-BASE THEORIES

### ARRHENIUS

(*Chemistry* 6th ed. page 654 / 7th ed. page 623)

The Arrhenius theory states that, in aqueous solution (water), acids produce hydrogen ions and bases produce hydroxide ions.

### BRØNSTED-LOWRY

(*Chemistry* 6th ed. pages 654–656 / 7th ed. pages 623–626)

The Brønsted-Lowry theory says that an acid is a proton ($H^+$) donor and a base is a proton acceptor.

In the reaction below, $HNO_3$ transfers a proton to $H_2O$ forming $H_3O^+$, the hydronium ion. $H_3O^+$ is the conjugate acid of $H_2O$ and $NO_3^-$ is the conjugate base of $HNO_3$. The formulas in a conjugate acid-base pair differ by one $H^+$.

$$HNO_3 + H_2O \longrightarrow H_3O^+ + NO_3^-$$

Acid     Base     Conjugate acid    Conjugate base

> EXAMPLE: Give the formulas for the conjugate base of $H_2SO_4$ and the conjugate acid of $CH_3NH_2$.
>
> ANSWER: $HSO_4^-$ is the conjugate base of $H_2SO_4$.
>
> $CH_3NH_3^+$ is the conjugate acid of $CH_3NH_2$.
>
> Note that each conjugate acid-base pair differs by 1 $H^+$:
>
> $HSO_4^- + 1H^+ = H_2SO_4$ ; $CH_3NH_3^+ = 1H^+ + CH_3NH_2$.

### LEWIS ACID-BASE MODEL

(*Chemistry* 6th ed. pages 696–699 / 7th ed. pages 663–665)

According to this theory, a Lewis acid is an electron pair acceptor and a Lewis base is an electron pair donor. Remember that this is a much broader definition than the other two, in that the acid-base interaction does not need to involve a transfer of a proton $H^+$.

## ACID AND BASE STRENGTH

(*Chemistry* 6th ed. pages 657–660, 676 / 7th ed. pages 626–629, 644–645)

The names and formulas of the six strong acids must be memorized. The six strong acids are $HCl$, $HBr$, $HI$, $HNO_3$, $H_2SO_4$, and $HClO_4$.

If an acid is not one of the six in the list, then for purposes of the AP exam, you can assume it is a weak acid. The chart below compares the dissociation of strong acids to weak acids.

| **Comparison of Strong and Weak Acids** | | | |
|---|---|---|---|
| Type of acid, HA | Reversibility of reaction | $K_a$ value | Ions existing when acid, HA, dissociates in $H_2O$ |
| Strong | Not reversible | $K_a$ value very large | $H^+$ and $A^-$, only. No HA present. |
| Weak | reversible | $K_a$ is small | $H^+$, $A^-$, and HA |

Weak acids exist in equilibrium with their ions in aqueous solution. The acid dissociation constant, $K_a$, measures the extent to which the acid dissociates in water:

$$HA(aq) + H_2O\ (l) \rightleftharpoons H_3O^+(aq) + A^-(aq)$$

The equilibrium expression for the reaction is

$$K_a = \frac{[H_3O^+][A^-]}{[HA]}$$

A table of $K_a$ values for monoprotic acids, containing one acidic hydrogen, appears in the appendix of *Chemistry* and many other textbooks and is worth studying. The larger the $K_a$ value, the stronger the acid.

EXAMPLE: List the acids in order of increasing strength: $HCN$, $HCl$, $HClO_2$, $HNO_2$.

ANSWER: $HCN$, $HNO_2$, $HClO_2$, and $HCl$. The first three weak acids are listed in order of increasing $K_a$ values. $HCl$ is stronger than all of the weak acids given. $HCl$ is a strong acid.

EXAMPLE: Arrange the following species in order of increasing base strength: $NO_2^-$, $ClO_2^-$, $CN^-$, $Cl^-$

ANSWER: $Cl^-$, $ClO_2^-$, $NO_2^-$, $CN^-$; The bases are listed in reverse order of their conjugate acids in the previous example because the stronger an acid, the weaker its conjugate base.

Strong bases include group I and II hydroxides such as $NaOH$. Weak bases such as $NH_3$, ammonia, are not group I or II hydroxides.

The base dissociation constant, $K_b$, measures the extent to which a base reacts with water. The reaction of a weak base, B, with water and its corresponding equilibrium expression is

$$B\,(aq) + HOH(l) \rightleftharpoons BH^+(aq) + OH^-(aq); \quad K_b = \frac{[BH^+]\,[OH^-]}{[B]}.$$

---

## AP tip

Writing the reaction for base dissociation can be tricky. Always remember to react the base with water. Use the Arrhenius and Brønsted-Lowry theories to help write the reaction. For example, the base must react with water to produce hydroxide ions according to the Arrhenius theory. The base accepts an $H^+$ ion according to the Brønsted-Lowry theory. Remember to check the charges of the reactants and products and be sure that the sums of the charges on both sides of the reaction are equal.

---

## THE EFFECT OF STRUCTURE ON ACID-BASE PROPERTIES

(*Chemistry* 6th ed. pages 693–695 / 7th ed. pages 661–662)

For binary acids, HX, the strength of the H–X bond and the polarity of the bond will determine the behavior of the acid. The polarity of the bonds in hydrogen halides become less polar going down a group. The strength of the H-F bond is what makes it a weak acid, whereas the rest of the hydrogen halides are strong acids: HI>HBr>HCl.

For a given series of oxoacids such as $HClO_4$, $HClO_3$, $HClO_2$, and $HClO$, the acid strength increases with increasing number of oxygens attached to the central atom. $HClO_4$ is a strong acid. The remainder of the oxoacids are listed in order of decreasing strength (decreasing number of oxygen atoms). The O–H bond becomes more polarized and weakened due to the electron density drawn toward the highly electronegative oxygen atoms.

## CALCULATING THE pH OF STRONG ACIDS AND BASES

(*Chemistry* 6th ed. pages 665–666, 677 / 7th ed. pages 634–635, 645–646)

The pH of a strong acid can be calculated directly from the hydrogen ion concentration,

$$pH = -\log [H^+]$$

$[H^+]$, the molar concentration of the hydrogen ion, is obtained from the molarity of the acid.

EXAMPLE: Calculate the pH of 0.010 M HCl.

ANSWER: The pH equals 2.00; $-\log (0.010) = 2.00$.

The pH of a strong base can be calculated from its hydroxide ion concentration.

$pOH = -\log [OH^-]$

$pH + pOH = 14.00$

The concentration of a strong acid or strong base can be determined from the solution's pH.

EXAMPLE: The pH of a $Sr(OH)_2$ solution is 13.50. Calculate the concentration of $Sr(OH)_2$.

ANSWER: $pOH = 14 - pH = 14 - 13.50 = 0.50$

$[OH^-] = $ inv log $(-pOH)$; inv log $(-0.50) = 0.32$ M $OH^-$

$$\frac{0.32 \text{ mol OH}}{1 \text{ L}} \times \frac{1 \text{ mol Sr(OH)}_2}{2 \text{ mol OH}^-} = 0.16 \text{ M Sr(OH)}_2$$

## AP tip

The number of significant figures in a pH measurement is equal to the number of decimal places in the pH. For example, 1.70 has 2 significant figures.

## CALCULATING THE pH OF WEAK ACIDS

(*Chemistry* 6th ed. pages 666–671 / 7th ed. pages 635–640)

The pH of a weak acid cannot be calculated directly from the concentration of the acid since all of the acid does not dissociate to form $H^+$. The equilibrium reaction of the acid must be considered.

EXAMPLE: Calculate the pH of 0.25 M HCN.

ANSWER: First, write the reaction of the acid with water. Use the Arrhenius and Brønsted-Lowry theories to help you write the products. Check that you have the correct charges on the products.

$HCN + H_2O \rightleftharpoons H_3O^+ + CN^-$

Second, set up an "ICE" chart as you did in the previous chapter for equilibrium problems.

| | $HCN$ | $+ H_2O \rightleftharpoons$ | $H_3O^+$ | $+ CN^-$ |
|---|---|---|---|---|
| I | 0.25 | | 0 | 0 |
| C | $-x$ | | $+x$ | $+x$ |
| E | $0.25 - x$ | | $x$ | $x$ |

Third, write the equilibrium expression for $K_a$ in the same manner as you did in the last chapter. Plug in the values from the equilibrium line of the ICE chart.

If $x$ is very small compared to the concentration of the acid you are subtracting it from, then you can assume $[HA - x]$ is approximately equal to [HA]. One way to determine if this approximation is valid is to compare the magnitude to $K$ to [HA]; if [HA] is greater than $K$ by a factor of $10^3$ or more then $x$ can be safely ignored. You can always assume that $x$ is small and then check the value of $x$ you calculate to see if [HA – $x$] is within 5% of [HA].

$$K_a = 6.2 \times 10^{-10} = \frac{[H_3O^+][CN^-]}{[HCN]}; \frac{x^2}{(0.25-x)} \cong \frac{x^2}{0.25}.$$

Fourth, solve for x which equals H₃O⁺.

$(6.2 \times 10^{-10})(0.25) = x^2 ; x = 1.2 \times 10^{-5}$ M.

Finally calculate the pH from the value of $x$, the H₃O⁺ concentration.

pH = –log $(1.2 \times 10^{-5})$ = 4.90

---

## AP tips

Always follow these steps when performing calculations involving the disassociation of weak acids or weak bases.
1. Write the reaction of the acid or base with water. (Use the acid-base theories and check charges.)
2. Set up ICE chart.
3. Write the equilibrium expression in terms of reactants and products (without numbers).
4. Solve for $x$, using the method of approximation. Test approximation.
5. Solve for pH. (Be careful for base equilibria, $x$ = OH⁻. You need to find pOH and then pH.)
6. Always remember that for weak acids and bases, at equilibrium, pH = –log [H₃O⁺].

---

## CALCULATING THE PERCENT DISSOCIATION

(*Chemistry* 6th ed. pages 672–676 / 7th ed. pages 641–644)

The percent dissociation of an acid (or a base) is the amount of the acid, HA, which has dissociated, $x$, divided by the acid's initial concentration, $HA_o$, multiplied by 100.

$$\frac{x}{[HA_o]} \times 100\%$$

When making assumptions in an equilibrium calculation, it is best to test the assumption by making sure that the percent dissociation is less than or equal to 5%. The test for the assumption is the same as the calculation for the percent dissociation.

EXAMPLE: The percent dissociation of an acid, HA, which is 0.100 M is 2.5 %. Calculate the $K_a$ of the acid.

ANSWER: x/0.100 M × 100% = 2.5% ; x = 2.5 × $10^{-3}$ M.

$$K_a = \frac{[H_3O^+][A^-]}{[HA]} \; ; K_a = \frac{(2.5 \times 10^{-3})^2}{(0.100 - 2.5 \times 10^{-3})} = 6.4 \times 10^{-5}.$$

## CALCULATING THE pH OF WEAK BASES

(*Chemistry* 6th ed. pages 678–682 / 7th ed. pages 646–650)

The calculations involving weak base equilibria are similar to the weak acid equilibria problems except that the equation is written for a base reacting with water and the calculation initially involves finding [OH⁻]. You will need to find the pOH and then the pH.

EXAMPLE: The pH of a 0.20 M solution of $H_2NNH_2$ is 11.38. Calculate $K_b$ for $H_2NNH_2$.

Write the reaction with water. Bases accept H⁺. Watch charges! One hint to help in writing the reaction is that the pH is 11.38. The basic pH indicates that OH⁻ must be one of the products.

Fill out the ICE Chart under the reaction.

$$H_2NNH_2 + HOH \rightleftharpoons OH^- + H_2NNH_3^+$$

|   | $H_2NNH_2$ | $OH^-$ | $H_2NNH_3^+$ |
|---|---|---|---|
| I | 0.20 | 0 | 0 |
| C | −x | +x | +x |
| E | 0.20 − x | x | x |

You are given the pH, but x equals [OH⁻].

Find pOH ; pH + pOH = 14.00 ; pOH = 14.00 − 11.38 = 2.62

pOH = −log [OH⁻] ; Find [OH⁻] = inv log (−pOH) ;

inv log (−2.62) = 2.4 × $10^{-3}$M. (You can also type this into your calculator as $10^{-pOH}$.)

Plug this value of x into the $K_b$ expression:

$K_b$ = [OH⁻] [$H_2NNH_3^+$] / [$H_2NNH_2$] = $x^2$ / (0.20 − x).

$K_b$ = ( 2.4 × $10^{-3}$)²/ 0.20 = 2.9 × $10^{-5}$

## POLYPROTIC ACIDS

(*Chemistry* 6th ed. pages 682–687 / 7th ed. pages 650–655)

Polyprotic acids can donate more than one proton, H⁺, and dissociate by losing 1 H⁺ at a time.

EXAMPLE: Calculate the [H⁺] of a 0.20 M solution.

Also determine the concentrations of $H_3AsO_4$ , $H_2AsO_4^-$, $HAsO_4^{2-}$, and $AsO_4^{3-}$.

For $H_3AsO_4$ $K_{a_1} = 5 \times 10^{-3}$, $K_{a_2} = 8 \times 10^{-8}$, $K_{a_3} = 6 \times 10^{-10}$.

|   | $H_3AsO_4$ + $H_2O$ | $\rightleftharpoons$ | $H_3O^+$ | $H_2AsO_4^-$ |
|---|---|---|---|---|
| I | 0.20 | | 0 | 0 |
| C | −x | | +x | +x |
| E | 0.20 − x | | x | x |

$H_3AsO_4 + H_3O^+ \rightleftharpoons H_3O^+ \quad H_2AsO_4^-$

$$K_{a_1} = \frac{[H_3O^+][H_2AsO_4^-]}{[H_3AsO_4]}$$

$$5 \times 10^{-3} = \frac{x^2}{0.20 - x}$$

Note: You cannot assume $x$ is small since the $K$ and 0.20 only differ by a factor of $10^2$.

$x = 3 \times 10^{-2}$ M (as a result of solving quadratics)

$[H_3O^+] = [H_2AsO_4^-] = 3 \times 10^{-2}$ M

$[H_3AsO_4] = 0.20 - 0.03 = 0.17$ M

Since $K_{a_3} \lll K_{a_2} \lll K_{a_1}$, very little of $H_2AsO_4^-$ and $HAsO_4^{2-}$ dissociates compared to $H_3AsO_4$, so $[H_3O^+]$ and $[H_2AsO_4^-]$ will not change very much by the $K_{a_2}$ dissociation, and we can use their concentrations to find the concentration of $HAsO_4^{2-}$.

|   | $H_2AsO_4^-$ + $H_2O$ | $\rightleftharpoons$ | $H_3O^+$ + | $HAsO_4^{2-}$ |
|---|---|---|---|---|
| I | $3 \times 10^{-2}$ | | $3 \times 10^{-2}$ | 0 |
| C | −x | | +x | +x |
| E | $3 \times 10^{-2}$ − x | | $3 \times 10^{-2}$ + x | x |

$$K_{a_2} = 8 \times 10^{-8} = \frac{(3 \times 10^{-2})[HAsO_4^{2-}]}{(3 \times 10^{-2})}$$

$[HAsO_4^{2-}] = 8 \times 10^{-8}$ M; the assumption that $K_{a_2}$ does not contribute significantly to $[H_3O^+]$ and $[H_2AsO_4^-]$ is good.

Repeat the process to find $[AsO_4^{3-}]$.

|   | $HAsO_4^{2-}$ + | $H_2O$ | $\rightleftharpoons$ | $H_3O^+$ + | $AsO_4^{3-}$ |
|---|---|---|---|---|---|
| I | $8 \times 10^{-8}$ | | | $3 \times 10^{-2}$ | 0 |
| C | −x | | | +x | +x |
| E | $8 \times 10^{-8}$ − x | | | $3 \times 10^{-2}$ + x | x |

$$K_{a_3} = 6 \times 10^{-10} = \frac{(3 \times 10^{-2})\,[AsO_4^{3-}]}{(8 \times 10^{-8})}$$

$[AsO_4^{3-}] = 2 \times 10^{-15}\,M$. Assumption that $x$ is small is good.

# ACID-BASE PROPERTIES OF SALTS

(*Chemistry* 6th ed. pages 687–693 / 7th ed. pages 655–660)

## PREDICTING THE pH OF SALTS

You are to determine if a salt is acidic, basic, or neutral by looking at its chemical formula. This process involves two steps as outlined in the flowcharts which follow. In the first flowchart, you will determine the acidity or basicity of the individual ions. Then, using the second chart and the results from the first flowchart, you can determine if the salt is acidic, basic, or neutral.

Determining the Approximate pH of Ions in a Salt

Determining the Approximate pH of a Salt

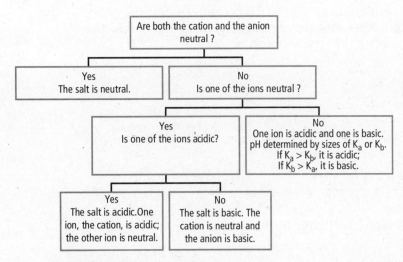

EXAMPLE: Determine whether an aqueous solution of $KC_2H_3O_2$ is acidic, basic, or neutral.

ANSWER: $KCH_3O_2$ is basic. Use the method of asking questions outlined above.

Split the salt into its cation, $K^+$, and its anion, $C_2H_3O_2^-$.

# CHAPTER 13

Is the cation a Group I metal? Yes; the cation does not affect the pH.

Is the anion a conjugate base of a strong acid? No; $C_2H_3O_2^-$ is the conjugate base of a weak acid, $HC_2H_3O_2$, which makes the solution basic.

Because we have a salt with a cation which doesn't affect the pH and a basic anion, an aqueous solution of the salt is basic.

## CALCULATING THE pH OF SALTS

To calculate the pH of a salt, first you must decide whether the salt is acidic, basic, or neutral. If the salt is basic, then its anion is the conjugate base of a weak acid. The anion will undergo hydrolysis. We will need to write an equation for the reaction of that ion with water to form the acid and $OH^-$ ions: $A^- + H_2O = HA + OH^-$. Then we will write the equilibrium expression

$$K_b = \frac{K_w}{K_a} = \frac{[HA][OH^-]}{[A^-]}.$$

If the salt is basic, then the hydrolysis reaction will produce an acid.

EXAMPLE: Determine the pH of a 0.100 M aqueous solution of NaCN. The $K_a$ for HCN is $5.8 \times 10^{-10}$.

ANSWER: In aqueous solutions, NaCN ionizes completely into $Na^+$ and $CN^-$. However, the cyanide ions ionize in water according to the following equation:

$$CN^- (aq) + H_2O(l) \rightleftharpoons HCN (aq) + OH^- (aq).$$

$Na^+$ ions do not affect the pH. Ions from Group IA and IIA never undergo hydrolysis because they are cations of strong bases. The equilibrium expression for the solution is

$$K_b = \frac{[OH^-][HCN]}{[CN^-]}$$

We need a value for $K_b$. Since we have $K_a$ for HCN, we can calculate the value of $K_b$ for $CN^-$.

$$K_b = \frac{1.0 \times 10^{-14}}{K_a} = \frac{1.0 \times 10^{-14}}{5.8 \times 10^{-10}}$$

Constructing a table for filling in the available information, we get

|   | $CN^-$ | $+ H_2O$ | $\rightleftharpoons$ | HCN $+$ | $OH^-$ |
|---|---|---|---|---|---|
| I | 0.100 | | | 0 | 0 |
| C | $-x$ | | | $+x$ | $+x$ |
| E | $0.100 - x$ | | | $x$ | $x$ |

Since $K$ is much smaller than 0.100, assume that $0.100 - x \approx 0.100$.

$$K_b = \frac{[OH^-][HCN]}{[CN^-]}; \quad 1.7 \times 10^{-5} = \frac{x^2}{0.100}$$

Solving for $x$, we have

$x = [OH^-] = 1.3 \times 10^{-3}$

$pOH = 2.89; \quad pH = 14.00 - pOH$

$pH = 11.11.$

# BUFFERS

(*Chemistry* 6th ed. pages 716–729 / 7th ed. pages 684–696)

Buffers resist changes in pH when acids or bases are added. The components of a buffer are summarized in the next table.

| Buffer | |
|---|---|
| **Components** | **Examples** |
| Weak acid + salt containing the conjugate base | HCN and NaCN |
| Weak base + salt containing the conjugate acid | $CH_3NH_2$ and $CH_3NH_3Cl$ |
| Weak acid + excess strong base<br>or<br>Weak base + excess strong acid | 2 mol of HCN + 1 mol NaOH react to yield<br>1 mol HCN and 1 mol NaCN<br>2 mol $NH_3$ + 1 mol HCl react to yield<br>1 mol $NH_3$ and 1 mol $NH_4Cl$ |

## CALCULATING THE pH OF A BUFFER

The calculation to find the pH of a buffer is similar to all equilibrium calculations EXCEPT there are now two initial concentrations, one for each part of the buffer pair.

EXAMPLE: Calculate the pH of a solution that is 0.60 M HF and 1.00 M KF. $K_a$ for HF is $7.2 \times 10^{-4}$.

ANSWER: First, write the reaction of the acid with water. (You are given $K_a$, and the buffer is a made of a weak acid and its conjugate base.)

$$HF + H_2O \rightleftharpoons H_3O^+ + F^-$$

Second, set up an ICE chart.

|  | HF | + H₂O ⇌ | H₃O⁺ | + F⁻ |
|---|---|---|---|---|
| I | 0.60 |  | 0 | 1.00 |
| C | −x |  | +x | +x |
| E | 0.60 − x |  | x | 1.00 + x |

Third, write the equilibrium expression for $K_a$.

Plug in the values from the equilibrium line of the ICE chart. Check to see if $x$ is small.

$$K_a = 7.2 \times 10^{-4} = \frac{[H_3O^+][F^-]}{[HF]} = \frac{x(1.00-x)}{(0.60-x)} = \frac{x(1.00)}{0.60}$$

Fourth, solve for $x$ which equals H₃O⁺.

$$K_a = 7.2 \times 10^{-4} = \frac{x(1.00)}{0.60}$$

$$x = [H_3O^+] = 4.3 \times 10^{-4} \text{ M}$$

$$pH = -\log [H_3O^+]$$

$$pH = -\log (4.3 \times 10^{-4}) = 3.37.$$

## PREPARATION OF A BUFFER

A buffer can be made from a weak acid and a salt containing its conjugate base or from a weak base and a salt containing its conjugate acid.

EXAMPLE: Calculate the mass of $NaC_2H_3O_2$ required to prepare a buffer of pH 4.55 when added to 0.500 L of 0.67 M acetic acid. (Assume no change in volume.) $K_a = 1.8 \times 10^{-5}$ for $HC_2H_3O_2$.

ANSWER: 1.8 g $NaC_2H_3O_2$

This problem can be solved using an ICE chart or by using the Henderson-Hasselbalch equation.

---
Henderson-Hasselbalch

pH = p$K_a$ + log ( [conjugate base] / [acid] )
---

$pK_a = -\log K_a$

Brackets [ ] represent the molar concentration of the substance.

$$pH = pKa + \log\left(\frac{[C_2H_3O_2^-]}{[HC_2H_3O_2]}\right)$$

$$4.55 = 4.74 + \log\left(\frac{[C_2H_3O_2^-]}{0.67 \text{ M}}\right)$$

$$-0.19 = \log\left(\frac{[C_2H_3O_2^-]}{0.67 \text{ M}}\right)$$

$$0.65 = \frac{[C_2H_3O_2^-]}{0.67 \text{ M}}$$

$$\text{mass NaC}_2\text{H}_3\text{O}_3 = \frac{0.44 \text{ mol } [C_2H_3O_2^-]}{L} \times \frac{1 \text{ mol NaC}_2\text{H}_3\text{O}_2}{1 \text{ mol } [C_2H_3O_2^-]} \times \frac{0.500 \text{ L} \times 82.0 \text{ g}}{1 \text{ mol NaC}_2\text{H}_3\text{O}_2}$$

## ADDITION OF STRONG ACID AND STRONG BASE TO A BUFFER

The next example summarizes the steps for this type of problem. The problem is solved in the same manner as the buffer problem on the previous page with HF and NaF, except for the first three steps which are new.

EXAMPLE: Calculate the pH when 100.0 mL of 0.50 M HCl are added to the buffer consisting of 20.0 g $HC_2H_3O_2$ and 18.0 g $NaC_2H_3O_2$ dissolved in 5.00 x $10^2$ mL of water.

ANSWER: Calculate the initial moles of weak acid and conjugate acid present in the buffer. Write these amounts under the reaction, above the ICE table.

|  | $HC_2H_3O_2$ + $H_2O$ ⇌ | $H_3O^+$ | $C_2H_3O_2^-$ |
|---|---|---|---|
| Initial moles | 0.33 | 0 | 0.21 |
| Add mol $H^+$ | +0.05 | 0 | –0.05 |
| New Initial Mol | 0.38 | 0 | 0.16 |
| I | 0.63 |  | 0.27 |
| C | –$x$ | +$x$ | +$x$ |
| E | 0.63 – $x$ | $x$ | 0.27 + $x$ |

Next, calculate the moles of strong acid, $H^+$, added. The added strong acid will react with the basic part of the buffer, $C_2H_3O_2^-$, to produce the other part of the buffer $HC_2H_3O_2$.

$$H^+ + C_2H_3O_2^- \longrightarrow HC_2H_3O_2$$

Make a new line under the initial moles called "Add mol $H^+$." Note: The minus sign shows the amount of conjugate base decreasing upon added $H^+$ and the plus sign shows the amount of acid increasing.

Calculate the new initial moles of weak acid and its conjugate base, after the conjugate base has reacted with the strong acid, by adding or subtracting the number of moles $H^+$ added.

Fill out the ICE chart, computing the initial concentration of weak acid and conjugate base by dividing the moles of weak acid and conjugate base by the total volume (volume of buffer + volume of added strong acid, in liters).

$HC_2H_3O_2$ = 0.38 mol / 0.600 L = 0.63 M

$C_2H_3O_2^-$ = 0.16 mol / 0.600 L = 0.27 M

Proceed to fill out the ICE chart and plug the values into the $K_a$ expression (or solve Henderson-Hasselbalch) as in a normal buffer problem.

$$K_a = \frac{[H^+][C_2H_3O_2]}{[HC_2H_3O_2]}$$

$$1.8 \times 10^{-5} = \frac{[H^+](0.27 + x)}{0.63 + x} \approx \frac{[H^+]0.27}{0.63}$$

$4.2 \times 10^{-5}$ M = $[H^+]$  pH = –log $[H^+]$ = 4.38

Alternatively, you can use the Henderson-Hasselbalch equation. Volume cancels.

pH = p$K_a$ + log [0.16 mol/0.38 mol] = 4.74 +(–0.38) = 4.36

## ADDITION OF STRONG BASE TO A BUFFER

The calculation for addition of a strong base to a buffer is similar to the previous problem.

EXAMPLE: Calculate the pH when 100.0 mL 0.50 M NaOH are added to the buffer consisting of 20.0 g of $HC_2H_3O_2$ and 18.0 g of $NaC_2H_3O_2$ dissolved in $5.00 \times 10^2$ mL of water.

ANSWER: Calculate the initial moles of weak acid and conjugate acid that are present in the buffer. Write these amounts under the reaction, above the ICE table.

|  | $HC_2H_3O_2$ + $H_2O$ | ⇌ | $H_3O^+$ | $C_2H_3O_2^-$ |
|---|---|---|---|---|
| Initial moles | 0.33 | | 0 | 0.21 |
| Add mol OH⁻ | –0.05 | | 0 | +0.05 |
| New initial mol | 0.28 | | 0 | 0.26 |
| I | 0.47 | | | 0.43 |
| C | –x | | +x | +x |
| E | 0.47 – x | | x | 0.43 + x |

Next, calculate the moles of strong base, OH⁻, added. The added strong base will react with the acidic part of the buffer, $HC_2H_3O_2$, to produce the other part of the buffer ($C_2H_3O_2^-$).

$$OH^- + HC_2H_3O_2 \longrightarrow HOH + C_2H_3O_2^-$$

Make a new line under the initial moles called "Add mol OH⁻."
Note, the minus sign shows the amount of acid decreasing upon
added OH⁻ and the plus sign shows the amount of conjugate
base increasing.

Calculate the new initial concentration of the acid and its
conjugate base and fill out the ICE chart as in the previous
problem.

The answer is:     $[H^+] = 2.0 \times 10^{-5} \, M$ ; $pH = 4.71$.

# TITRATION

(*Chemistry* 6th ed. pages 729–745 / 7th ed. pages 696–711)

A titration can be used to determine the concentration of an unknown
solution. The calculations involve stochiometric principles.

The table below summarizes the other types of problems involving
acid-base titrations. Examples of each type of titration problem follow
the table.

## Characteristics of Titrations

| Titrations | Before equivalence point | At equivalence point | After equivalence point |
|---|---|---|---|
| Strong acid titrated with a strong base | Use excess $[H^+]$ ion to calculate the pH  pH < 7 | pH = 7 | Use excess $[OH^-]$ ion to calculate the pH  pH > 7 |
| Species in solution which affect the pH (in addition to $H_2O$) | $H^+$ | $H_2O$ | $OH^-$ |
| Weak acid titrated with a strong base | Weak acid + conjugate base (a buffer problem)  pH < 7 | Salt hydrolysis problem  pH > 7 | Use excess $OH^-$ ion to calculate the pH  pH > 7 |
| Species in solution which affect the pH (in addition to $H_2O$) | HA, A⁻ | A⁻ | OH⁻ |
| Weak base titrated with a strong acid | Weak base + conjugate acid  (a buffer problem)  pH > 7 | Salt hydrolysis  pH < 7 | Excess $H^+$ ion  pH < 7 |
| Species in solution which affect the pH (in addition to $H_2O$) | B, HB⁺ | HB⁺ | H⁺ |

## TITRATION OF A STRONG ACID WITH A STRONG BASE

(*Chemistry* 6th ed. pages 730–733 / 7th ed. pages 696–699)

At any point during the titration of a strong acid with a strong base (or a strong base with a strong acid), the pH can be calculated from the molarity of the ion present in excess after the complete reaction.

At the equivalence point of any type of titration,

$$\text{moles } H^+ = \text{moles } OH^-.$$

For a strong acid-strong base titration, the pH equals 7.

EXAMPLE: Consider the titration of 40.0 mL of 0.200 M HBr by 0.100 M KOH. Calculate the pH of the resulting solution when the following volumes of KOH have been added.

a. 10.0 mL       b. 80.0 mL       c. 100.0 mL

ANSWER: First, using the volume and molarity of the HBr, calculate the moles of $H^+$ to be titrated. The resulting moles of $H^+$ appears in the first column in the following table.

Next, for the titration in question, calculate the moles of $OH^-$ added from the volume and concentration of KOH. The resulting moles of $OH^-$ used in each titration appear in the second column of the table.

Calculate the moles of ion in excess, $H^+$ or $OH^-$. The results for each titration appear in column four in the table below.

Using the total volume, in column five of the table, and the moles of ion in excess in column four, determine the molarity of the ion in excess. The molarity of the ion in excess for each of the three titrations is in column six.

It is the molarity of the ion in excess which determines the pH. In part b of the example, neither ion is in excess. The pH equals 7 because the titration is at the equivalence point.

Be careful, in part c of the example, the excess ion is the $OH^-$ ion. To calculate the pH, you must determine pOH and then the pH.

| Problem | Mol $H^+$ | Mol $OH^-$ | Mol Ion in Excess | Total Volume (When acid and base react) | Molarity Ion In Excess | pH |
|---|---|---|---|---|---|---|
| a | $8.00 \times 10^{-3}$ | $1.00 \times 10^{-3}$ | $7.00 \times 10^{-3}$ ($H^+$) | 0.0500 L | 0.140 M ($H^+$) | 0.854 |
| b | $8.00 \times 10^{-3}$ | $8.00 \times 10^{-3}$ | None Equivalence point | Not needed | None | 7.00 |
| c | $8.00 \times 10^{-3}$ | $10.0 \times 10^{-3}$ | $2.0 \times 10^{-3}$ ($OH^-$) | 0.140 L | 0.014 M ($OH^-$) | 12.15 |

## AP tip

When you are calculating the pH during a titration, be careful to take the negative log of the concentration of $H^+$, not just the moles of $H^+$ which is a common mistake. And if the excess ion or the value of $x$ is $OH^-$, be sure that you calculate the pOH and then the pH.

## TITRATION OF A WEAK ACID WITH A STRONG BASE

There are three characteristic types of calculations for the titration of a weak acid with a strong base: before the equivalence point, at the equivalence point, and after the equivalence point.

> EXAMPLE: A 25.0 mL sample of 0.100 M $HC_3H_5O_2$ is titrated with the 0.100 M NaOH. Calculate the pH when the following volumes of 0.100 M NaOH are added: $K_a = 1.3 \times 10^{-5}$ for $HC_3H_5O_2$.
>
> a) 8.0 mL        b) 12.5 mL      c) 25.0 mL              d) 30.0 mL

## CALCULATION OF THE pH BEFORE THE EQUIVALENCE POINT

Part a of the example above involves the titration before the equivalence point. Perform the stoichiometry and record the results in a table under the reaction. The first three lines in the table on the next page are explained below.

Calculate the initial moles of the weak acid present in the sample.

$0.0250 \text{ L} \times 0.100 \text{ mol /L} = 2.50 \times 10^{-3} \text{ mol } HC_3H_5O_2$

Calculate the moles of strong base, $OH^-$, added.

$0.0080 \text{ L} \times 0.100 \text{ mo/L} = 8.0 \times 10^{-4} \text{ mol}$

Calculate the new initial moles of weak acid and its conjugate base present after the $OH^-$ has reacted.
    This problem is now identical to a "buffer" problem.
    Make a new line under the initial moles called "mol $OH^-$ added." Note the minus sign to show the amount of acid decreasing upon added $OH^-$ and the plus sign to show the amount of conjugate base increasing.
    Calculate the new initial moles of weak acid and its conjugate base, after the weak acid has reacted with the strong base, by subtracting or adding the number of moles $OH^-$ added.

| | $HC_3H_5O_2$ + $H_2O$ | $\rightleftharpoons$ | $H_3O^+$ + | $C_3H_5O_2^-$ |
|---|---|---|---|---|
| Initial moles | $2.50 \times 10^{-3}$ | | 0 | 0 |
| Add mol OH⁻ | $-8.0 \times 10^{-4}$ | | 0 | $+8.0 \times 10^{-4}$ |
| New Initial Mol | $1.70 \times 10^{-3}$ | | 0 | $8.0 \times 10^{-4}$ mol |
| I | $1.70 \times 10^{-3}$ / 0.0330 L = 0.0515 M | | | $8.0 \times 10^{-4}$ / 0.0330 L = 0.0242 M |
| C | $-x$ | | $+x$ | $+x$ |
| E | $0.0515 - x$ | | $x$ | $0.0242 + x$ |

The stoichiometric calculations are described on the previous page. You can now perform the equilibrium calculation.

To begin filling out the ICE chart, compute the initial concentration of weak acid and conjugate base by dividing the moles of weak acid and conjugate base by the total volume (volume of acid titrated + volume of added strong base, in liters).

$$K_a = \frac{[H^+]\,[C_3H_5O_2^-]}{[HC_3H_5O_2]}$$

$$1.3 \times 10^{-5} = \frac{[H^+]\,(0.0242 + x)}{0.0515 + x} \approx \frac{[H^+]0.0242}{0.0515}$$

$2.77 \times 10^{-5}$ M = $[H^+]$;  pH = $-\log [H^+]$ = 4.56

Alternatively, you can use the Henderson-Hasselbalch equation.

## HALFWAY TO THE EQUIVALENCE POINT

In part b of the example on the previous page, a 25.00 mL sample of 0.100 M $HC_3H_5O_2$ is titrated with 12.5 of 0.100 M NaOH. $K_a$ is $1.3 \times 10^{-5}$ for $HC_3H_5O_2$.

Half of the acid being titrated is neutralized. This point is halfway to the equivalence point.

Since half the acid is neutralized and half remains, [HA] = [A⁻].

$$K_a = \frac{[H^+]\,\cancel{[A^-]}}{\cancel{[HA]}}$$

$$pK_a = pH$$

$$pK_a = -\log K_a = -\log (1.3 \times 10^{-5}) = 4.89$$

## AT THE EQUIVALENCE POINT

In part c of the example on the previous page, a 25.00 mL sample of 0.100 M $HC_3H_5O_2$ is titrated with 25.0 mL of 0.100 M NaOH.

$K_a$ is $1.3 \times 10^{-5}$ for $HC_3H_5O_2$.

Calculate the initial moles of the weak acid in the sample.

Calculate the moles of strong base, OH⁻, added.

$0.0250 \text{ L} \times 0.100 \text{ mol/L} = 2.50 \times 10^{-3} \text{ mol OH}^- = \text{mol C}_3\text{H}_5\text{O}_2^-$

There is no weak acid present once the strong base added completely reacts. All that is present is $A^-$ ($C_3H_5O_2^-$), the conjugate base of weak acid, HA ($HC_3H_5O_2$).

$$\text{HA} + \text{NaOH} \longrightarrow \text{NaA} + \text{HOH}$$

The problem from this point on is a salt hydrolysis problem.

$$C_3H_5O_2^- + H_2O \rightleftharpoons OH^- + HC_3H_5O_2$$

I    $2.50 \times 10^{-3}/0.0500 \text{ L}$

     0.0500 M          0     0

C    $-x$                $+x$    $+x$

E    $0.0500 - x$         $x$     $x$

$$K_b = \frac{K_w}{K_a} = \frac{1.0 \times 10^{-14}}{1.3 \times 10^{-5}} = 7.7 \times 10^{-10}$$

$$K_b = \frac{[OH^-][HC_3H_5O_2]}{[C_3H_5O_2^-]} = \frac{x^2}{0.0500 - x} \approx \frac{x^2}{0.0500}$$

$$x = [(7.7 \times 10^{-10})(0.0500)]^{1/2} = 6.2 \times 10^{-6} \text{ M OH}^-$$

$$\text{pOH} = 5.21 \text{ ; pH} = 14.00 - 5.21 = 8.79$$

## AFTER THE EQUIVALENCE POINT

The calculations involved here are identical to those in a strong acid-strong base titration.

In part d of the example on the previous page, a 25.00 mL sample of 0.100 M $HC_3H_5O_2$ is titrated with 30.0 mL of 0.100 M NaOH. $K_a$ is $1.3 \times 10^{-5}$ for $HC_3H_5O_2$.

$0.0300 \text{ L} \times 0.100 \text{ mol/L} = 0.00300 \text{ mol NaOH}$

$0.02500 \text{ L} \times 0.100 \text{ mol/L} = 0.002500 \text{ mol HC}_3\text{H}_5\text{O}_2$

Excess $OH^- = 0.00300 \text{ mol} - 0.002500 \text{ mol}/0.0500 \text{ L} = 0.010 \text{ M OH}^-$

$\text{pOH} = 2.00 \text{ ; pH} = 14 - \text{pOH} = 12.00$

## TITRATION OF A WEAK BASE WITH A STRONG ACID

The calculations for a weak base-strong acid titration are very similar to the weak acid-strong base titration.

The differences are these:
- The pH before the equivalence point is greater than 7.
- At the equivalence point, the pH is less than 7 since the salt formed is acidic.
- After the equivalence point, the pH is determined by the molarity of the excess $H^+$.

## CHARACTERISTICS OF TITRATION CURVES

It is important for you to be able to sketch and identify the key points on titration curves for the AP Exam. The next example will help you to review titration curves for the different types of titrations discussed in this section.

EXAMPLE: The questions that follow refer to the following three titration curves. All solutions are equimolar.

Figure 1:

Figure 2:

Figure 3:

(a) Which titration curve represents a strong acid titrated by a strong base? Explain your answer.
(b) Which titration curve represents a weak acid titrated by a strong base? Explain your answer.
(c) Which titration curve represents a weak base titrated by a strong acid? Explain your answer.
(d) For each lettered part of each curve, A through L, identify the major species which affects the pH.

ANSWER: (a) Figure 1 represents a strong acid titrated by a strong base. The pH is very low when no base has been added. The pH at the equivalence point is 7.00.

ANSWER: (b) Figure 2 represents a weak acid titrated by a strong base. The pH is acidic, but slightly higher than the titration curve in figure 1. The equivalence point occurs at pH 9, signifying the presence of the conjugate base of the weak acid being titrated.

ANSWER: (c) Figure 3 represents a weak base titrated by a strong acid. When no acid has been added, the pH is basic.

ANSWER: (d) The points are labeled as follows:

Figure 1:

A  strong acid, $H^+$

B strong acid, $H^+$

C water

D strong base, $OH^-$

Figure 2:

E  weak acid, HA

F weak acid, HA and its conjugate base, A$^-$

G basic ion, A$^-$

H strong base, OH$^-$

Figure 3:

I weak base, B

J weak base, B , and its conjugate acid, HB$^+$

K acidic ion HB$^+$

L strong acid H$^+$

## Multiple-Choice Questions

Calculators may not be used on this part of the exam.

1.  The calculation of concentration and pH for weak acids is more complex than for strong acids due to
    (A) the incomplete ionization of weak acids.
    (B) the low $K_a$ value for strong acids.
    (C) the more complex atomic structures of strong acids.
    (D) the low percent ionization of strong acids.
    (E) the inconsistent $K_b$ value for strong acids.

    **Answer:  A**  Strong acids ionize 100%, therefore the [H$^+$] is equal to the initial concentration of the strong acid. This is not the case for weak acids. Since not all of the acid ionizes, it is necessary to determine both how much of the acid forms H$^+$ and how much is left in molecular form (*Chemistry* 6th ed. pages 665–670 / 7th ed. pages 634–639).

2.  The general reaction of an acid dissolving in water may be shown as

    HA(*aq*) + HOH(*l*) ⇌ H$_3$O$^+$(*aq*) + A$^-$(*aq*).

    A conjugate acid base pair for this reaction is
    (A) HA and HOH.
    (B) HA and A$^-$.
    (C) HOH and A$^-$.
    (D) H$_3$O$^+$ and A$^-$.
    (E) HA and H$_3$O$^+$.

    **Answer:  B**  The conjugate acid differs from its conjugate base by a proton, (H$^+$).  The acid form has the proton (HA in this example) and the base form has lost the proton (A$^-$ in this case) (*Chemistry* 6th ed. page 655 / 7th ed. page 624).

3.  Strong acids are those which
    (A) have an equilibrium lying far to the left.
    (B) yield a weak conjugate base when reacting with water.
    (C) have a conjugate base which is a stronger base than water.
    (D) readily remove the H⁺ ions from water.
    (E) are only slightly dissociated (ionized) at equilibrium.

ANSWER: **B** Strong acids give up their protons (H⁺ ions) easily. If their conjugate base form were strong, then the acid formed would hold strongly to the H⁺, which would not be a characteristic of a strong acid (*Chemistry* 6th ed. pages 657–659 / 7th ed. pages 626–628).

4.  When calculating the pOH of a hydrofluoric acid solution ($K_a = 7.2 \times 10^{-4}$) from its concentration, the contribution of water ionizing ($K_w = 1.0 \times 10^{-14}$) is usually ignored because
    (A) hydrofluoric acid is such a weak acid.
    (B) hydrofluoric acid can dissolve glass.
    (C) the ionization of water provides relatively few H⁺ ions.
    (D) the [OH⁻] for pure water is unknown.
    (E) the conjugate base of HF is such a strong base.

ANSWER: **C** By comparing the $K_a$ for HF with $K_w$ (for water), you can see that even though hydrofluoric acid is a very weak acid it is a much stronger acid than is water, so much so that the H⁺ ion contribution of water may be ignored (*Chemistry* 6th ed. pages 666–669 / 7th ed. pages 635–638).

5.  The percent dissociation (percent ionization) for weak acids
    (A) is always the same for a given acid, no matter what the concentration.
    (B) usually increases as the acid becomes more concentrated.
    (C) compares the amount of acid that has dissociated at equilibrium with the initial concentration of the acid.
    (D) may only be used to express the dissociation of weak acids.
    (E) has no meaning for polyprotic acids.

ANSWER: **C** The percent dissociation does change with the concentration of the acid (see sample exercise 14.10, page 672–673, 6th edition / page 641–642, 7th edition). For example, 1.0 M acetic acid does not ionize nearly as much as 0.10 M acetic acid. The term applies to the amount of acid dissociated/the initial concentration, expressed as a percent (*Chemistry* 6th ed. pages 673–674 / 7th ed. pages 641–642).

6.  The [OH⁻] of a certain aqueous solution is $1.0 \times 10^{-5}$M. The pH of this same solution must be
    (A) $1.0 \times 10^{-14}$.
    (B) 5.00.
    (C) 7.00.
    (D) 9.00.
    (E) 12.00.

ANSWER: **D** Referring to pH + pOH = 14.00 at 25°C, if [OH⁻] = $1 \times 10^{-5}$M, pOH = 5.0, pH = 14.00 – 5.0 = 9.0 (*Chemistry* 6th ed. pages 661–664 / 7th ed. pages 630–633).

7. In many calculations for the pH of a weak acid from the concentration of the acid, an assumption is made that often takes the form $[HA]_o - x = [HA]_o$.
   This
   (A) is valid because $x$ is very small compared to the initial concentration of the weak acid.
   (B) is valid because the concentration of the acid changes by such large amounts.
   (C) is valid because actual value of $x$ cannot be known.
   (D) is valid because pH is not dependent upon the concentration of the weak acid.
   (E) approximation is always shown to be valid and so need not be checked.

ANSWER: A We can expect $x$ to be very small (as indicated by the low value for $K_a$ for weak acids), so that the concentration of the weak acid does not significantly change from its initial value. As the final step in such problems, you must determine that the change in the initial [HA] is less than 5% for the approximation to be considered valid (*Chemistry* 6th ed. pages 669–672 / 7th ed. pages 638–641).

8. HA is a weak acid which is 4.0% dissociated at 0.100M. Determine the $K_a$ for this acid.
   (A) 0.0040
   (B) 0.00016
   (C) 0.040
   (D) 1.6
   (E) 16 5

ANSWER: B

$[HA]_o = 0.100M$

$[H^+] = [A^-] = 0.004$ M

$K_a = [H^+][A^-] / [HA] = (0.004 \times 0.004) / 0.100 =$

(seen more simply for our purposes here)

$$(4 \times 10^{-3})(4 \times 10^{-3}) / (10^{-1}) = 1.6 \times 10^{-4}$$

Note that although the actual value for [HA] is 0.100 – 0.004 = 0.096M, this approximation is within the 5% rule, and students do not have calculators for this part of the test (*Chemistry* 6th ed. pages 672–675 / 7th ed. pages 641–643).

9. Sulfur trioxide is an acidic oxide due to the
   (A) high electronegativity of sulfur in the O–S bond, forming strong covalent bonds.
   (B) low electronegativity of sulfur in the O–S bond, forming strong covalent bonds.
   (C) high electronegativity of sulfur in the O–S bond, forming strong ionic bonds.
   (D) low electronegativity of sulfur in the O–S bond, forming strong ionic bonds.
   (E) attraction for the H–O bonds by the $H^+$ ion in water.

ANSWER: **A** As oxides dissolve and form hydrates, one must compare the electronegativity difference of the X–O bond to the H–O bond. If the electronegativity difference between atoms X and O is small, this bond is primarily covalent, which means that the O–H bond is easily broken to form acidic solutions (*Chemistry* 6th ed. pages 693–696 / 7th ed. pages 661–663).

10. Ionic substances known as salts can form acidic, basic, and neutral solutions when dissolved in water. When dissolved in water
    (A) $KNO_3$ forms a basic solution.
    (B) NaCl forms an acidic solution.
    (C) $NaNO_3$ forms an acidic solution.
    (D) NaF forms a basic solution.
    (E) $KClO_4$ forms an acidic solution.

ANSWER: **D** When NaF is dissolved in water the $Na^+$ has little affinity for $H^+$ but the $F^-$, the anion of a weak acid, does have an attraction for the $H^+$ in water, leaving an excess of $OH^-$ in solution. This kind of reaction with water is known as a hydrolysis reaction and, in this case, forms a basic solution (*Chemistry* 6th ed. pages 685–689 / 7th ed. pages 654–657).

## FREE-RESPONSE QUESTIONS

Calculators may be used for this part of the exam.

1. The $K_w$ for water at 25°C is $1.0 \times 10^{-14}$, but is $1.0 \times 10^{-13}$ at 60°C.
   (a) Give the chemical equation for the autoionization of water.
   (b) Determine the $[OH^-]$ for water at 60°C.
   (c) Determine the pH of water at 60°C.
   (d) Is the reaction of the autoionization of water endothermic or exothermic? Support your answer with data and explanation.

ANSWER 1(a): $H_2O(l) + H_2O(l) \rightleftharpoons H_3O^+(aq) + OH^-(aq)$
(*Chemistry* 6th ed. pages 660–661 / 7th ed. pages 629–630)

ANSWER 1(b): Since $[H_3O^+] = [OH^-]$

and $K_w = [H_3O^+][OH^-] = 1.0 \times 10^{-13}$

$[OH^-] = \sqrt{1.0 \times 10^{-13}}\ 1.0 \times 10^{-13} = 3.2 \times 10^{-7}\,M$

(*Chemistry* 6th ed. pages 660–662 / 7th ed. pages 629–631)

ANSWER 1(c): pH = –log [$H_3O^+$] = –log 3.2 × $10^{-7}$ = <u>6.49</u>. (*Chemistry* 6th ed. pages 661–664 / 7th ed. pages 630–633)

ANSWER 1(d): Since raising the temperature from 25°C to 60°C causes the $K_w$ to increase, more product must be forming. If raising the temperature causes the reaction to shift to the right (toward products), then the reaction must be endothermic. (You could also approach this as a bond breaking reaction; bond breaking is always endothermic) (*Chemistry* 6th ed. pages 662–664 / 7th ed. pages 631–633).

2.  Phosphoric acid, $H_3PO_4$, is a triprotic acid.
    (a) Show the three equations involved in the dissociation of this substance.
    (b) Illustrate how these three equations might be combined to show the complete dissociation of phosphoric acid.
    (c) If a 7.0 M $H_3PO_4$ solution dissociated, calculate the pH of the solution.
    $K_{a_1}$ = 7.5 × $10^{-3}$  $K_{a_2}$ = 6.2 × $10^{-8}$  $K_{a_3}$ = 4.8 × $10^{-13}$.
    (d) Determine the concentration for the ions $H_2PO_4^-$, $HPO_4^{2-}$, and $PO_4^{3-}(aq)$ in the dissociated 7.0 M $H_3PO_4$ solution.
    (e) Determine the pOH for this same 7.0 M $H_3PO_4$ solution.

ANSWER 2(a):  $H_3PO_4(aq) \rightleftharpoons H^+(aq) + H_2PO_4^-(aq)$

$H_2PO_4^-(aq) \rightleftharpoons H^+(aq) + HPO_4^{2-}(aq)$

$HPO_4^{2-}(aq) \rightleftharpoons H^+(aq) + PO_4^{3-}(aq)$

ANSWER 2(b):  $H_3PO_4(aq) \rightleftharpoons H^+(aq) + \cancel{H_2PO_4^-}(aq)$

$\cancel{H_2PO_4^-}(aq) \rightleftharpoons H^+(aq) + \cancel{HPO_4^{2-}}(aq)$

$\cancel{HPO_4^{2-}}(aq) \rightleftharpoons H^+(aq) + PO_4^{3-}(aq)$

---

$H_3PO_4(aq) \rightleftharpoons 3H^+(aq) + PO_4^{3-}(aq)$

ANSWER 2(c): Note that only the first dissociation forms significant amounts of $H^+$, so use:

$K_{a_1}$ = [$H^+$] [$H_2PO_4^-$] / [$H_3PO_4$] = 7.5 × $10^{-3}$

If $x$ = [$H^+$] = [$H_2PO_4^-$]

Assume [$H_3PO_4$]$_0$ = 7.0 – $x$ ~ 7.0 M.

($x$) ($x$) / 7.0 = 7.35 × $10^{-3}$

$x$ = $\sqrt{(7.35 \times 10^{-3} \times 7.0)}$ = 0.23 M = [$H^+$]

pH = –log [$H^+$] = –log 0.23 = <u>0.64</u>

Note: Check to see that $(7.0 - x) = (7.0 - 0.23)$ is about equal to 7.0, (this is within the allowable 5%, as the rule suggests, since 5% of 7.0 is 0.35), therefore the assumption is acceptable.

**ANSWER (2d):** From part (c), $[H^+] = \underline{0.23M} = [H_2PO_4^-]$.

In $H_2PO_4^-(aq) \rightleftharpoons H^+(aq) + HPO_4^{2-}(aq)$   $K_a = 6.2 \times 10^{-8}$

$[H^+][HPO_4^{2-}] / [H_2PO_4] = (0.23)[HPO_4^{2-}] / 0.23 = 6.2 \times 10^{-8}$

$[HPO_4^{2-}] = \underline{6.2 \times 10^{-8} M}$

For $[PO_4^{3-}]$:

$HPO_4^{2-}(aq) \rightleftharpoons H^+(aq) + PO_4^{3-}(aq)$  $K_a = 4.8 \times 10^{-13}$

$[H^+] = 0.23 M$  (from above)

$[HPO_4^{2-}] = 6.2 \times 10^{-8}M$  (also, from above)

$$\frac{[H^+][PO_4^{3-}]}{[HPO_4^{2-}]} = \frac{(0.23)[PO_4^{3-}]}{6.2 \times 10^{-8}} = 4.8 \times 10^{-13}$$

$$\left[PO_4^{3-}\right] = \frac{(4.8 \times 10^{-13})(6.2 \times 10^{-8})}{(0.23)} = \underline{1.3 \times 10^{-19} M}$$

**ANSWER:** From part c, pH = 0.64.

pH + pOH = 14.00

pH = 14.00 − 0.64 = $\underline{13.36}$

(*Chemistry* 6th ed. pages 682–687, especially sample exercises 14.15, page 683 and 14.16, page 685 / 7th ed. pages 650–655, especially sample exercises 14.15, pages 651–652 and 14.16, page 653)

# 14

# SOLUBILITY

In this chapter, you will apply basic equilibrium concepts involving solids dissolving to form aqueous solutions.

You should be able to
- Write balanced equations for the dissolution of a salt and its corresponding solubility product expression.
- Predict the relative solubilities of salts which dissolve to give the same number of ions from their $K_{sp}$ values.
- Calculate the $K_{sp}$ value from the solubility of a salt and also calculate the solubility of the salt in units of mol/L or g/L from the given $K_{sp}$ value.
- Predict the effect of a common ion on the solubility of a salt and perform calculations.
- Perform calculations to predict if a precipitate will form when two solutions are mixed.
- Do problems involving selective precipitation.
- Perform calculations involving complex ions and solubility.
- Use qualitative analysis to separate a mixture of ions.

## AP tip

Solubility problems usually appear in the first free-response question. Awareness of the types of problems outlined above and the methods used to solve them will lead you to succeed in this topic on the AP exam.

## SOLUBILITY PRODUCT

(*Chemistry* 6th ed. pages 751–752, 756 / 7th ed. pages 717–718, 721)

You already have memorized the solubility rules and know which salts are soluble. For slightly soluble or insoluble salts, an equilibrium exists between the solid and its aqueous ions. For example, lead(II) chloride dissolves in water as follows:

$$PbCl_2(s) \rightleftharpoons Pb^{2+}(aq) + 2\,Cl^-(aq)$$

At first, when the salt is added to the water, there are no ions present. As the solid dissolves, the concentration of the ions increases. A simultaneous competing process is the reverse of the dissolution, that is, the reforming of the solid called crystallization. At some point, the maximum amount of dissolution is achieved, which is called the saturation point. However, remember that on a molecular level, a dynamic equilibrium exists between dissolved solute and undisssolved solid. (The rate of dissolution equals the rate of crystallization.) The solution is saturated when no more solid dissolves and equilibrium is reached.

The equilibrium expression for the dissolution of lead(II) chloride is

$$K_{sp} = [Pb^{2+}][Cl^-]^2.$$

The constant, $K_{sp}$, is the solubility product constant. For salts producing the same number of ions, the $K_{sp}$ value can be used to measure the extent to which the solid dissolves. The larger the $K_{sp}$ value, the more soluble the salt.

**EXAMPLE:** Given the following salts and their $K_{sp}$ values, which salt is the most soluble? Which salt is the least soluble?

| Formula | $K_{sp}$ |
|---|---|
| $NiCO_3$ | $1.4 \times 10^{-7}$ |
| $MnS$ | $2.3 \times 10^{-13}$ |
| $CaSO_4$ | $6.1 \times 10^{-5}$ |

The most soluble salt is the salt with the largest $K_{sp}$ value, $CaSO_4$.

The least soluble salt is the salt with the lowest value of $K_{sp}$, $MnS$. You are able to compare the $K_{sp}$ values to determine the relative solubilities of the salts because they all produce the same number of ions.

# CALCULATIONS INVOLVING SOLUBILITY

(*Chemistry* 6th ed. pages 752–756 / 7th ed. pages 718–721)

The solubility of a salt is the amount of salt that will dissolve in 1 liter of water. The solubility of a salt can be given in units of mol/L or g/L. The solubility of a salt can be used to determine the $K_{sp}$ value for the salt.

## CALCULATING $K_{SP}$ FROM SOLUBILITY

EXAMPLE: The solubility of $Pb_3(PO_4)_2$ is $6.2 \times 10^{-12}$ M. Calculate the $K_{sp}$ value for the solid.

$K_{sp}$ equals $9.9 \times 10^{-55}$.

Step 1: Write the reaction for the dissolution of the solid.

$Pb_3(PO_4)_2 (s) \rightleftharpoons 3 Pb^{2+} (aq) + 2 PO_4^{3-} (aq)$.

Step 2: Underneath the reaction, make an ICE chart.

|  | $Pb_3(PO_4)_2(s)$ $\rightleftharpoons$ | $3 Pb^{2+} (aq) +$ | $2 PO_4^{3-} (aq)$ |
|---|---|---|---|
| Initial |  | 0 | 0 |
| Change |  | +3x | +2x |
| Equilibrium |  | 3x | 2x |

In the ICE chart above, $x$ represents $x$ mol/L of $Pb_3(PO_4)_2(s)$ dissolving to reach equilibrium which equals $6.2 \times 10^{-12}$ M.

For every 1 mol per liter of $Pb_3(PO_4)_2$ which dissolves, 3 moles per liter of $Pb^{2+}$ and 2 moles per liter of $PO_4^{3-}$ form.

$3x$ is the mol/L of $Pb^{2+} (aq)$ produced when the solid, $Pb_3(PO_4)_2$, dissolves.

$2x$ is the mol/L of $PO_4^{3-} (aq)$ produced when the solid, $Pb_3(PO_4)_2$, dissolves.

Step 3: Write the equilibrium expression for the reaction and plug in the values from the equilibrium line of the ICE chart.

$K_{sp} = [Pb^{2+}]^3 [PO_4^{3-}]^2 = (3x)^3 (2x)^2 = 108 x^5$.

The value of $x$ is the solubility of $Pb_3(PO_4)_2$, which equals $6.2 \times 10^{-12}$ M.

$K_{sp} = 108 (6.2 \times 10^{-12})^5 = 9.9 \times 10^{-55}$.

Alternate solution:

$$6.2 \times 10^{-12} \frac{mol\ Pb_3(PO_4)_2}{1\ L} \times \frac{3\ mol\ Pb^{2+}}{1\ mol\ Pb_3(PO_4)_2} = 1.9 \times 10^{-11} M\ Pb^{2+}.$$

$$6.2 \times 10^{-12} \frac{\text{mol Pb}_3(\text{PO}_4)_2}{1 \text{ L}} \times \frac{2 \text{ mol Pb}_4{}^{3-}}{1 \text{ mol Pb}_3(\text{PO}_4)_2} = 1.2 \times 10^{-11} \text{M PO}_4{}^{3-}$$

Plug these values into the $K_{sp}$ expression and solve for $K_{sp}$.

$K_{sp} = [\text{Pb}^{2+}]^3 [\text{PO}_4{}^{3-}]^2 = (1.9 \times 10^{-11})^3 (1.2 \times 10^{-11})^2 = 9.9 \times 10^{-55}$.

## CALCULATING SOLUBILITY FROM $K_{SP}$

If you are given a group of salts which do not all have the same cation to anion ratio and asked which is more soluble, you must perform a calculation to determine the solubility of each salt.

EXAMPLE: Given the two salts in the table below, which is more soluble? Show calculations to support your answer.

| | $K_{sp}$ |
|---|---|
| $FeC_2O_4$ | $2.1 \times 10^{-7}$ |
| $Cu(IO_4)_2$ | $1.4 \times 10^{-7}$ |

You cannot directly compare the $K_{sp}$ values to predict which is more soluble because the salts dissolve to produce a different number of ions. $FeC_2O_4$ dissolves to produce two ions and $Cu(IO_4)_2$ produces three ions. You must calculate the solubility for each salt from its $K_{sp}$ value.

Step 1: Write the reaction for the dissolution of the solid.

$FeC_2O_4 (s) \rightleftharpoons Fe^{2+} (aq) + C_2O_4{}^{2-} (aq)$.

Step 2: Underneath the reaction, make an ICE chart.

| | $FeC_2O_4 (s)$ $\rightleftharpoons$ | $Fe^{2+} (aq)$ + | $C_2O_4{}^{2-} (aq)$ |
|---|---|---|---|
| Initial | | 0 | 0 |
| Change | | $+x$ | $+x$ |
| Equilibrium | | $x$ | $x$ |

Step 3: Write the equilibrium expression for $K_{sp}$, plug in the equilibrium line, and solve for $x$.

$K_{sp} = [Fe^{2+}][C_2O_4{}^{2-}]$ ; $2.1 \times 10^{-7} = x^2$ ; $x = 4.6 \times 10^{-4}$ mol $Fe^{2+}$/L.

$$4.6 \times 10^{-4} \text{mol Fe}^{2+} \times \frac{1 \text{ mol FeC}_2\text{O}_4}{1 \text{ L}} = 4.6 \times 10^{-4} \text{mol FeC}_2\text{O}_4/\text{L}.$$

Repeat the same calculations for the next salt.

Step 1: Write the reaction for the dissolution of the solid.

$Cu(IO_4)_2 (s) \rightleftharpoons Cu^{2+} (aq) + 2 IO_4{}^- (aq)$.

Step 2: Underneath the reaction, make an ICE chart.

$$Cu(IO_4)_2 \, (s) \rightleftharpoons Cu^{2+}(aq) + 2\, IO_4^-(aq)$$

| | | $Cu^{2+}$ | $2\,IO_4^-$ |
|---|---|---|---|
| Initial | | 0 | 0 |
| Change | | $+x$ | $+2x$ |
| Equilibrium | | $x$ | $2x$ |

Step 3: Write the equilibrium expression for $K_{sp}$, plug in the equilibrium line, and solve for $x$.

$K_{sp} = [Cu^{2+}][IO_4^-]^2; \ 1.4 \times 10^{-7} = x(2x)^2; \ 4x^3 = 1.4 \times 10^{-7};$

$x = 3.3 \times 10^{-3} \, mol/L.$

$$3.3 \times 10^{-3} \, mol \, Cu^{2+} \times \frac{1 \, mol \, Cu(IO_4)_2}{1 \, L} = \frac{3.3 \times 10^{-3} \, mol \, Cu(IO_4)_2}{L}$$

$3.3 \times 10^{-3}$ mol $Cu(IO_4)_2$ /L is greater than $4.6 \times 10^{-4}$ mol $FeC_2O_4$ /L.

Therefore $Cu(IO_4)_2$ is more soluble even though $K_{sp}$ for $FeC_2O_4$ is slightly larger than $K_{sp}$ for $Cu(IO_4)_2$.

## COMMON ION EFFECT

(*Chemistry* 6th ed. pages 757–760 / 7th ed. pages 722–724)

When a salt is dissolved in water containing a common ion, its solubility is decreased.

For example, consider the solubility equilibrium of silver sulfate.

$$Ag_2SO_4(s) \rightleftharpoons 2\, Ag^+(aq) + SO_4^{2-}(aq).$$

When silver sulfate is dissolved in 0.100 $M$ $AgNO_3$, the $Ag^+$ ion from silver nitrate causes the equilibrium to shift to the left, decreasing the solubility of silver sulfate.

**EXAMPLE:** Calculate the molar solubility of $Ag_2SO_4$ in 0.10 $M$ $AgNO_3$.

$K_{sp}$ for $Ag_2SO_4(s) = 1.2 \times 10^{-5}$.

Fill out the ICE chart. Be sure to include the initial concentration of the ions from the soluble salt, $AgNO_3$.

$$Ag_2SO_4(s) \rightleftharpoons 2\, Ag^+(aq) + SO_4^{2-}(aq)$$

| | | $2\,Ag^+$ | $SO_4^{2-}$ |
|---|---|---|---|
| Initial | | 0.10 | 0 |
| Change | | $+2x$ | $+x$ |
| Equilibrium | | $0.10 + 2x$ | $x$ |

Plug in the equilibrium line into the expression:

$K_{sp} = [Ag^+]^2 [SO_4^{2-}] = (0.10 + 2x)^2(x).$

Assume that $0.10 + 2x$ is about 0.10, since $K_{sp}$ is small you can assume that the change ($2x$) from the initial concentration (0.10) is negligible.

$K_{sp} = 1.2 \times 10^{-5} = (0.10)^2(x)$ ; $x = 1.2 \times 10^{-3}$ mol $SO_4^{2-}$/L.

$$1.2 \times 10^{-3} \frac{\text{mol}}{\text{L}} \ SO_4^{2-} \times \frac{1 \text{ mol } Ag_2SO_4}{1 \text{ mol } SO_4^{2-}} = 1.2 \times 10^{-3} \frac{\text{mol}}{\text{L}} \ Ag_2SO_4.$$

The solubility of silver sulfate in 0.100 $M$ silver nitrate, $1.2 \times 10^{-3}$ mol/L, is less than the solubility of silver sulfate in pure water, $1.4 \times 10^{-2}$ mol/L.

## pH AND SOLUBILITY

Chromium(III) hydroxide dissolves according to the equilibrium

$$Cr(OH)_3 \ (s) \ \rightleftharpoons \ Cr^{3+}(aq) + 3 \ OH^-(aq).$$

An increase in pH, caused by the addition of $OH^-$ ions, will shift the equilibrium to the left, decreasing the solubility of $Cr(OH)_3$.

A decrease in pH, caused by the addition of $H^+$ ions, will shift the equilibrium to the right, increasing the solubility of $Cr(OH)_3$. The $H^+$ ions remove the $OH^-$ ions from the solution.

A salt with the general formula, MX, will show increased solubility in acidic solution if the anion, $X^-$, is an effective base (if HX is a weak acid). Common anions that make effective bases include $S^{2-}$, $OH^-$ and $CO_3^{2-}$.

EXAMPLE: Calculate the solubility of $Fe(OH)_3$ in a solution with a pH equal to 5.0. $K_{sp} = 4.0 \times 10^{-38}$.

Fill out an ICE chart. Calculate the initial hydroxide concentration from the pH.

pOH = 14 –pH = 14.0 – 5.0 = 9.0 ; $[OH^-] = 1.0 \times 10^{-9}$ $M$.

|  | $Fe(OH)_3(s)$ $\rightleftharpoons$ | $Fe^{3+}(aq)$ + | $3OH^-(aq)$ |
|---|---|---|---|
| Initial |  | 0 | $1.0 \times 10^{-9}$ |
| Change |  | $x$ | $+3x$ |
| Equilibrium |  | $x$ | $1.0 \times 10^{-9}$ |

$K_{sp} = 4.0 \times 10^{-38} = [Fe^{3+}][ OH^-]^3 = x(1.0 \times 10^{-9})^3$

Because the pH is 5.0, the equilibrium [OH] must be equal to $1.0 \times 10^{-9}$.

$x = 4.0 \times 10^{-11} M$.

# PRECIPITATE FORMATION

(*Chemistry* 6th ed. pages 760–763 / 7th ed. pages 724–727)

A precipitate may or may not form when two solutions are mixed, depending on the concentrations of the ions involved in the formation of the solid.

## ION PRODUCT

The ion product, $Q$, is written in the same way as the $K_{sp}$ expression. For lead(II) chloride, $Q = [Pb^{2+}][Cl^-]^2$

Calculation of the value, $Q$, involves the use of the initial concentrations of the solutions mixed, $[Pb^{2+}]_0$ and $[Cl^-]_0$, instead of the equilibrium concentrations.

A comparison of the value of $Q$ to $K_{sp}$ determines if a precipitate is formed.

$Q > K_{sp}$: precipitation occurs

$Q < K_{sp}$: no precipitation occurs

$Q = K_{sp}$: the solution is saturated

EXAMPLE: Will a precipitate form when 100.0 mL of $4.0 \times 10^{-4} M$ $Mg(NO_3)_2$ is added to 100.0 mL of $2.0 \times 10^{-4} M$ NaOH?

Step 1: Determine the identity of the precipitate formed.

$Mg(OH)_2$ is the possible precipitate. $NaNO_3$ is always soluble.

Step 2: Determine the concentration of the ions after they are mixed and before any reaction occurs.

Determine the moles of concentration of each solute present. Be sure to divide by the total volume of the two solutions mixed.

$[Mg^{2+}]_0 = (0.1000 \text{ L} \times 4.0 \times 10^{-4} \text{ mol/ L}) / 0.2000 \text{ L} = 2.0 \times 10^{-4} M.$

$[OH^-]_0 = (0.1000 \text{ L} \times 2.0 \times 10^{-4} \text{ mol/ L}) / 0.2000 \text{ L} = 1.0 \times 10^{-4} M.$

Step 3: Write the ion product expression, calculate its value, and compare it to $K_{sp}$, which equals $8.8 \times 10^{-12}$.

$Q = [Mg^{2+}][OH^-]^2 = (2.0 \times 10^{-4} M)(1.0 \times 10^{-4} M)^2 = 2.0 \times 10^{-12}$

Since $Q < K_{sp}$, no precipitate will form.

# SELECTIVE PRECIPITATION

(*Chemistry* 6th ed. pages 763–764 / 7th ed. pages 727–729)

A reagent is added to a mixture of metal ions, thus forming a precipitate. One metal ion will precipitate before the other, allowing the mixture to be separated.

EXAMPLE: A solution contains $0.25 M$ $Ni(NO_3)_2$ and $0.25 M$ $Cu(NO_3)_2$. A solution of $Na_2CO_3$ is slowly added to this solution.

a) Will $NiCO_3$ ($K_{sp} = 1.4 \times 10^{-7}$) or $CuCO_3$ ($K_{sp} = 2.5 \times 10^{-10}$) precipitate first?

b) Calculate the concentration of $CO_3^{2-}$ necessary to begin the precipitation of each salt.

c) Determine the concentration of $Cu^{2+}$ when $NiCO_3$ begins to precipitate.

ANSWERS:

a)  $CuCO_3$ will precipitate first because its $K_{sp}$ is smaller.

b)  For $CuCO_3$, precipitation begins when

$[CO_3^{2-}] = K_{sp}/[Cu^{2+}] = 2.5 \times 10^{-10}/0.25\ M = 1.0 \times 10^{-9}\ M.$

For $NiCO_3$, precipitation begins when

$[CO_3^{2-}] = K_{sp}/[Ni^{2+}] = 1.4 \times 10^{-7}/0.25\ M = 5.6 \times 10^{-7}\ M.$

c)  The concentration of $[Cu^{2+}]$ when $NiCO_3$ begins to precipitate

$[Cu^{2+}] = K_{sp}/[CO_3^{2-}] = 2.5 \times 10^{-10}/5.6 \times 10^{-7} M = 4.5 \times 10^{-4}\ M.$

## COMPLEX ION EQUILIBRIA

(*Chemistry* 6th ed. pages 766–771 / 7th ed. pages 731–736)

A complex ion consists of a metal ion surrounded by ligands which are Lewis bases such as $H_2O$, $OH^-$, $NH_3$, $Cl^-$ and $CN^-$.

The coordination number is the number of ligands which attach to the transition metal ion. Common coordination numbers include 4 for $Cu^{2+}$ and $Co^{2+}$ and 2 for $Ag^+$.

The ligands attach one at a time to the metal ion; each step has a formation constant, $K_{f_i}$:

$$Ag^+ + S_2O_3^{2-} \rightleftharpoons Ag(S_2O_3)^- \qquad K_{f_1} = 7.4 \times 10^8$$

$$Ag(S_2O_3)^- + S_2O_3^{2-} \rightleftharpoons Ag(S_2O_3)_2^{3-} \qquad K_{f_2} = 3.9 \times 10^4$$

Formation of a complex ion causes a precipitate to dissolve. The equilibrium, $AgBr(s) \rightleftharpoons Ag^+ + Br^-$ is not affected by the addition of $H^+$. But the concentration of $Ag^+$ can be lowered by the addition of excess $S_2O_3^{2-}$ forming the complex ion, $Ag(S_2O_3)_2^{3-}$.

> EXAMPLE: Calculate the mass of AgBr that can dissolve in 1.00 L of 0.500 $M$ $Na_2S_2O_3$. $K_{sp}$ for AgBr = $5.0 \times 10^{-13}$.

> Step1: Determine the overall reaction by adding up the reactions for the solubility equilibrium and the stepwise formation of the complex ion. Remember, to get the equilibrium constant for the overall reaction, $K$, multiply together the $K$ for each step.

| | | | | |
|---|---|---|---|---|
| $AgBr(s)$ | $\rightleftharpoons$ | $Ag^+(aq)$ + | $Br^-(aq)$ | $K_{sp} = 5.0 \times 10^{-13}$ |
| $Ag^+$ + $S_2O_3^{2-}$ | $\rightleftharpoons$ | $Ag(S_2O_3)^-$ | | $K_{f_1} = 7.4 \times 10^8$ |
| $Ag(S_2O_3)^-$ + $S_2O_3^{2-}$ | $\rightleftharpoons$ | $Ag(S_2O_3)_2^{3-}$ | | $K_{f_2} = 3.9 \times 10^4$ |

---

| | | | | |
|---|---|---|---|---|
| $AgBr(s)$ | $2\,S_2O_3^{2-}$ | $\rightleftharpoons$ | $Ag(S_2O_3)_2^{3-}$ + $Br^-$ | $K' = 14.4$ |

Step 2: Complete an ICE chart with the overall reaction:

| | $AgBr(s)$ | $2\,S_2O_3^{2-}$ | $\rightleftharpoons$ | $Ag(S_2O_3)_2^{3-}$ + | $Br^-$ |
|---|---|---|---|---|---|
| Initial | | 0.500 | | 0 | 0 |
| Change | | $-2x$ | | $+x$ | $+x$ |
| Equilibrium | | $0.500 - 2x$ | | $x$ | $x$ |

$K' = 14.4 = [Ag(S_2O_3)_2^{3-}][Br^-]/[S_2O_3^{2-}]^2$

$= x^2 / (0.500 - 2x)^2$ ; take the square root of both sides

$3.79 = x / (0.500 - 2x)$

$x = 1.90 - 7.58x$ ; $x = 0.221\ M\ Br^- = 0.221\ M\ AgBr(s)$

$1.00\ L \times \dfrac{0.221\ \text{mol AgBr}}{1\ L} \times \dfrac{187.8\ g}{1\ \text{mol}} = 41.5\ g\ AgBr = 42\ g\ AgBr$

# QUALITATIVE ANALYSIS

(*Chemistry* 6th ed. pages 764–766 / 7th ed. pages 729–731)

Qualitative analysis involves separating a mixture of cations or anions based on their solubilities. Cations can be separated into five major groups based on their solubilities: Group I: insoluble chlorides; Group II: sulfides soluble in acidic solution; Group II: sulfides insoluble in basic solution; Group IV: insoluble carbonates; and Group V: alkali metal and ammonium ions. Each of these groups can be treated further to separate and identify the individual ions.

EXAMPLE: Separate a mixture containing the Group I cations: $Ag^+$, $Hg_2^{2+}$ and $Pb^{2+}$.

The flowchart below shows the separation of Group I ions using qualitative analysis.

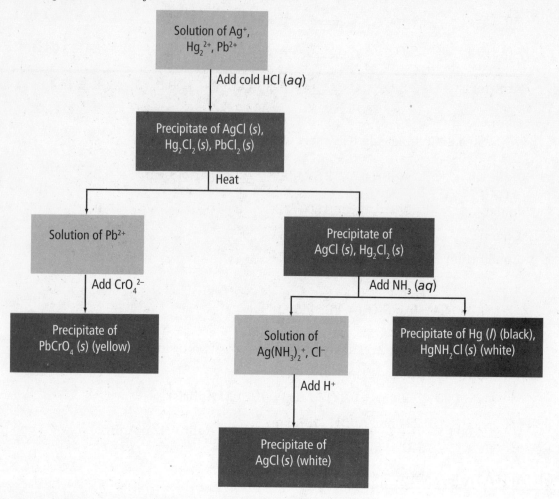

## Multiple-Choice Questions

No calculators are to be used in this section.

1. Temperature is often given with $K_{sp}$ values because
   (A) the solubility of solids always increases with increasing temperature.
   (B) the solubility of solids varies with temperature changes.
   (C) solubility changes with temperature but $K_{sp}$ values do not.
   (D) $K_{sp}$ varies with temperature even though concentrations do not.
   (E) the number of ions varies with the kind of salt that is dissolving.

**Answer:   B** The solubility of ionic salts is related to temperature. Though the solubility most often increases with increasing temperature, this is not always the case. Since $K_{sp}$ is calculated from solubility, this value is temperature dependent, as are all equilibrium constants (*Chemistry* 6th ed. pages 752–754 / 7th ed. pages 718–719).

2. The $K_{sp}$ expression for silver phosphate is
    (A) $K_{sp} = [Ag^+][PO_4^{3-}]$.
    (B) $K_{sp} = [Ag^+]^2[PO_4^{3-}]$.
    (C) $K_{sp} = [Ag^+]^3[PO_4^{3-}]$.
    (D) $K_{sp} = [Ag^+][PO_4^{3-}]^3$.
    (E) $K_{sp} = [Ag^+][PO_4^{3-}] / [Ag_3PO_4]$.

ANSWER: C Begin by writing the equation for dissolving the ionic salt, often called the solubility equilibrium reaction. Recall that $K_{sp}$ is just a special case of the equilibrium constant, which is [products]/[reactants], each raised to a power expressed by its coefficients, and that solids are not shown in the expression. In this case:

$Ag_3PO_4(s) \rightleftharpoons 3Ag^+ (aq) + PO_4^{3-} (aq)$     $K_{sp} = [Ag^+]^3[PO_4^{3-}]$.

(*Chemistry* 6th ed. pages 751–752 / 7th ed. pages 717–718).

3. The solubility of nickel(II) hydroxide, $K_{sp} = 1.6 \times 10^{-16}$, is about
    (A) $\sqrt[3]{4.0 \times 10^{-17}}$.
    (B) $\sqrt[3]{\left(\dfrac{1.6 \times 10^{-16}}{2}\right)}$.
    (C) $\sqrt{1.6 \times 10^{-16}}$.
    (D) $\sqrt{(1.6 \times 10^{-16} \times 3)}$.
    (E) $\sqrt{(1.6 \times 10^{-16} \times 4)}$.

ANSWER: A First show the solubility equilibrium expression:

$Ni(OH)_2(s) \rightleftharpoons Ni^{2+}(aq) + 2OH^-(aq)$.

If $x$ = mol/L of solid dissolving, the $[Ni^{2+}] = x$ and $[OH^-] = 2x$.

From $K_{sp} = [Ni^{2+}][OH^-]^2 = (x)(2x)^2 = 4x^3$,

$$\text{solubility} = x = \sqrt[3]{\left(\frac{K_{sp}}{4}\right)}$$

(*Chemistry* 6th ed. pages 755–756 / 7th ed. pages 720–721)

4. Determine the solubility of lead(II) fluoride, $K_{sp} = 4.0 \times 10^{-8}$, in a 0.0040 $M$ lead(II) nitrate solution.
    (A) $2.0 \times 10^{-3} M$
    (B) $2.0 \times 10^{-2} M$
    (C) $\sqrt{4.0 \times 10^{-5}} M$
    (D) $\dfrac{\sqrt{1.0 \times 10^{-5}}}{2} M$
    (E) $(4.0 \times 10^{-8} / 4.0 \times 10^{-3}) M$

ANSWER: D The solubility equilibrium expression is $PbF_2(s) \rightleftharpoons Pb^{2+}(aq) + 2F^-(aq)$.

$K_{sp} = [Pb^{2+}][F^-]^2 = 4.0 \times 10^{-8}$.

$[Pb^{2+}] = 4.0 \times 10^{-3} M$.

Since each mole of $PbF_2(s)$ that dissolves produces two moles of $F^-(aq)$, the solubility of lead(II) fluoride = $[F^-] / 2$.

$$[F^-] = \sqrt{\frac{K_{sp}}{[Pb^{2+}]}} = \sqrt{\left(\frac{4.0 \times 10^{-8}}{4.0 \times 10^{-3}}\right)} = \sqrt{1.0 \times 10^{-5}}\ M.$$

So the solubility of the $PbF_2$ will equal one-half that amount (*Chemistry* 6th ed. pages 757–759 / 7th ed. pages 722–724).

5. The solubility of magnesium hydroxide ($K_{sp} = 8.0 \times 10^{-12}$) in a buffered solution of pH = 11.00 would be
   (A) $1.0 \times 10^{-11} M$.
   (B) $8.0 \times 10^{10} M$.
   (C) $8.0 \times 10^{-12} M$.
   (D) $8.0 \times 10^{-11} M$.
   (E) $8.0 \times 10^{-6} M$.

ANSWER: E Begin by writing the solubility equilibrium expression:

$Mg(OH)_2(s) \rightleftharpoons Mg^{2+}(aq) + 2OH^-(aq)$.

With a pH=11.00, pOH = 14.00–11.00 = 3.00, and $[OH^-] = 1.0 \times 10^{-3} M$.

$K_{sp} = [Mg^{2+}][OH^-]^2 = 8.0 \times 10^{-12}$

Solubility = $[Mg^{2+}] = K_{sp} / [OH^-]^2 = 8.0 \times 10^{-12} / (1.0 \times 10^{-3})^2 = 8.0 \times 10^{-6} M$.

(*Chemistry* 6th ed. pages 759–760 / 7th ed. page 724)

6. The solubility of silver sulfide is $8.0 \times 10^{-17}$M. Determine the $K_{sp}$ of this salt.
   (A) $6.4 \times 10^{-50}$
   (B) $(8.0 \times 10^{-7})(16.0 \times 10^{-17})$
   (C) $(8.0 \times 10^{-7})(16.0 \times 10^{-17})^2$
   (D) $(8.0 \times 10^{-7})(4.0 \times 10^{-17})$
   (E) $(8.0 \times 10^{-7})(4.0 \times 10^{-17})^2$

ANSWER: C The equation for the solubility reaction is

$Ag_2S(s) \rightleftharpoons 2Ag^+(aq) + S^{2-}(aq)$.

$K_{sp} = [Ag^+]^2[S^{2-}]$.

Keeping in mind that there are two silver ions formed for each sulfide ion formed, $K_{sp} = (16. \times 10^{-7})^2 (8.0 \times 10^{-7})$.

(*Chemistry* 6th ed. pages 665–668 / 7th ed. pages 634–637)

7. Equal volumes of $1.6 \times 10^{-5}\,M$ KCl and $1.6 \times 10^{-5}\,M$ AgNO$_3$ are mixed. The $K_{sp}$ for silver chloride is $1.6 \times 10^{-10}$. As these two solutions intermingle,
   (A) a precipitate of AgCl forms.
   (B) there is no precipitate formed.
   (C) NaCl will precipitate.
   (D) AgNO$_3$ will precipitate.
   (E) the [Na$^+$] will become 0.020 $M$.

ANSWER: **B** When the equal volumes are mixed but before there is any reaction, $[Ag^+] = 0.80 \times 10^{-5}\,M$ and $[Cl^-] = 0.80 \times 10^{-5}\,M$.

So the ion product, $Q$, $= (0.80 \times 10^{-5})(0.80 \times 10^{-5}) = 6.4 \times 10^{-11}$.

Since the $K_{sp}$ value of $1.6 \times 10^{-10}$ is not exceeded by $Q$, a precipitate of AgCl will not form (*Chemistry* 6th ed. pages 669–670 / 7th ed. pages 638–639).

8. A saturated solution of the strong base, MOH, has a pH of 11.00. Determine the $K_{sp}$ for this base.
   (A) $1.0 \times 10^{-6}$
   (B) $1.0 \times 10^{-11}$
   (C) $5.0 \times 10^{-11}$
   (D) $2.5 \times 10^{-12}$
   (E) $1.0 \times 10^{-22}$

ANSWER: **A** The reaction: MOH($s$) $\rightleftharpoons$ M$^+$($aq$) + OH$^-$($aq$). If pH = 11.00, then $[OH^-] = 1.0 \times 10^{-3}\,M$, which is also the value for $[M^+]$. $K_{sp} = [M^+][OH^-] = (1.0 \times 10^{-3})(1.0 \times 10^{-3}) = 1.0 \times 10^{-6}$ (*Chemistry* 6th ed. pages 673–676 / 7th ed. pages 642–644).

9. Reactions like Ag$^+$($aq$) + 2NH$_3$($aq$) $\rightleftharpoons$ Ag(NH$_3$)$_2$$^+$(aq) show the formation of a complex ion. If the formation constant, $K_f$, is given,
   (A) the relative stability of the complex ion is known.
   (B) the stability of the Lewis acid is predictable.
   (C) the stability of the metal cation is predictable.
   (D) the pH effects are also indicated.
   (E) the stability of many complex ions may be calculated from these data.

ANSWER: **A** If the $K_f$ value is very large, then the complex ion is quite stable in solution, so this is a good predictor of stability for this complex ion. Since $K_f$ values (formation constants) are given for specific complex ions, this only allows questions about the one equilibrium system. (Note: the Lewis acid in this reaction is the metal cation.) (*Chemistry* 6th ed. pages 678–679 / 7th ed. pages 647–648)

10. The $K_{sp}$ of the metal hydroxide, MOH, is $1.0 \times 10^{-8}$. What is the pH of a saturated solution of MOH?
    (A) pH = 4.00.
    (B) pH = 8.00.
    (C) pH = 10.00.
    (D) pH = 12.00.
    (E) pH = 14.00.

ANSWER: **C** First determine the solubility of MOH, which will lead to $[OH^-]$.

The reaction MOH(s) $\rightleftharpoons$ M$^+$(aq) + OH$^-$(aq).

$K_{sp}$ = [M$^+$][OH$^-$] = $1.0 \times 10^{-8}$, and [M$^+$] = [OH$^-$] = the solubility of MOH.

The solubility then is $\sqrt{1.0 \times 10^{-8}}$ = $1.0 \times 10^{-4}$ M = [OH$^-$].

Therefore pOH = $-\log[OH^-]$ = $-\log(1.0 \times 10^{-4})$ = 4.00 from pH + pOH = 14.00, pH = 14.00 − 4.00 = 10.00 (*Chemistry* 6th ed. pages 673–676 / 7th ed. pages 642–644).

## FREE-RESPONSE QUESTIONS

1. Lead(II) chromate has a $K_{sp}$ of $2.0 \times 10^{-16}$. Exactly 4.0 mL of 0.0040 $M$ lead(II) nitrate is mixed with 2.0 mL of 0.00020 M sodium chromate.
   (a) Give the solubility equilibrium reaction.
   (b) Show the $K_{sp}$ expression for this reaction.
   (c) Will a precipitate form? Support your answer.
   (d) What would be the effect on the solubility equilibrium system if a concentrated potassium chromate solution is added dropwise?
   (e) Which is more soluble, AgCl, $K_{sp}$ $2 \times 10^{-10}$, or Ag$_2$CO$_3$, $K_{sp}$ = $8 \times 10^{-12}$?

   Support your answer with calculations.

ANSWER 1(a): PbCrO$_4$(s) $\rightleftharpoons$ Pb$^{2+}$(aq) + CrO$_4^{2-}$(aq)

ANSWER 1(b): $K_{sp}$ = [Pb$^{2+}$][CrO$_4^{2-}$]

ANSWER 1(c):

$$Q = \left[\left(\frac{0.0040M}{0.0060L}\right)0.0040L\right]\left[\left(\frac{0.00020M}{0.0060L}\right)0.0020L\right] = 1.8 \times 10^{-7}.$$

Since the $K_{sp}$ ($2.0 \times 10^{-16}$) has been exceeded, a precipitate does form.

ANSWER 1(d): This would cause an equilibrium shift to the left, and more solid PbCrO$_4$ will form.

**ANSWER 1(e):** With a 1:1 salt, like AgCl, solubility = $\sqrt{K_{sp}}$, while with a 2:1 salt, solubility = $\sqrt[3]{\left(\dfrac{K_{sp}}{4}\right)}$.

Using the above gives solubility of $1 \times 10^{-5}$ M for AgCl and $2 \times 10^{-4}M$ for PbCrO$_4$, so it can be seen that lead(II) chromate is more soluble even though it has the lower $K_{sp}$ value. ($K_{sp}$ can only be used for a direct comparison of solubility if a 1:1 salt is compared to a 1:1 salt, or if a 2:1 salt is compared to a 2:1 salt, etc.) (*Chemistry* 6th ed. pages 669–670 / 7th ed. pages 638–639)

2.
- (a) Compare the solubility of solid Mg(OH)$_2$ in water, with a solution of high pH, and with a solution of low pH.
- (b) Apply similar reasoning and explain why copper(II) sulfide is more soluble in acidic solution than in water.
- (c) In the qualitative analysis method of selective precipitation, Group 3 Cations (Co$^{2+}$, Fe$^{2+}$, Mn$^{2+}$, Ni$^{2+}$, and Zn$^{2+}$) are removed from an acidified solution by the addition of sodium hydroxide solution, following the removal of Group 2 Cations with H$_2$S. Explain how solid sulfides of these ions (not hydroxides) are formed.
- (d) Given

$$Ag^+ + S_2O_3^{2-} \rightleftharpoons Ag(S_2O_3)^- \qquad K = 7.4 \times 10^8$$

$$Ag(S_2O_3)^- + S_2O_3^{2-} \rightleftharpoons Ag(S_2O_3)_2^{3-} \quad K = 4.0 \times 10^4$$

Calculate [Ag(S$_2$O$_3$)$_2$$^{3-}$] when 0.500 L of 2.0 M Na$_2$S$_2$O$_3$ is mixed with 0.500 L of 0.040 M AgNO$_3$.

**ANSWER 2(a):** First consider the solubility equilibrium reaction:

$$Mg(OH)_2(s) \rightleftharpoons Mg^{2+}(aq) + 2OH^-(aq).$$

If pH < 7, H$^+$ reacts with OH$^-$ to form water; this causes a shift in the solubility equilibrium toward the right, increasing the solubility of Mg(OH)$_2$.

If pH > 7, the OH$^-$ ions cause a shift in the solubility equilibrium to the left, forming more solid, decreasing the solubility of magnesium hydroxide.

Summary: Mg(OH)$_2$ will be most soluble in acidic solution, less in water, and even less in basic solution.

**ANSWER 2(b):** Again, first consider the solubility equilibrium reaction,

$$CuS(s) \rightleftharpoons Cu^{2+}(aq) + S^{2-}(aq)$$

and the reactions in acidic solution,

$$S^{2-}(aq) + H^+(aq) \rightleftharpoons HS^-(aq)$$

$$HS^-(aq) + H^+(aq) \rightleftharpoons H_2S(aq).$$

The reaction of the $S^{2-}$ with $H^+$ (and the $HS^-$ with $H^+$) cause the solution equilibrium to shift to the right allowing more of the CuS to dissolve.

ANSWER 2(c): When the system is acidified, the reaction with metal ions, $M^{2+}$, is $H_2S\ (aq) + M^{2+}(aq) \rightleftharpoons MS(s) + 2H^+(aq)$. The addition of the acid has kept this equilibrium shifted to the left. Adding a base forms water, (from $H^+ + OH^- \longrightarrow H_2O$) shifting the equilibrium toward the right, forming insoluble sulfides.

ANSWER 2(d): 0.020 $M$

With the large excess of $S_2O_3^{2-}$ ions these two reactions go to completion (i.e. equilibrium is far toward the products). This means that essentially all of the $Ag^+$ goes to form the complex ion, $Ag(S_2O_3)_2^{3-}$; hence

$$[Ag+] = \frac{(0.040\ M \times 0.500\ L)}{(0.500\ L + 0.500\ L)} =$$

$[Ag(S_2O_3)_2^{3-}]$ after reaction = 0.0200 $M$.

(*Chemistry* 6th ed. pages 759–769 / 7th ed. pages 724–734)

# 15

# ELECTROCHEMISTRY

Electrochemistry is the study of the interchange of electrical and chemical energy. There are two types of electrochemical cells, galvanic and electrolytic. In galvanic cells, spontaneous oxidation-reduction (redox) reactions generate electric current. In electrolytic cells, a nonspontaneous chemical reaction occurs with the application of an electric current.

You should be able to
- Identify and compare the two types of electrochemical cells: galvanic and electrolytic.
- Draw and label a galvanic cell, including labeling the electrodes, the flow of electrons, and the flow of ions.
- Write half-reactions and determine which reaction occurs at the anode and which reaction occurs at the cathode.
- Give the line notation for a galvanic cell or write a balanced redox reaction from the given line notation.
- Calculate the cell potential for a galvanic cell and an electrolytic cell.
- Determine if a reaction is spontaneous from its cell potential.
- Calculate the cell potential under nonstandard conditions when the solutions are not 1M. This involves the use of the Nernst equation.
- Determine the strengths of oxidizing agents and reducing agents.
- Draw and label an electrolytic cell.
- Determine the reactions which occur at the anode and the cathode during electrolysis.
- Perform stoichiometric calculations involving electrolysis.

## GALVANIC CELLS

### COMPONENTS OF THE GALVANIC CELL

(*Chemistry* 6th ed. pages 828–830 / 7th ed. pages 791–793)

In a galvanic cell, a spontaneous chemical reaction is used to produce electrical energy. The current produced by the cell is measured in volts by a voltmeter. One compartment of the galvanic cell contains the anode, where oxidation, a loss of electrons, occurs. The other compartment contains the cathode where reduction, a gain of electrons, occurs. The electrodes, the anode, and the cathode are immersed in respective solutions containing metal cations. The two compartments of a galvanic cell are connected by a salt bridge or a porous disk. This is shown in the figure to the right.

The oxidizing agent is the substance being reduced at the cathode. The reducing agent is the substance being oxidized at the anode.

### LINE NOTATION

(*Chemistry* 6th ed. pages 835–836 / 7th ed. pages 798–799)

A galvanic cell can be abbreviated with line notation.

reactant /product ‖ reactant/product
anode reaction      cathode reaction

The salt bridge is indicated by the symbol ‖.

EXAMPLE: Give the correct line notation for the galvanic cell pictured below.

ANSWER: $Zn/Zn^{2+}\|H^+/H_2$

## AP tips

Here are some mnemonic devices to help you remember some facts about electrochemistry and redox reactions.

- "LEO" goes "GER" means Loss of Electrons is Oxidation and Gain of Electrons is Reduction.

- To recall what happens at the anode and the cathode: RedCat and AnOx means reduction occurs at the cathode and oxidation occurs at the anode.

- To know the migration of ions toward the electrodes for both types of cells, "CAT" ions move to the "CAT" ode and "AN" ions move to the "AN" ode.

You will be provided with a table of standard reduction potentials on the AP test.

You should be able to sketch a galvanic cell and label the electrodes, the flow of electrons, and the flow of ions.

## STANDARD REDUCTION POTENTIALS

(*Chemistry* 6th ed. pages 830–834 / 7th ed. pages 794–797)

The cell potential, $E_{cell}$, is the potential of the cell to do work on its surroundings by driving an electric current through a wire. By definition, a potential of 1 volt is produced when 1 joule of energy moves 1 coulomb of electric charge across a potential. The magnitude of the cell potential is a measure of the driving force behind an electrochemical reaction. Sometimes it is referred to as the electromotive force or emf. Tables of reduction potentials give standard voltages for reduction half-reactions measured at standard conditions of 1 atmosphere, 1 molar solution, and 25°C.

## CALCULATING THE CELL POTENTIAL OF A GALVANIC CELL

(*Chemistry* 6th ed. pages 834–835 / 7th ed. pages 797–798)

The reaction occurring in a galvanic cell can be broken down into an oxidation half-reaction and a reduction half-reaction. Using the Table of Standard Reduction Potentials on page 30 of this book, you can calculate the cell potential of the overall reaction.

EXAMPLE: Consider a galvanic cell based on the reaction:

$Al + Ni^{2+} \longrightarrow Al^{3+} + Ni$

Give the balanced cell reaction and calculate the cell potential, $E°_{cell}$, for the reaction.

Step 1: Write the oxidation and reduction half-reactions.

Oxidation: $Al \longrightarrow Al^{3+} + 3\,e^-$  Reduction: $Ni^{2+} + 2\,e^- \longrightarrow Ni$

Step 2: For the reduction half-reaction, look up the potential in the table.

$Ni^{2+} + 2e^- \longrightarrow Ni \ E°_{red} = -0.23$ V.

Step 3: For the oxidation half-reaction, $E°_{ox} = -E°_{red}$.

The oxidation half-reaction must be the reverse of the reduction reaction that you find in the table for the value of $E°_{red}$.

Oxidation: $Al \longrightarrow Al^{3+} + 3e^- \ E°_{ox} = -E°_{red} = -(-1.66V) = +1.66$ V.

Step 4: The cell potential for the overall reaction is equal to the sum of the reduction potential, $E°_{red}$, and the oxidation potential, $E°_{ox}$.

$E°_{cell} = E°_{ox} + E°_{red}; E°_{cell} = -0.23$ V $+ 1.66$ V $= 1.43$ V.

To obtain the balanced cell reaction, you must make sure that the electrons lost equal the electrons gained. When multiplying the half-reactions through by a coefficient, do not change the value of $E°$.

3 ($Ni^{2+} + 2e^- \longrightarrow Ni$)        $E°_{red} = -0.23$ V

2 ($Al \longrightarrow Al^{3+} + 3 e^-$)        $E°_{ox} = + 1.66$ V

$3 Ni^{2+} + 2 Al \longrightarrow 3 Ni + 2 Al^{3+} \ E°_{cell} = 1.43$ V

## SPONTANEOUS REACTIONS

(*Chemistry* 6th ed. pages 837–840 / 7th ed. pages 800–803)

Gibbs free energy, $\Delta G°$, can be calculated from the cell potential, $E°_{cell}$.
$$\Delta G° = -nFE°_{cell}$$
Faraday's constant, $F$, has a value of 96,485 C/mol e⁻.
The number of moles of electrons transferred in a redox reaction is represented by $n$.
A spontaneous reaction is one that has a negative value for $\Delta G°$ or a positive value for $E°_{cell}$.

EXAMPLE: Will 1 $M$ HCl dissolve silver metal and form $Ag^+$ solution?

ANSWER: Write the half reactions and calculate $E°_{cell}$.

$2H^+ + 2 e^- \longrightarrow 2H_2$        $E°_{red} = 0.00$ V

$2Ag \longrightarrow 2Ag^+ + 2e^-$        $E°_{ox} = -0.80$ V

$2H^+ + 2 Ag \longrightarrow H_2 + 2 Ag^+$    $E°_{cell} = -0.80$ V

The negative value for $E°_{cell}$ indicates that the reaction will not occur.

You may be asked if an element or ionic species is capable of reducing another element or ion. To determine if the reaction will occur, write the half-reactions and calculate the cell potential as in the previous example.

EXAMPLE: Bromine, $Br_2$, can oxidize iodide, $I^-$, to iodine, $I_2$. However, $Br_2$ cannot oxidize chloride, $Cl^-$, to chlorine, $Cl_2$. Explain why the first reaction occurs yet, the second one does not.

Begin by writing the appropriate half-reactions. Then calculate the cell potential for the overall reaction.

First the reaction in which $Br_2$ oxidizes $I^-$:

$$Br_2 + 2e^- \longrightarrow 2Br^- \qquad E^°_{red} = 1.09 \text{ V}$$

$$2I^- \longrightarrow I_2 + 2e^- \qquad E^°_{ox} = -0.54 \text{ V}$$

$$Br_2 + 2I^- \longrightarrow 2Br^- + I_2 \qquad E^°_{cell} = 0.55 \text{ V}$$

This reaction occurs; $E^°_{cell}$ is positive.

$$Br_2 + 2e^- \longrightarrow 2Br^- \qquad E^°_{red} = 1.09 \text{ V}$$

$$2Cl^- \longrightarrow Cl_2 + 2e^- \qquad E^°_{ox} = -1.36 \text{ V}$$

$$Br_2 + 2Cl^- \longrightarrow 2Br^- + Cl_2 \qquad E^°_{cell} = -0.27 \text{ V}$$

This reaction does not occur; $E^°_{cell}$ is negative

## CELL DEPENDENCE ON CONCENTRATION

(*Chemistry* 6th ed. pages 841–844 / 7th ed. pages 803–806)

The galvanic cell represented by the reaction
$$3 Ni^{2+} + 2 Al \longrightarrow 3 Ni + 2 Al^{3+}$$
has a cell potential, $E^°_{cell}$, equal to 1.43 V under standard conditions (all solutions are 1 molar).

Increasing the concentration of $Ni^{2+}$ will shift the reaction to the right by Le Châtelier's principle, increasing the driving force on the electrons and increasing the cell potential.

The relationship between the cell potential and concentrations at 25°C is given by the Nernst equation:

$$E_{cell} = E^°_{cell} - (0.0591/n) \log Q$$

The cell potential, $E_{cell}$, is for nonstandard conditions.
The moles of electrons transferred are represented by $n$.
The mass action quotient is represented by $Q$.

At equilibrium, $E_{cell}$ equals 0.

EXAMPLE: Calculate the cell potential for the reaction

$$3 Ni^{2+} + 2 Al \rightarrow 3 Ni + 2 Al^{3+}$$

in which $[Al^{3+}] = 2.00 \ M$ and $[Ni^{2+}] = 0.750 \ M$.

You already know that $E^°_{cell} = 1.43$ V and the number of moles of electrons transferred, $n$, equals 6.

$Q = [Al^{3+}]^2 / [Ni^{2+}]^3 = (2.00)^3/(0.750)^2 = 14.2$

$E_{cell} = 1.43\ V - (0.0591/6\ )\ \log 14.2 = 1.42\ V$

## DETERMINING THE STRENGTH OF OXIDIZING AND REDUCING AGENTS

(*Chemistry* 6th ed. pages 857–860 / 7th ed. pages 819–820)

You may be asked to list atoms or ions in order of increasing strength as reducing agents or oxidizing agents.

For a substance to be oxidized, it must lose electrons and another substance must gain electrons because oxidation and reduction always occur together. The substance that causes another substance to be oxidized is called an oxidizing agent. An oxidizing agent is reduced; it is the reactant in the reduction half-reaction. The larger (more positive) $E°_{red}$, the stronger the oxidizing agent.

A reducing agent is oxidized; it is the reactant in the oxidation half-reaction. The larger (more positive) the $E°_{ox}$, the stronger the reducing agent.

> EXAMPLE: Classify each of the following as an oxidizing agent, reducing agent, or both. Within each list, arrange in order of increasing strength as oxidizing agents and reducing agents.
>
> $Br_2$, Mg, $Fe^{2+}$, $I_2$, $Cl^-$, $Cu^{2+}$
>
> Oxidizing agents : $Fe^{2+} < Cu^{2+} < I_2 < Br_2$
>
> Reducing agents: $Cl^- < Fe^{2+} < Mg$

To be an oxidizing agent, a substance must be capable of gaining electrons or being reduced. Of the species listed, Mg and $Cl^-$ are the only ones listed that cannot have a lower oxidation state. For the oxidizing agents listed above, the respective reduction potentials are –0.44 V, 0.16 V or 0.34 V for $Cu^{2+}$ (which can be reduced to $Cu^0$ or $Cu^+$), 0.54 V, and 1.09 V. The more positive the cell potential, the stronger the oxidizing agent.

Reducing agents must be capable of being oxidized to a higher oxidation state. $Cl^-$ and Mg can go to $Cl^0$ and $Mg^{2+}$. $Fe^{2+}$ can exist as $Fe^{3+}$ or $Fe^0$ so it can act as an oxidizing agent or reducing agent. For the reducing agents listed above, the corresponding oxidation potentials are –1.36 V, –0.77 V, +2.37 V. The more positive the cell potential, the stronger the reducing agent.

## ELECTROLYTIC CELLS

(*Chemistry* 6th ed. pages 855–857 / 7th ed. pages 816–818)

In an electrolytic cell, a nonspontaneous reaction is made to occur by forcing an electric current through the cell. In an earlier example, it was shown that the following reaction is spontaneous:

$$3\ Ni^{2+} + 2Al \longrightarrow 3\ Ni + 2\ Al^{3+}$$

The reverse of this reaction: $3\ Ni + 2Al^{3+} \longrightarrow 3\ Ni^{2+} + 2Al$ is nonspontaneous and can be made to occur by the addition of an external power source. This electrolytic cell can be set up with two compartments just like the galvanic cell, with the replacement of a power supply for the voltmeter.

In the process of electroplating, the electrolytic cell can also be set up using only one compartment, as shown in the figure below.

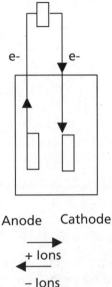

For example, if an object is to be plated with copper, make it the cathode and immerse it into a copper(II) sulfate solution. At the cathode, the reaction that will occur and deposit copper onto the object is $Cu^{2+} + 2e^- \longrightarrow Cu$. The anode can also be made of copper. The oxidation of copper occurs at this anode.

## REACTIONS THAT OCCUR IN AN ELECTROLYTIC CELL

To determine which reaction occurs at the anode and the cathode during electrolysis, you must consider all possible reactions and their reduction and oxidation potentials. If the reaction takes place in an aqueous solution, the oxidation and reduction of water must be considered.

EXAMPLE: A solution of copper (II) sulfate is electrolyzed. Calculate the cell potential of the reaction, $E^\circ_{cell}$.

|         | Possible reactions[1]                                          | Cell potential, $E^\circ$(V) |
|---------|---------------------------------------------------------------|----------------------|
| Cathode | $Cu^{2+} + 2e^- \longrightarrow Cu$                           | 0.34                 |
|         | $SO_4^{2-} + 4H^+ + 2e^- \longrightarrow$ $H_2SO_3 + H_2O$    | 0.20                 |
|         | $2H_2O + 2e^- \longrightarrow H_2 + 2OH^-$                     | −0.83                |
| Anode   | $Cu \longrightarrow Cu^{2+} + 2e^-$                           | −0.34                |
|         | $2H_2O \longrightarrow O_2 + 4H^+ + 4e^-$                     | −1.23                |

[1]Note: There is no half-reaction for the oxidation of $SO_4^{2-}$. S in $SO_4^{2-}$ is in its highest oxidation state, 6+, and cannot be oxidized further.

For each electrode, the reaction with the more positive potential will occur. At the cathode, $Cu^{2+}$ will be reduced. At the anode, Cu will be oxidized.

$$Cu^{2+} + 2e^- \longrightarrow Cu$$

$$Cu \longrightarrow Cu^{2+} + 2e^-$$

## AP tip

Frequently, the electrodes are inert for electrolysis. For example, during the electrolysis of KI ($aq$) $K^+$, $I^-$, and $H_2O$ are the only species present. Only $I^-$ and $H_2O$ are present to be oxidized at anode. Note: In aqueous KI, there is no K($s$) to be oxidized.

## STOICHIOMETRY OF ELECTROLYTIC PROCESSES

(*Chemistry* 6th ed. pages 856–857 / 7th ed. page 818)

In this section, you will review how much chemical change occurs with the flow of a given current for a specified time. You might be asked how much metal was plated (formed) or how long an electroplating process will take or how much current is required to produce a specified amount of metal over a period of time.

Some units to be familiar with include A, amperes; 1 A = 1 C/s; coulombs, C; Faraday's constant is 96,485 C = 1 mol $e^-$.

> **EXAMPLE:** A current of 10.0 A is passed through a solution containing $M^{2+}$ for 30.0 min. It produces 5.94 g of metal, $M$. Determine the identity of metal, $M$.
>
> 10.0 A = 10.0 C/s × 30.0 min × 60 s /min = $1.80 \times 10^4$ C
>
> $1.80 \times 10^4$ C × 1 mol $e^-$ / 96, 485 C × 1 mol $M$ / 2 mol $e^-$
>
> = $9.33 \times 10^{-2}$ mol $M$
>
> Molar mass of $M$ = g $M$ / mol $M$ = 5.94 g $M$ / $9.33 \times 10^{-2}$ mol $M$ = 63.7 g/mol is Cu.

This can also be shown in one step:

$$5.94 \text{ g M} \times \frac{2 \text{ mole } e^-}{1 \text{ mol } M} \times \frac{96485 \text{ C}}{10.0 \text{ C}} \times \frac{1 \text{ s}}{\text{mole } e^-} \times \frac{1 \text{ min}}{60 \text{ s}} \times \frac{1}{30.0 \text{ min}} = 63.7 \text{ g/mol}$$

## COMPARISON OF GALVANIC AND ELECTROLYTIC CELLS

Galvanic and electrolytic cells have a few features in common. For both types of cells, reduction always occurs at the cathode and oxidation at the anode. In an electrolytic cell, electrons travel from the battery to the cathode. In both cases, electrons travel in the wire, but you wouldn't say the electrons travel from the anode to the cathode in an electrolytic cell. Positive ions are always attracted to the cathode whether the cell is electrolytic or galvanic. The table below compares the galvanic cell to the electrolytic cell.

|  | Galvanic | Electrolytic |
|---|---|---|
| Sign of the cathode | + | − |
| Sign of the anode | − | + |
| Ions attracted to the cathode | Cations (+) | Cations (+) |
| Ions attracted to the anode | Anions (−) | Anions(−) |
| Sign of $E^\circ_{cell}$ | + | − |
| Spontaneity | Spontaneous | Nonspontaneous |

In both types of cells, the + ions or cations move toward the cathode because there is an excess of negative ions at the cathode caused by the reduction of + ions in solution. Likewise, oxidation at the anode produces + ions, so negative ions or anions in the salt bridge must move to the anode to maintain electrical neutrality.

## MULTIPLE-CHOICE QUESTIONS

No calculators are to be used on this part of the exam.

1. In the reaction $MnO_4^- + 5Fe^{2+} + 8H^+ \longrightarrow Mn^{2+} + 5Fe^{3+} + 4H_2O$,
   (A) $Fe^{3+}$ is oxidized and $Mn^{2+}$ is reduced.
   (B) $H^+$ is oxidized and $MnO_4^-$ is reduced.
   (C) $Fe^{2+}$ is oxidized and $MnO_4^-$ is reduced.
   (D) electrons are transferred from $Mn^{2+}$ to $Fe^{2+}$.
   (E) electrons are transferred from $H^+$ to $H_2O$.

ANSWER: C  In changing from $Fe^{2+}$ to $Fe^{3+}$, iron(II) ions lose an electron, which is oxidation; $MnO_4^-$ changing to $Mn^{2+}$ is a gain of electrons, which is reduction (*Chemistry* 6th ed. page 828 / 7th ed. page 791).

2. The $E^\circ$ for $Mg \longrightarrow Mg^{2+} + 2e^-$ is +2.37 volts.
   The $E^\circ$ for $2 Mg^{2+} + 4e^- \longrightarrow 2 Mg$ is
   (A) 2.37 volts.
   (B) −2.37 volts.
   (C) 4.74 volts.
   (D) −4.74 volts.
   (E) zero.

ANSWER: B  When the half-reaction is reversed, the sign of the potential is also reversed. However, the $E^\circ$ does not change. The potential is an intensive property, which means that it is not dependent upon how many times the reaction takes place (*Chemistry* 6th ed. pages 833–834 / 7th ed. pages 796–797).

3. The function of a salt bridge (or porous barrier) in an electrochemical cell (galvanic cell) is to allow ions to flow
   (A) though the wire from reducing agent to oxidizing agent.
   (B) to encourage charge building up on both sides of the cell.
   (C) to keep the net charge on each side at zero.
   (D) from the oxidizing side of the cell to the reducing side.
   (E) from the reducing agent side of the cell to the oxidizing agent side.

**ANSWER: C** While electrons flow through the wire from reducing agent to oxidizing agent, it is ions which flow from one side of the cell through the salt bridge to the other side to keep the net charge at zero. If this were not done, a large amount of energy would be required to 'force' electrons to move from a positive environment into a negative environment; with such charge differences due to the charge buildup on the two sides the cell would not function; the current would be zero (*Chemistry* 6th ed. pages 828–830 / 7th ed. pages 791–793).

4. Given:     $Ag^+ + 1e^- \longrightarrow Ag$         +0.80 V

   $Al^{3+} + 3e^- \longrightarrow Al$         –1.66 V

   The $E°$ for $Al + 3Ag^+ \longrightarrow Al^{3+} + 3Ag$, when arranged as a galvanic cell, is
   (A) –2.46 V.
   (B) –0.74 V.
   (C) +0.74 V.
   (D) +0.86 V.
   (E) +2.46 V.

**ANSWER: E** Reverse the $Al^{3+} \longrightarrow Al$ to give = +1.66 V, and add to the potential for $Ag^+ \longrightarrow Ag$ of +0.80 V to give +2.46 V (*Chemistry* 6th ed. pages 830–835 / 7th ed. pages 794–798).

5. Given: $Mg^{2+} + 2e^- \longrightarrow Mg$         –2.37 V

   $Fe^{3+} + 1e^- \longrightarrow Fe^{2+}$         + 0.77V.

   When the reaction $2Mg + Fe^{3+} \longrightarrow 2Mg^{2+} + Fe^{2+}$ comes to equilibrium, the $E°_{cell}$ value (cell potential) becomes
   (A) 3.91 V.
   (B) 3.14 V.
   (C) 1.60 V.
   (D) 0.83 V.
   (E) 0.00 V.

**ANSWER: E** When the cell reaction reaches equilibrium, no electrons flow and $E°_{cell}$ = zero (*Chemistry* 6th ed. pages 835–839, 843 / 7th ed. pages 798–802, 805).

6. For the reaction $Cu^{2+}(aq) + Zn(s) \longrightarrow Zn^{2+}(aq) + Cu(s)$ $E° = +1.10$ V. How many faradays are transferred as 1.00 mole of zinc is consumed?
   (A) $96{,}500 \times 2.00 \times 6.023 \times 10^{23}$
   (B) 96,500
   (C) $96{,}500 \times 2$
   (D) 2.00
   (E) zero

ANSWER: D The charge on one mole of electrons is called the faraday, ($F$, named for the Englishman Michael Faraday). In this case, two moles of electrons are transferred, hence 2.00 faradays of charge are transferred. (One faraday is 96,500 coulombs of charge, the charge on one mole of electrons.) (*Chemistry* 6th ed. pages 837–839 / 7th ed. pages 800–802)

7. The galvanic cell based on $Ag^+ + 1e^- \longrightarrow Ag$ ($E° = +0.80$ V) and $Fe^{3+} + 1e^- \longrightarrow Fe^{2+}$ ($E° = +0.77$ V)
   (A) has electrons lost by $Ag^+$ and gained by $Fe^{2+}$.
   (B) has the mass of the silver electrode decreasing as this reaction proceeds.
   (C) has an overall cell potential difference (voltage) of +1.57 V.
   (D) has electrons flowing from $Fe^{2+}$ to $Ag^+$.
   (E) has the compartment containing the $Ag^+$ as the anode.

ANSWER: D The overall equation for this cell reaction is $Fe^{2+}(aq) + Ag^+(aq) \longrightarrow Fe^{3+}(aq) + Ag(s)$ in which $Fe^{2+}$ is oxidized (loses electrons) and is therefore the anode. Because $Fe^{2+}$ loses electrons, they flow from $Fe^{2+}$ to $Ag^+$, which is reduced to $Ag(s)$, and the mass of the silver electrode increases (*Chemistry* 6th ed. pages 835–837 / 7th ed. pages 798–800).

8. Cathodic protection is a method often used to protect iron. Selecting from the list of four metals, Mg, Zn, Sn, and Cu, a chemist could protect iron with
   (A) Zn or Mg.
   (B) Sn or Cu.
   (C) Zn or Ag.
   (D) Cu or Mg.
   (E) Zn or Cu.

ANSWER: A Look for metals which are better reducing agents than iron so that the metal furnishes the electrons. Such metals have an $E°$ more negative than iron (*Chemistry* 6th ed. pages 853–854 / 7th ed. pages 814–815).

9. The Nernst equation is helpful in calculating $E°_{cell}$ values for electrochemical cells in which
   (A) the temperature is not 0°C.
   (B) the temperature is not 25°C.
   (C) the concentrations of the solutions are not all 1.00 $M$.
   (D) equilibrium has been reached.
   (E) the value of $K_C$ is very large.

ANSWER: C The Nernst equation gives the relationship between cell potential and the concentration of the cell components: $E_{cell} = E°_{cell} - (0.0591/n) \log Q$, at 25°C, where $n$ is the number of moles of electrons transferred (*Chemistry* 6th ed. pages 853–854 / 7th ed. pages 814–815).

10. Copper may be used for electroplating, with a half-reaction of $Cu^{2+} + 2e^- \longrightarrow Cu$. If a current of 10.0 A is applied to a $Cu^{2+}$ solution for 60.0 minutes, the mass of copper plated out can be calculated as
    (A) 10.0 × 3600. × 1 / 96,500 × 2.00 / 1 × 63.5.
    (B) 10.00 × 3600. × 6.02 × $10^{23}$ × 1 / 2.00 × 63.5.
    (C) 96,500 × 1 / 10.0 × 1 / 2.00 × 1 / 3600. × 63.5.
    (D) 96,500 × 1/ 10.0 × 1 / 2.00 × 1 / 3600. × 1 / 63.5.
    (E) 96,500 × 10.0 × 3600. × 2.00 × 1 / 63.5.

ANSWER: A Adding units to the style of responses where a setup is given is often very helpful. In this case it might better look like: (10.0C/s)(3600/s)(1 mole e⁻/96,500 C)(2.00 mole e⁻/1 mole Cu)(63.5g/mol Cu).
(*Chemistry* 6th ed. pages 855–857 / 7th ed. pages 816–818)

## FREE-RESPONSE QUESTIONS

Calculators may be used on this part of the exam, as well as Tables and Equations.

1. (a) Which occurs more often in nature, oxidation or reduction? Discuss your answer.
   (b) The half-reaction for the hydrogen electrode, $2H^+ + 2e^- \longrightarrow H_2$, has been given the $E°$ value of zero. Explain why such a standard is necessary.
   (c) The system $Mg(s) + Cu^{2+}(aq) \longrightarrow Mg^{2+}(aq) + Cu(s)$ has a value of $E°_{cell}$ of +2.71 V. What would be the $E°_{cell}$ value if $[Mg^{2+}] = 1.00\ M$ and $[Cu^{2+}] = 0.050\ M$ with no other changes in conditions?

ANSWER: 1(a) They occur equally; for every electron lost by one substance an electron must be gained by another substance (*Chemistry* 6th ed. page 833 / 7th ed. page 796).

ANSWER: 1(b) Because the potential of any half-reaction cannot be measured directly, some standard is required to rank the potentials of all half-reactions. Using the hydrogen half-reaction as an arbitrary standard allows such calculations. When we say, for example, that $Zn^{2+} + 2e^- \rightarrow Zn$ has a potential of –0.76 V, we really mean with respect to the potential for the hydrogen half-reaction. A good comparison might be assigning sea level as the zero measure for altitude and then comparing mountain heights and ocean depths to this standard marker (*Chemistry* 6th ed. page 831 / 7th ed. page 794).

**ANSWER: 1(c)** From the Nernst equation, $E_{cell} = E°_{cell} -(0.0591/ n)$ log $Q$, the value of $Q$ is $[Mg^{2+}] / [Cu^{2+}] = 1.00 / 0.050 = 20$. Since log $Q$ increases (log 20. = 1.3), from $E_{cell} = E°_{cell} -(0.0591/ n)$ log 20., $E_{cell}$ decreases by 0.038, to 2.67 volts (*Chemistry* 6th ed. pages 853–854 / 7th ed. pages 814–815).

2. Consider the following galvanic cell:  $Mg(s) | Mg^{2+}(aq) \| Al^{3+}(aq) | Al(s)$

   Standard Reduction Potentials :  $Mg^{2+} +2e^- \longrightarrow Mg$  –2.37 V

   $Al^{3+} +3e^- \longrightarrow Al$  –1.66 V

   (a) Give the equation for the spontaneous cell reaction which produces charge flow.
   (b) Indicate the oxidizing agent and the reducing agent in this reaction.  Label the anode and the cathode electrodes.
   (c) Calculate the $E°_{cell}$.
   (d) How will the potential difference (voltage) of the galvanic cell $Cu(s) + 2Ag^+(aq) \longrightarrow Cu^{2+}(aq) + 2Ag(s)$ be affected by the addition of NaCl(aq)?  Explain.
   (e) How does increasing the size of the anode affect the voltage? Justify your response.
   (f) The electrolysis of aqueous NaCl does not produce metallic sodium.  Explain.

**ANSWER: 2(a)**  $3Mg(s) + 2Al^{3+}(aq) \longrightarrow 3Mg^{2+}(aq) + 2Al(s)$. (*Chemistry* 6th ed. pages 833–844 / 7th ed. pages 796–806)

**ANSWER: 2(b)** Mg loses $e^-$, therefore is oxidized; it is the reducing agent and is the site of oxidation called the anode. In like manner, $Al^{3+}$, which is gaining these same electrons, is the oxidizing agent, and the Al electrode is the site of reduction, called the cathode (*Chemistry* 6th ed. pages 833–844 / 7th ed. pages 796–806).

**ANSWER: 2(c)**

> +2.37V
>
> <u>–1.66V</u>
> +0.71 V

(*Chemistry* 6th ed. pages 833–844 / 7th ed. pages 796–806)

**ANSWER: 2(d)** The added $Cl^-(aq)$ will react with the $Ag^+(aq)$, precipitating as AgCl(s).  This lowers the silver ion concentration, which shifts the equilibrium to the left, making the silver ions less available to react with the solid copper, and <u>lowers</u> the voltage.  This could also be supported via the Nernst equation, where it would have been seen that $Q$ ($[Cu^{2+}] / [Ag^+]^2$ ) is increasing, and the overall cell voltage is thus decreasing (*Chemistry* 6th ed. pages 833–844 / 7th ed. pages 796–806).

ANSWER: 2(e) If the size of the anode (site of oxidation) is changed there will be no effect on the voltage. The oxidation and reduction potentials are determined by the nature of the metal and not the amount of the metal. You might be able to "run" the cell longer by changing the size of certain electrodes, but would not produce more voltage (*Chemistry* 6th ed. pages 833–844 / 7th ed. pages 796–806).

ANSWER: 2(f) When the standard reduction potentials for the two possible reactions are compared,

$$Na^+ + 1e^- \longrightarrow Na \qquad\qquad E° = -2.71 \text{ V}$$

$$2H_2O + 2e^- \longrightarrow H_2 + 2OH^- \qquad E° = -0.83 \text{ V},$$

it is evident that water is the better competitor for electrons of these two (E° for water is the more positive value (*Chemistry* 6th ed. pages 833–844 / 7th ed. pages 796–806).

# 16

# NUCLEAR CHEMISTRY

This chapter deals with nuclear transformations and the kinetics of radioactive decay.

In this chapter, you should be able to

▪ Identify the products of nuclear transformation processes.

▪ Interconvert between the half-life of a nuclide and the amount of that nuclide remaining after a specific amount of time.

▪ Be familiar with critical mass, fusion/fission, mass defect, binding energy, nucleon, strong nuclear force, differences in penetrating power, and the ionizing ability of alpha, beta, and gamma radiation.

▪ Use a graph to predict the type of decay that will occur.

## RADIOACTIVE DECAY

Radioactive decay occurs when a nucleus undergoes decomposition to form a different nucleus and additional particles. Alpha-particle and gamma-ray production involve a change in the mass number of the decaying nucleus. The mass number (the sum of the protons and neutrons) of the decaying nucleus does not change during $\beta$-particle production.

### ALPHA-PARTICLE PRODUCTION

(*Chemistry* 6th ed. page 880 / 7th ed. pages 843–844)

An alpha particle, $\alpha$ particle, is a helium nucleus: $_2^4\text{He}$

Heavy radioactive nuclides, isotopes, undergo decay to form alpha particles.

EXAMPLE: Write a balanced equation for the decay of uranium-235 by α– particle production.

$$^{235}_{92}U \longrightarrow ^4_2He + ^{231}_{90}Th$$

The product can be identified by knowing that the total of the mass number and the atomic number must be the same on both sides of the equation. Thus the mass number for the product must be 235 – 4 = 231. The atomic number for the product must be 92 – 2 = 90 which is the atomic number for thorium.

## GAMMA-RAY PRODUCTION

(*Chemistry* 6th ed. page 880 / 7th ed. pages 843–844)

A gamma ray, γ, is a high-energy photon often produced during nuclear decay. A nucleus with excess energy can relax to its ground state by emitting a gamma ray: $^{235}_{92}U \longrightarrow ^4_2He + ^{231}_{90}Th + ^0_0\gamma$

## BETA-PARTICLE PRODUCTION

(*Chemistry* 6th ed. page 880 / 7th ed. pages 843–844)

A beta particle, β particle, is an electron represented in a nuclear reaction by the symbol: $^0_{-1}e$

The β particle has a mass number 0. Its mass is very small compared to a proton or neutron. The mass number and atomic number must be conserved as seen in alpha-particle production.

EXAMPLE: Write a balanced equation for the decay of carbon-14 by β-particle production.

$$^{14}_6C \longrightarrow ^0_{-1}e + ^{14}_7N$$

Thus the mass number for the product must be 14 – 0 = 14. The atomic number for the product must be 6 – (–1)= 7 which is the atomic number for nitrogen.

## POSITRON PRODUCTION

(*Chemistry* 6th ed. page 881 / 7th ed. page 845)

A positron is a particle which has the same mass as an electron but opposite charge.

EXAMPLE: Write a balanced equation for the decay of bismuth-83 by positron production.

$$^{205}_{83}Bi \longrightarrow ^0_{+1}e + ^{205}_{82}Pb$$

Thus the mass number for the product must be 205 – 0 = 205. The atomic number for the product must be 83 – 1 = 82, which is the atomic number for lead.

## ELECTRON CAPTURE

(*Chemistry* 6th ed. page 881 / 7th ed. page 845)

Electron capture occurs when a nucleus captures an inner orbital electron.

$$^{241}_{96}\text{Cm} + ^{0}_{-1}\text{e} \longrightarrow ^{241}_{95}\text{Am}$$

# THE KINETICS OF RADIOACTIVE DECAY

(*Chemistry* 6th ed. pages 883–886 / 7th ed. pages 846–849)

The decay of nuclides is represented by $\ln\left(\dfrac{N}{N_0}\right) = -kt$

where  $N_o$ = the original mass of nuclides
  $N$ = the mass remaining at time $t$
  $k$ = the first-order rate constant
  $t$ = time.

## HALF-LIFE

(*Chemistry* 6th ed. page 884 / 7th ed. page 847)

The half-life, $t_{1/2}$, is the time required for the number of nuclides to reach half of their original value. It can be determined by the equation:

$$t_{1/2} = 0.693/k.$$

EXAMPLE: Iodine-131, used in the diagnosis and treatment of thyroid disease, has a half-life of 8.1 days. If a patient with thyroid disease consumes a sample containing 10 μg of iodine-131, how long will it take for the amount of iodine-131 to decrease to 1/100 of the original amount?

Using the equation:  $\ln(N/N_o) = -kt$,

substitute $0.010\ N_o$ for $N$,  because $N = 0.010\ N_o$.

Also for $k$, substitute 0.693 / 8.1 days because $t_{1/2}$ = 0.693/k and $t_{1/2}$ = 8.1 days.

Solving for $t$, you get

$\ln(0.010) = -\,0.693t/8.1$ days

$t = 54$ days.

## Multiple-choice Questions

1. Isotopes of the same element are nuclides with
   (A) the same number of protons and the same atomic number (Z).
   (B) the same number of protons and the same number of neutrons.
   (C) the same mass number (A) and the same number of electrons.
   (D) the same mass number (A) and the same number of protons.
   (E) the same sum of protons and neutrons as well as the same mass number (A).

Answer: A "Isotope" is the term given to a group of nuclides that have the same atomic number, therefore the same number of protons (individual atoms are known as nuclides). For example, chlorine-35 and chlorine-37 both have an atomic number of 17, with 17 protons in the nucleus, but differ in mass due to a different number of neutrons (18 vs. 20) (*Chemistry* 6th ed. pages 877–878 / 7th ed. pages 840–841).

2. When carbon-14 undergoes radioactive decay, it forms one nitrogen-14 and one
   (A) carbon-13.
   (B) carbon-12.
   (C) nitrogen-13.
   (D) nitrogen-12.
   (E) beta particle.

Answer: E Perhaps this is most easily seen as $^{14}_{6}\text{C} \longrightarrow ^{14}_{7}\text{N} + ^{0}_{-1}\text{e}$ .

Note that both Z and A values (Z represents atomic number and A stands for atomic mass) are conserved in nuclear reactions; that is, they give the same sum on both sides of the equation (*Chemistry* 6th ed. pages 880–882 / 7th ed. pages 843–845).

3. When potassium-40 undergoes decay by electron capture, the product is
   (A) $^{39}\text{K}$
   (B) $^{40}\text{K}$
   (C) $^{41}\text{K}$
   (D) $^{39}\text{Ar}$
   (E) $^{40}\text{Ar}$

Answer: E The nuclear reaction is $^{40}_{19}\text{K} + ^{0}_{-1}\text{e} \longrightarrow ^{40}_{18}\text{Ar}$

This will only make sense to you if you understand what is meant by electron capture (*Chemistry* 6th ed. page 881 / 7th ed. page 845).

4. When aluminum undergoes bombardment by an alpha particle, the products are phosphorus-30 and
   (A) a proton.
   (B) an electron.
   (C) a neutron.
   (D) a helium nucleus.
   (E) a 2+ nucleus.

ANSWER: C The nuclear reaction is $^{27}_{13}Al + ^4_2He \longrightarrow ^{30}_{15}P + ^1_0n$.

Note that $^1_0n$ is the symbol for a neutron, with a mass = 1 and a charge = 0 (*Chemistry* 6th ed. page 880 / 7th ed. pages 843–844).

5. The atomic number is decreased by one by
   (A) electron capture only.
   (B) positron emission only.
   (C) both electron capture and positron emission.
   (D) both gamma and beta emission.
   (E) proton bombardment.

ANSWER: C The process of electron capture involves an inner orbital electron being captured by the nucleus, causing the atomic number to decrease by one. Positron production involves emission of a particle which has the same mass as an electron but the opposite charge (*Chemistry* 6th ed. pages 880–881 / 7th ed. pages 843–844).

6. When $^{214}_{82}Pb$ undergoes decay producing two beta particles, the other product is
   (A) $^{214}_{82}Pb$.
   (B) $^{214}_{84}Po$.
   (C) $^{214}_{80}Pb$.
   (D) $^{212}_{84}Po$.
   (E) $^{214}_{82}Po$.

ANSWER: B This time be sure that you understand that the beta particle is $^0_{-1}e$, and that you form two of them: $^{214}_{82}Pb \longrightarrow ^0_{-1}e + ^0_{-1}e + ^{214}_{84}Po$. (*Chemistry* 6th ed. page 880 / 7th ed. pages 843–844)

7. Referring to the graph which considers neutron to proton ratios, you can conclude that when the neutron to proton ratio is too low
   (A) positron emission is likely.
   (B) beta emission is likely.
   (C) positron will be absorbed by the nucleus.
   (D) alpha particle production is expected.
   (E) electron capture is expected.

ANSWER: **A** If the neutron/proton is too low, this suggests too many protons, which is an unstable condition. This condition can be relieved by positron production. You will find more on this topic in free-response question #1 (*Chemistry* 6th ed. pages 878–880 / 7th ed. pages 841–843).

8. The number of particles of a given nuclide decays from 120 units to 15 units in 60. minutes, indicating a half-life of
   (A) 10. minutes.
   (B) 15. minutes.
   (C) 20. minutes.
   (D) 30. minutes.
   (E) 60. minutes.

ANSWER: **C** The change in the number of nuclide particles from 120 to 60 to 30 to 15 units suggests a change through three half-lives. Since three half-lives took 60. minutes, one half-life would be 60./3 = 20. minutes (*Chemistry* 6th ed. pages 883–884 / 7th ed. pages 846–847).

9. The number of neutrons emitted when U-235 is bombarded with a neutron, forming Sm-160 and Zn-72, is
   (A) 1.
   (B) 2.
   (C) 3.
   (D) 4.
   (E) 5.

ANSWER: **D**

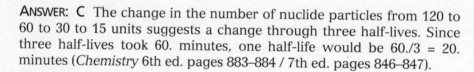

$^{235}_{92}U + ^{1}_{0}n \longrightarrow ^{160}_{62}Sm + ^{72}_{30}Zn + ?^{1}_{0}n$. Since the total mass on the right equals the total mass on the left = 236 and Sm – 160 + Zn – 72 contribute a mass of (160 + 72) = 232, four neutrons are necessary to balance this expression. Note that a neutron, usually shown as $^{1}_{0}n$, has a mass of one and a charge of zero (*Chemistry* 6th ed. pages 880–882 / 7th ed. pages 843–845).

10. U-238 emits an alpha particle to produce a nucleus which then decays by emitting a beta particle. The mass number and atomic number of the two successive products are
    (A) 90-234 and 91-235.
    (B) 90-234 and 91-234.
    (C) 91-233 and 91-234.
    (D) 90-234 and 92-238.
    (E) 92-234 and 89-234.

ANSWER: B

$$^{238}_{92}U \longrightarrow {}^{4}_{2}He + {}^{234}_{90}X$$

$$^{234}_{90}X \longrightarrow {}^{0}_{-1}e + {}^{234}_{91}Y$$

(*Chemistry* 6th ed. pages 880–882 / 7th ed. pages 843–845)

# FREE-RESPONSE QUESTIONS

1. Refer to the graph in multiple-choice question #7: the graph of the number of neutrons plotted vs. the number of protons shows the "zone of stability."
   (a) What is meant by the "zone of stability?
   (b) What are the characteristics of those nuclides found in the "zone of stability"?
   (c) Discuss ways in which such a graph is helpful to the nuclear chemist.

ANSWER 1(a): This is the area in which stable nuclides reside.

ANSWER 1(b): As light nuclides, the number of neutrons and protons are about equal. With more massive nuclides the number of neutrons must exceed the number of protons. Nuclides with an even number of both neutrons and protons seem to have a higher incidence of stability (you could also note the so called "magic numbers" of 2, 8, 20, 28, 50, etc. for the number of protons in very stable nuclides).

ANSWER 1(c): The comparison of the stability of a given nuclide to the neutron to proton ratio helps to predict the nature of decay that unstable substances will most likely experience. Either beta emission or positron emission usually leads to the formation of a more stable substance (*Chemistry* 6th ed. pages 878–879 / 7th ed. pages 841–843).

2. Contrast nuclear reactions with ordinary chemical reactions. Include in your discussion or outline
   (i) the effects of temperature, pressure, concentration, and the addition of a catalyst;
   (ii) the relative amounts of energy involved; and
   (iii) what is changed or converted in each type of reaction.

ANSWER 2i: Chemical Reactions: The rate of reaction is influenced by changes in temperature, pressure, concentration, and by the addition of a suitable catalyst.

Nuclear Reactions: These are not influenced by changes in pressure, temperature, concentration, or the addition of a catalyst. (Note: Concentration may affect the amount of available critical mass in certain cases).

Both types of reactions can be very fast, very slow, or somewhere between these extremes.

**Answer 2ii:** Chemical Reactions: These usually involve a process that gains or loses around 400 kJ or less.

Nuclear Reactions: These can involve very large amounts of energy, perhaps as much as $10^{10}$ kJ.

**Answer 2iii:** Chemical Reactions: These reactions involve gaining, losing, or sharing electrons. The outermost electron configurations are most important.

Nuclear Reactions: As the name suggests, these involve rearrangement of the nucleus of the atom. Often a different nucleus is formed through a process called radioactive decay. The relative number of protons and neutrons is especially important (*Chemistry* 6th ed. pages 877–881 / 7th ed. pages 840–844).

# Part III

## Practice Tests

Noble gases 18 8A

Halogens → 17 7A

metals → ← nonmetals

Alkaline earth metals → 2 2A

Alkali metals

Transition metals

| Group | 1 1A | 2 2A | 3 | 4 | 5 | 6 | 7 | 8 | 9 | 10 | 11 | 12 | 13 3A | 14 4A | 15 5A | 16 6A | 17 7A | 18 8A |
|---|---|---|---|---|---|---|---|---|---|---|---|---|---|---|---|---|---|---|
| | 1 H 1.008 | | | | | | | | | | | | | | | | | 2 He 4.003 |
| | 3 Li 6.941 | 4 Be 9.012 | | | | | | | | | | | 5 B 10.81 | 6 C 12.01 | 7 N 14.01 | 8 O 16.00 | 9 F 19.00 | 10 Ne 20.18 |
| | 11 Na 22.99 | 12 Mg 24.31 | | | | | | | | | | | 13 Al 26.98 | 14 Si 28.09 | 15 P 30.97 | 16 S 32.07 | 17 Cl 35.45 | 18 Ar 39.95 |
| | 19 K 39.10 | 20 Ca 40.08 | 21 Sc 44.96 | 22 Ti 47.88 | 23 V 50.94 | 24 Cr 52.00 | 25 Mn 54.94 | 26 Fe 55.85 | 27 Co 58.93 | 28 Ni 58.69 | 29 Cu 63.55 | 30 Zn 65.38 | 31 Ga 69.72 | 32 Ge 72.59 | 33 As 74.92 | 34 Se 78.96 | 35 Br 79.90 | 36 Kr 83.80 |
| | 37 Rb 85.47 | 38 Sr 87.62 | 39 Y 88.91 | 40 Zr 91.22 | 41 Nb 92.91 | 42 Mo 95.94 | 43 Tc (98) | 44 Ru 101.1 | 45 Rh 102.9 | 46 Pd 106.4 | 47 Ag 107.9 | 48 Cd 112.4 | 49 In 114.8 | 50 Sn 118.7 | 51 Sb 121.8 | 52 Te 127.6 | 53 I 126.9 | 54 Xe 131.3 |
| | 55 Cs 132.9 | 56 Ba 137.3 | 57 La* 138.9 | 72 Hf 178.5 | 73 Ta 180.9 | 74 W 183.9 | 75 Re 186.2 | 76 Os 190.2 | 77 Ir 192.2 | 78 Pt 195.1 | 79 Au 197.0 | 80 Hg 200.6 | 81 Tl 204.4 | 82 Pb 207.2 | 83 Bi 209.0 | 84 Po (209) | 85 At (210) | 86 Rn (222) |
| | 87 Fr (223) | 88 Ra 226 | 89 Ac† (227) | 104 Rf | 105 Db | 106 Sg | 107 Bh | 108 Hs | 109 Mt | 110 Ds | 111 Uuu | 112 Uub | | 114 Uuq | | | | |

*Lanthanides

| 58 Ce 140.1 | 59 Pr 140.9 | 60 Nd 144.2 | 61 Pm (145) | 62 Sm 150.4 | 63 Eu 152.0 | 64 Gd 157.3 | 65 Tb 158.9 | 66 Dy 162.5 | 67 Ho 164.9 | 68 Er 167.3 | 69 Tm 168.9 | 70 Yb 173.0 | 71 Lu 175.0 |
|---|---|---|---|---|---|---|---|---|---|---|---|---|---|

†Actinides

| 90 Th 232.0 | 91 Pa (231) | 92 U 238.0 | 93 Np (237) | 94 Pu (244) | 95 Am (243) | 96 Cm (247) | 97 Bk (247) | 98 Cf (251) | 99 Es (252) | 100 Fm (257) | 101 Md (258) | 102 No (259) | 103 Lr (260) |
|---|---|---|---|---|---|---|---|---|---|---|---|---|---|

Group numbers 1–18 represent the system recommended by the International Union of Pure and Applied Chemistry.

# Practice Test 1

These questions are representative of the AP Chemistry examination, but keep in mind that it is impossible to predict exactly how well you will do on the actual exam. The first section of this test is 45% of your total test grade. Time yourself so that you finish this part in 90 minutes. Remember that on the actual examination, there is a guess factor; scoring is the number of questions answered correctly minus one-fourth the number wrong.

## AP CHEMISTRY EXAMINATION
### Section I: Multiple-Choice Questions
### Time: 90 minutes
### Number of Questions: 75

No calculators are to be used in this section; no tables (except periodic table) permitted.

## Part A

Directions: The following questions consist of five lettered headings followed by a listing of numbered phrases. For each phrase you are to select the one heading that is most closely related to it. Headings may be used once, more than once, or not at all.

Questions 1–4:
(A)     0.10 M $Ca(NO_3)_2$
(B)     0.10  M $NaC_2H_3O_2$
(C)     0.10 M $CuSO_4$
(D)     0.10 M $HC_2H_3O_2$
(E)     0.10 M $C_2H_5OH$

1.  Which of the above would be expected to have the lowest pH?

2.  Which of the above would be expected to be colored?

3.  Which of the above would be expected to have the lowest freezing temperature?

4.  Which of the above would be expected to have the highest pH?

Questions 5–7
(A)     $XeF_4$
(B)     $PCl_4^-$
(C)     Zn
(D)     Fe
(E)     $F_2O$

5.  Which of the above would be expected to have a tetrahedral shape?

6.  Which of the above would be expected to form a protective coating in the galvanizing of steel?

7.  Which of the above would be expected to have a bent (angular) shape?

**GO ON TO NEXT PAGE**

# Part B

<u>Directions:</u> Each of the questions or incomplete statements below is followed by five suggested answers or completions. Select the one that is best in each case.

8. Ideal gases differ from real gases in which of the following ways?
   I.   Ideal gases have zero molecular volume.
   II.  Ideal gases have no intermolecular forces.
   III. $PV = nRT$ describes both ideal and real gases at all temperatures and pressures.
   (A) I only
   (B) II only
   (C) III only
   (D) I and II only
   (E) I, II, and III

9. The VSEPR model predicts that the structure of $XeF_4$ is square planar. This is due to
   (A) the tendency of four electron pairs always to form a square.
   (B) the tendency of four electron pairs always to form a tetrahedron.
   (C) bond angles of 90°.
   (D) the high degree of repulsion between the two lone pairs.
   (E) a high nuclear charge.

10. Which of the following salts will form a basic solution in water solution?
    (A) KCl
    (B) $Na_2SO_4$
    (C) $CuCl_2$
    (D) $Na_2CO_3$
    (E) $NH_4NO_3$

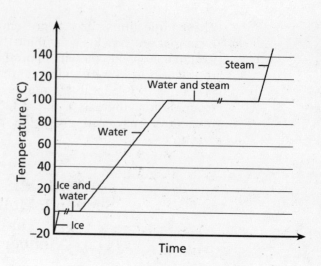

11. A heating curve for water (ice → water → steam) shows that the slope in the section for water heating is different than the slope for steam heating. This difference in slope occurs because
    (A) it takes more energy to vaporize water than to melt ice.
    (B) little energy is needed to cool steam at 100°C.
    (C) little energy is needed to cool ice at 0°C.
    (D) water and steam have different molar heat capacities.
    (E) raising the temperature of ice is not involved.

12. The phase diagram for water is not typical of most substances. Carbon dioxide has a phase diagram more typical of most substances. The main nontypical feature of the phase diagram for water is
(A) the critical point is at a higher temperature than the triple point.
(B) the triple point is below a pressure of one atm.
(C) the solid/liquid line extends indefinitely.
(D) water can exist in all three physical states.
(E) the solid/liquid line has a negative slope.

13. A sample of helium gas has a volume of 5.6 L at STP. The number of moles of He present are
(A) 0.25 mole.
(B) 0.50 mole.
(C) $0.25 \times 6.02 \times 10^{23}$.
(D) $0.50 \times 6.02 \times 10^{23}$.
(E) $4.0 \times 6.02 \times 10^{23}$.

14. The $K_{sp}$ expression for lead (II) bromide is
(A) $[Pb^{2+}][Br^-]$.
(B) $[Pb^{2+}][2Br^-]$.
(C) $[Pb^{2+}]^2[Br^-]$.
(D) $[Pb^{2+}][Br^-]^2$.
(E) $[Pb^{2+}]^2[Br^-]^2$.

15. Grinding an ionic salt into a fine powder accomplishes which of the following:
I. increases its solubility.
II. increases the value of the $K_{sp}$.
III. increases the rate of dissolving.
(A) I only
(B) II only
(C) III only
(D) I and II only
(E) I, II, and III

16. It is possible to predict if two ionic solutions will form a precipitate by noting
(A) that the ion product ($Q$) is a high value.
(B) that the ion product ($Q$) is a low value.
(C) that the ion product ($Q$) exceeds the $K_{sp}$ value.
(D) that the solubility product exceeds the ion product.
(E) temperature changes as a solid forms.

17. A buffered solution is one which can be described by which of the following:
I. a solution which resists pH change.
II. may contain a weak acid and its salt.
III. contains species which can react with both $H^+$ and $OH^-$.
(A) I only
(B) II only
(C) III only
(D) I and II only
(E) I, II, and III

18. Entropy is greater in which of the following:
I. solid salt than salt solution.
II. ice formed on a cold window than water vapor.
III. graphite rather than diamond.
(A) I only
(B) II only
(C) III only
(D) I and II only
(E) I, II, and III

19. The vapor pressure of a heavier substance (MM) will be lower than that of a lighter substance, all other things being equal, because
(A) of weaker intermolecular forces.
(B) the molecules are more polarizable.
(C) hydrogen "bonding" isn't considered.
(D) vapor pressure is a constant for a given substance.
(E) the enthalpy of vaporization is lower.

**GO ON TO NEXT PAGE**

20. For the reaction $O_2(g) \rightarrow 2O(g)$
    (A) $\Delta H$ would be +, $\Delta S$ would be +.
    (B) $\Delta H$ would be –, $\Delta S$ would be –.
    (C) $\Delta H$ would be +, $\Delta S$ would be –.
    (D) $\Delta H$ would be –, $\Delta S$ would be +.
    (E) would be spontaneous at low temperatures.

21. The purpose of the salt bridge in an electrochemical cell is
    (A) to make possible electron flow through the external circuit.
    (B) to provide for ion flow through the cell.
    (C) to provide for oxidation at the cathode.
    (D) to provide for reduction at the anode.
    (E) to reach equilibrium sooner.

22. A concentration cell of Ag / Ag$^+$ has Ag$^+$ of 1.0 M on the right side of the cell and 0.100 M Ag$^+$ on the left side.
    (A) Electrons would flow (external circuit) from left to right.
    (B) The Ag on the left side is the cathode.
    (C) The Ag on the right side will decrease in mass.
    (D) The solution will gradually turn light blue.
    (E) Electrons would flow internally in the cell from left to right.

23. In the half-reaction, $AgBrO_3 \rightarrow Ag + BrO_3^-$, the number of electrons
    (A) lost is 1.
    (B) lost is 2.
    (C) gained is 1.
    (D) gained is 2.
    (E) gained is 3.

24. Of the following salts, which is insoluble?
    (A) $Na_2S$
    (B) $Pb(NO_3)_2$
    (C) $KCl$
    (D) $CaCl_2$
    (E) all of the above are soluble

25. To neutralize a solution of NaOH and use the least volume of acid, you should select
    I.   1.0M HCl.
    II.  1.0 M $HC_2H_3O_2$.
    III. 1.0 M $H_2SO_4$.
    (A) I only.
    (B) II only.
    (C) III only.
    (D) I and II use equal and lesser volumes.
    (E) I, II, and III use the same volumes of acids.

26. The concentration of ions in 1.2 M aluminum nitrate solution is
    (A) 0.40 $M$.
    (B) 1.2 $M$.
    (C) 2.4 $M$.
    (D) 3.6 $M$.
    (E) 4.8 $M$.

27. The gas with the most rapid effusion rate is
    (A) $NH_3(g)$.
    (B) $HCl(g)$.
    (C) $Cl_2(g)$.
    (D) $CH_4(g)$.
    (E) $C_3H_8(g)$.

28. For which of the following is the sign of the enthalpy change different from the other four?
    (A) $2KClO_3(s) \rightarrow 2KCl(s) + 3O_2(g)$
    (B) $H_2O(l) \rightarrow H_2O(s)$
    (C) $Br_2(g) \rightarrow 2Br(g)$
    (D) $2HgO(s) \rightarrow 2Hg(l) + O_2(g)$
    (E) $Cl(g) + e^- \rightarrow Cl^-(g)$

29. Which of the following pairs illustrates the Law of Multiple Proportions?
(A) KCl, KBr
(B) $P_2O_5$, $N_2O_5$
(C) NO, $NO_2$
(D) $CO_2$, $SiO_2$
(E) $NH_4F$, $NH_4I$

30. The geometry of carbon disulfide, $CS_2$, can be described by the VSEPR model as
(A) bent or angular.
(B) linear.
(C) trigonal planar.
(D) trigonal pyramid.
(E) tetrahedral.

31. A hydrocarbon, $C_4H_{10}$, has a critical temperature above room temperature. This means that
(A) it can never be liquified.
(B) the intermolecular forces are strong enough to form a liquid when pressure is applied.
(C) the critical pressure is also high.
(D) it will exist as a liquid at room conditions.
(E) it is extremely flammable.

32. Instantaneous reaction rate
I. may not be constant over time.
II. can be calculated from slope of a line tangent to a concentration vs. time curve.
III. always has a positive value.
(A) I only
(B) II only
(C) III only
(D) I and II are both valid statements
(E) I, II, and III are all valid statements

33. One function of a catalyst is to
(A) shift the equilibrium toward products.
(B) lower the activation energy.
(C) increase the value of $\Delta H$.
(D) lower the value of the rate constant.
(E) make the reverse reaction less spontaneous.

34. A 1.0 M solution of which of the following is orange?
(A) KCl
(B) $K_2Cr_2O_7$
(C) $K_2CrO_4$
(D) $CuSO_4$
(E) $FeCrO_4$

35. Two solutions are mixed: 1.0 L of 0.20 M NaOH and 1.0 L of 0.10 M $H_3PO_4$. After reaction, the ions in the largest concentration are $Na^+$ and
(A) $OH^-$.
(B) $H^+$, $(H_3O^+)$.
(C) $PO_4^{3-}$.
(D) $HPO_4^{2-}$.
(E) $H_2PO_4^-$.

36. The hybridization of carbon in acetylene, $C_2H_2$, a triple bonded substance is
(A) sp hybrid orbitals.
(B) $sp^2$ hybrid orbitals.
(C) $sp^3$ hybrid orbitals.
(D) $dsp^3$ hybrid orbitals
(E) $d^2sp^3$ hybrid orbitals.

37. $2Al + 6HCl \rightarrow 3H_2 + 2AlCl_3$
Aluminum reacts with hydrochloric acid, as indicated in the equation above, to produce hydrogen gas. The $H_2$ produced was then collected by water displacement at 27°C (V.P. $H_2O$ = 21 torr) and a barometric pressure of 757 torr. If 0.555 L of gas is collected, the partial pressure of the hydrogen is
(A) $0.555 \times (273+27)$.
(B) $0.555 \times 757$.
(C) $757 - 27$.
(D) $757 - 21$.
(E) $0.555 \times 300 / 273$.

38. A saturated solution of magnesium hydroxide has added to it dropwise a solution of pH = 4. This results in
(A) a precipitate forming.
(B) much bubbling and frothing occurring.
(C) magnesium metal precipitates.
(D) the solution becoming increasingly basic.
(E) water forming.

**GO ON TO NEXT PAGE**

39. The volume of base at the equivalence point on the titration curve of an acid with a base is determined by
    (A) the strength of the acid.
    (B) the relative strength of the acid and the base.
    (C) the strength of the base used.
    (D) the $K_a$ value for the acid.
    (E) the amount of acid and base added at that point.

40. Knowing only the temperature and the $K_{sp}$ value will not provide the relative solubility of ionic salts because
    (A) different salts sometimes produce different number of ions when dissolving.
    (B) solubility does not consider lattice energy effects.
    (C) not all ionic salts are soluble.
    (D) the two ions formed are of different sizes.
    (E) the two ions formed have different charges.

41. The $K_a$ for acetic acid, $HC_2H_3O_2$, is $1.8 \times 10^{-5}$ at 25°C. In a 0.10 M solution of potassium acetate, which of the following are true?
    I.   $[HC_2H_3O_2] = [OH^-]$
    II.  $[H^+] = [OH^-]$
    III. $[C_2H_3O_2^-] < 0.10$ M.
    (A) I only
    (B) II only
    (C) III only
    (D) I and II
    (E) I, II, and III

42. The kinetic molecular theory indicates that temperature is best considered
    (A) the kinetic energy of gas particles.
    (B) the random motion of gas particles.
    (C) a function of the Celsius scale.
    (D) equal to the Kelvin scale.
    (E) an indicator of the average kinetic energy of gas particles.

43. Silver chromate solid exists in equilibrium with $Ag^+$ and $CrO_4^{2-}$:

    $Ag_2CrO_4(s) \rightleftharpoons 2Ag^+(aq) + CrO_4^{2-}(aq)$.

    If silver nitrate is added to this solution
    (A) a precipitate of $Ag_2CrO_4$ forms.
    (B) the solution gets increasingly blue.
    (C) the concentration of $CrO_4^{2-}$ increases.
    (D) the $[Ag_2CrO_4]$ increases.
    (E) the common ion effect does not apply.

44. 4.00 mmol of a weak acid is mixed with 2.00 mmol of NaOH and the pH is determined to be 5.00. The value of $K_a$ for this weak acid is
    (A) $4.00 / 2.00 \times 5.00$.
    (B) $4.00 / 2.00 \times 10^5$.
    (C) $2.00 / 4.00 \times 10^{-5}$.
    (D) $\log (10 \times 10^{-5})$.
    (E) $1.0 \times 10^{-5}$.

45. For which of the following equations is the sign of entropy as shown?
    (A) $Na(s) + 1/2\ Cl(g) \rightarrow NaCl(s)$ is (+).
    (B) $H_2(g) + 3N_2(g) \rightarrow 2NH_3(g)$ is (+)
    (C) $Hg(l) \rightarrow Hg(g)$ is (+).
    (D) $CO_2(s) \rightarrow CO_2(g)$ is (−).
    (E) $C_6H_{12}O_6(aq) \rightarrow C_6H_{12}O_6(s)$ is (+).

46. At what temperature will a process with $\Delta H = -10$ kJ and $\Delta S = -100$ J/K be spontaneous?
    (A) any temperature under 100K.
    (B) any temperature over 100K.
    (C) any temperature under 10 K.
    (D) any temperature under 0.10K.
    (E) any temperature over −100K.

47. The $\Delta G$ for $S(s) + O_2(g) \rightarrow SO_2(g)$, from $S(s) + 1\ 1/2\ O_2(g) \rightarrow SO_3(g)$ $\Delta G = -370$ kJ and $O_2(g) + 2SO_2(g) \rightarrow 2SO_3(g)$ $\Delta G = -140$ kJ is
    (A) +510 kJ.
    (B) −510 kJ.
    (C) +300 kJ.
    (D) −300 kJ.
    (E) −230 kJ.

48. The process with the greatest rate of reaction would be
(A) $\Delta G = -30$ kJ.
(B) $\Delta G = -60$ kJ.
(C) $\Delta G = +30$ kJ.
(D) $\Delta G = +60$ kJ.
(E) impossible to predict from only $\Delta G$ values.

49. The $E^o = -0.60$ V for $FeS(s) \rightarrow Fe^{2+}(aq) + S^{-2}(aq)$. This indicates that
(A) FeS has very low solubility.
(B) $K_{sp}$ for FeS is very large.
(C) $S^{-2}$ ions are unstable.
(D) $Fe^{2+}$ ions are unstable.
(E) both $S^{2-}$ and $Fe^{2+}$ ions are unstable.

50. The net equation for an electrochemical cell made with $Al^{3+}(aq)$ and $Mg(s)$ as reactants is
(A) $Al(s) + Mg(s) \rightarrow Al^{3+}(aq) + Mg^{2-}(aq)$.
(B) $Al^{3+}(aq) + Mg(s) \rightarrow Al(s) + Mg^{2-}(aq)$.
(C) $3Al^{3+}(aq) + 2Mg(s) \rightarrow 2Mg^{2+}(aq) + 3Al^{3+}(aq)$.
(D) $2Al^{3+}(aq) + Mg(s) \rightarrow 2Al(s) + Mg^{2+}(aq)$.
(E) $2Al^{3+}(aq) + 3Mg(s) \rightarrow 3Mg^{+2}(aq) + 2Al(s)$.

51. An aqueous lead (II) chloride solution is to undergo electrolysis to deposit the lead. To determine how long this process will take, you must also be given
(A) the mass of $PbCl_2$ and the current available.
(B) the number of moles of electrons transferred and the voltage available.
(C) the number of coulombs required to deposit one mol of lead.
(D) the half reaction for the lead ion/lead pair and the voltage.
(E) the mass of the $PbCl_2$ and the voltage measured.

Questions 52 and 53 refer to the electrochemical cell:
$$Zn(s) \mid Zn^{2+}(aq) \parallel Ag^+(aq) \mid Ag(s).$$

52. The anode for this cell is the
(A) Zn.
(B) $Zn^{2+}$.
(C) $Ag^+$.
(D) Ag.
(E) electrode undergoing reduction.

53. As this cell functions
(A) electrons flow from the Zn electrode to the Ag electrode.
(B) ions flow from the $Zn^{2+}$ side to the $Ag^+$ side.
(C) the mass of the Ag electrode decreases.
(D) the $[Ag^+]$ increases.
(E) $E^o_{cell} = 0$.

54. To calculate the molarity of a solution, it is necessary to know
I.   the mass of the solute
II.  molar mass of the solute
III. volume of water added
IV.  total volume of the solution.
(A) I only
(B) II only
(C) III only
(D) I, II, and III only
(E) I, II, and IV only

**GO ON TO NEXT PAGE**

55. When 1.0 L of 1.0 $M$ lead (II) nitrate is added to 1.0 L of 1.0 $M$ potassium nitrate the concentration of ions in solution is
    (A) 1.0 $M$ $Pb^{2+}$, 1.0 $M$ $NO_3^-$, 1.0$M$ $K^+$.
    (B) 1.0 $M$ $Pb^{2+}$, 2.0 $M$ $NO_3^-$, 1.0$M$ $K^+$.
    (C) 0.50 $M$ $Pb^{2+}$, 2.0 $M$ $NO_3^-$, 0.50$M$ $K^+$.
    (D) 0.50 $M$ $Pb^{2+}$, 1.5 M $NO_3^-$, 0.50$M$ $K^+$.
    (E) 0.50 $M$ $Pb^{2+}$, 0.50 $M$ $NO_3^-$, 0.50$M$ $K^+$.

56. Naturally occurring copper is composed of two isotopes, $^{63}Cu$ and $^{65}Cu$. If copper is 69.1% $^{63}Cu$ and 30.9% $^{65}Cu$, the average atomic mass could be calculated as
    (A) (30.9 + 69.1)/2.
    (B) (63 + 65)/2.
    (C) ($65 \times 30.9$) + ($63 \times 69.1$).
    (D) ($0.309 \times 65$) + $0.691 \times 63$).
    (E) [($0.309 \times 65$) + $0.691 \times 63$)]/2.

57. The number of atoms in a sliver of calcium of mass 5.234 mg may be calculated as
    (A) (5.234/1000) $\times$ (1/40.08) $\times$ ($6.022 \times 10^{23}$).
    (B) ($5.234 \times 40.08$) $\times$ ($6.022 \times 10^{23}$).
    (C) (40.08/5.234) $\times$ ($6.022 \times 10^{23}$).
    (D) ($6.022 \times 10^{23}$) $\times$ 40.08 $\times$ 1000.
    (E) 5.234 $\times$ ($6.022 \times 10^{23}$/40.08).

58. The Arrhenius definition of bases is
    (A) a substance that produces $H^+$ in water solution.
    (B) a substance that produces protons in water solution.
    (C) a substance that produces $OH^-$ in water solution.
    (D) a substance that acts as a proton donor in any solution.
    (E) a substance that acts as a proton acceptor in any solution.

59. A weak electrolyte
    (A) forms little electrical energy.
    (B) is only slightly soluble.
    (C) is never a base.
    (D) forms few ions when dissolving.
    (E) all of the above apply.

60. Rate laws can be written
    (A) only from experimental data.
    (B) only for first-order reactions.
    (C) directly from the coefficients of the reactants.
    (D) as a ratio of the coefficients of reactants to coefficients of products.
    (E) only if the rate constant is known.

61. The solubility of AgCl in HCl
    (A) is low, as $Cl^-$ ions cause an equilibrium shift toward the solid form.
    (B) is low, as $Ag^+$ has already reacted with $Cl^-$.
    (C) is high, as HCl readily ionizes.
    (D) is high, as AgCl can then form.
    (E) is high, as $H_2$ forms, then escapes.

Questions 62–63 refer to the reaction: A + 2B $\longrightarrow$ C + 2D, and these data:

| Exp. No. | Initial[A] | Initial [B] | Initial Rate $\Delta[A]/\Delta t$ in mol/L•s. |
|----------|-----------|-------------|------------------------------------------------|
| 1 | 0.100M | 0.050 M | $7.00 \times 10^{-6}$ |
| 2 | 0.200 M | 0.050 M | $14.00 \times 10^{-6}$ |
| 3 | 0.100 M | 0.100 M | $7.00 \times 10^{-6}$ |

62. The rate law for this reaction is
(A) Rate = k[A][B].
(B) Rate = k[A][B]$^2$.
(C) Rate = k[A].
(D) Rate = k[B].
(E) Rate = k[B]$^2$.

63. For Experiment 2, the rate of appearance of product D is
(A) $3.50 \times 10^{-6}$ mol/L•s.
(B) $7.00 \times 10^{-6}$ mol/L•s.
(C) $14.0 \times 10^{-6}$ mol/L•s.
(D) $28.0 \times 10^{-6}$ mol/L•s.
(E) cannot be determined from these data.

64. The equations $\Delta G^\circ = -RT \ln K$ and $\Delta G^\circ = H^\circ - TS^\circ$ can be written as $\ln K = -\Delta H^\circ/RT + \Delta S^\circ/R$. This means that if ln $k$ is plotted versus $1/T$, the results will be linear and the slope will be
(A) $\Delta H^\circ/R$.
(B) $-\Delta H^\circ/R$.
(C) $\Delta S^\circ/R$.
(D) $-\Delta S^\circ/R$.
(E) ln $K$.

65. In all cases, a reaction is spontaneous if
(A) $\Delta S < 0$ and $\Delta H > 0$.
(B) $\Delta S > 0$ and $\Delta H > 0$.
(C) $\Delta S = \Delta H$.
(D) $\Delta S > 0$ and $\Delta H < 0$.
(E) $\Delta S < 0$ and $\Delta H < 0$.

66. $PbSO_4(s) + H_2O (l) \rightarrow PbO_2(s) + Pb(s) + H^+(aq) + SO_4^{2-}(aq)$. In the above reaction the oxidizing agent is
(A) Pb.
(B) $PbO_2$.
(C) $H_2O$.
(D) $SO_4^{2-}$.
(E) $PbSO_4$.

67. The atomic mass for carbon given in the Periodic Table is 12.01, rather than 12.00, because
(A) carbon is the basis for all relative masses.
(B) natural occurring forms of $^{12}C$, $^{13}C$, and $^{14}C$ exist.
(C) all isotopes of carbon have six protons.
(D) carbon has many atoms of mass 12.01 a.m.u.
(E) a mole of carbon is $6.022 \times 10^{23}$ atoms.

68. The iron shaft on a sailboat engine may be protected from corrosion by using a zinc collar. The zinc is called
(A) the sacrificial anode.
(B) the sacrificial cathode.
(C) surface reducer.
(D) catalytic protector.
(E) a galvanized metal.

69. The half-reaction with a standard reduction potential of zero is
(A) $H^+(aq) + 1e^- \rightarrow H(g)$.
(B) $H^+(aq) + 2e^- \rightarrow H_2(g)$.
(C) $2H^+(aq)(1.0M) + 2e^- \rightarrow H_2(g)(1.0$ atm).
(D) $Zn(s) - 2e^- \rightarrow Zn^{2+}(aq)(1M)$.
(E) $Zn^{2+}(1M) + 2e^- \rightarrow Zn(s)$.

70. For the reaction X$\rightarrow$ Y, the rate for this second-order reaction has been determined to be $6.0 \times 10^{-4}$ mol/L•s. when the value of [X] = 0.22 $M$. If the concentration of X is 0.66 $M$, the rate, in mol/L•s., will be
(A) $6.0 \times 10^{-4}$.
(B) $(6.0 \times 10^{-4})/9$.
(C) $(6.0 \times 10^{-4}) \times 3$.
(D) $(6.0 \times 10^{-4})^2$.
(E) $(6.0 \times 10^{-4}) \times 9$.

**GO ON TO NEXT PAGE**

71. The phrase "like dissolves like" refers to
    (A) molecules of like electronegativity differences.
    (B) molecules of like polarity.
    (C) molecules of like sharing of electron pairs within a bond.
    (D) compounds soluble in each other.
    (E) all of the above.

72. The oxidation state assigned to nitrogen in $NO_3^-$ is
    (A) +1.
    (B) +2.
    (C) +5.
    (D) +6.
    (E) −1.

73. The 1.0 $M$ solution which provides the fewest ions in solution is
    (A) $CH_3COOH$.
    (B) $Ca(NO_3)_2$.
    (C) $NaNO_3$.
    (D) $NaCl$.
    (E) $H_2SO_4$.

74. One liter quantities of 1.0 $M$ dilute acids would usually be prepared in the laboratory, to be used as stock solutions, using
    (A) a buret.
    (B) a balance.
    (C) a test tube.
    (D) a volumetric flask.
    (E) all of the above.

75. Consider two solutions, Solution X with a pH of 7 and Solution Y with a pH of 9. The hydrogen ion ratio would be
    (A) X/Y = 7/9.
    (B) X/Y = 2/1.
    (C) X/Y = 1/3.
    (D) X/Y = 100/1.
    (E) X/Y = 1/100.

# A5.5 Standard Reduction Potentials at 25°C (298 K) for Many Common Half-Reactions

| Half-Reaction | e° (V) | Half-Reaction | e° (V) |
|---|---|---|---|
| $F_2 + 2e^- \rightarrow 2F^-$ | 2.87 | $Cu^+ + e^- \rightarrow Cu$ | 0.52 |
| $Ag^{2+} + e^- \rightarrow Ag^+$ | 1.99 | $O_2 + 2H_2O + 4e^- \rightarrow 4OH^-$ | 0.40 |
| $Co^{3+} + e^- \rightarrow Co^{2+}$ | 1.82 | $Cu^{2+} + 2e^- \rightarrow Cu$ | 0.34 |
| $H_2O_2 + 2H^+ + 2e^- \rightarrow 2H_2O$ | 1.78 | $Hg_2Cl_2 + 2e^- \rightarrow 2Hg + 2Cl^-$ | 0.34 |
| $Ce^{4+} + e^- \rightarrow Ce^{3+}$ | 1.70 | $AgCl + e^- \rightarrow Ag + Cl^-$ | 0.22 |
| $PbO_2 + 4H^+ + SO_4^{2-} + 2e^-$ | | $SO_4^{2-} + 4H^+ + 2e^- \rightarrow H_2SO_3 + H_2O$ | 0.20 |
| $\qquad \rightarrow PbSO_4 + 2H_2O$ | 1.69 | $Cu^{2+} + e^- \rightarrow Cu^+$ | 0.16 |
| $MnO_4^- + 4H^+ + 3e^- \rightarrow MnO_2 + 2H_2O$ | 1.68 | $2H^+ + 2e^- \rightarrow H_2$ | 0.00 |
| $2e^- + 2H^+ + IO_4^- \rightarrow IO_3^- + H_2O$ | 1.60 | $Fe^{3+} + 3e^- \rightarrow Fe$ | −0.036 |
| $MnO_4^- + 8H^+ + 5e^- \rightarrow Mn^{2+} + 4H_2O$ | 1.51 | $Pb^{2+} + 2e^- \rightarrow Pb$ | −0.13 |
| $Au^{3+} + 3e^- \rightarrow Au$ | 1.50 | $Sn^{2+} + 2e^- \rightarrow Sn$ | −0.14 |
| $PbO_2 + 4H^+ + 2e^- \rightarrow Pb^{2+} + 2H_2O$ | 1.46 | $Ni^{2+} + 2e^- \rightarrow Ni$ | −0.23 |
| $Cl_2 + 2e^- \rightarrow 2Cl^-$ | 1.36 | $PbSO_4 + 2e^- \rightarrow Pb + SO_4^{2-}$ | −0.35 |
| $Cr_2O_7^{2-} + 14H^+ + 6e^- \rightarrow 2Cr^{3+} + 7H_2O$ | 1.33 | $Cd^{2+} + 2e^- \rightarrow Cd$ | −0.40 |
| $O_2 + 4H^+ + 4e^- \rightarrow 2H_2O$ | 1.23 | $Fe^{2+} + 2e^- \rightarrow Fe$ | −0.44 |
| $MnO_2 + 4H^+ + 2e^- \rightarrow Mn^{2+} + 2H_2O$ | 1.21 | $Cr^{3+} + e^- \rightarrow Cr^{2+}$ | −0.50 |
| $IO_3^- + 6H^+ + 5e^- \rightarrow \frac{1}{2}I_2 + 3H_2O$ | 1.20 | $Cr^{3+} + 3e^- \rightarrow Cr$ | −0.73 |
| $Br_2 + 2e^- \rightarrow 2Br^-$ | 1.09 | $Zn^{2+} + 2e^- \rightarrow Zn$ | −0.76 |
| $VO_2^+ + 2H^+ + e^- \rightarrow VO^{2+} + H_2O$ | 1.00 | $2H_2O + 2e^- \rightarrow H_2 + 2OH^-$ | −0.83 |
| $AuCl_4^- + 3e^- \rightarrow Au + 4Cl^-$ | 0.99 | $Mn^{2+} + 2e^- \rightarrow Mn$ | −1.18 |
| $NO_3^- + 4H^+ + 3e^- \rightarrow NO + 2H_2O$ | 0.96 | $Al^{3+} + 3e^- \rightarrow Al$ | −1.66 |
| $ClO_2 + e^- \rightarrow ClO_2^-$ | 0.954 | $H_2 + 2e^- \rightarrow 2H^-$ | −2.23 |
| $2Hg^{2+} + 2e^- \rightarrow Hg_2^{2+}$ | 0.91 | $Mg^{2+} + 2e^- \rightarrow Mg$ | −2.37 |
| $Ag^+ + e^- \rightarrow Ag$ | 0.80 | $La^{3+} + 3e^- \rightarrow La$ | −2.37 |
| $Hg_2^{2+} + 2e^- \rightarrow 2Hg$ | 0.80 | $Na^+ + e^- \rightarrow Na$ | −2.71 |
| $Fe^{3+} + e^- \rightarrow Fe^{2+}$ | 0.77 | $Ca^{2+} + 2e^- \rightarrow Ca$ | −2.76 |
| $O_2 + 2H^+ + 2e^- \rightarrow H_2O_2$ | 0.68 | $Ba^{2+} + 2e^- \rightarrow Ba$ | −2.90 |
| $MnO_4^- + e^- \rightarrow MnO_4^{2-}$ | 0.56 | $K^+ + e^- \rightarrow K$ | −2.92 |
| $I_2 + 2e^- \rightarrow 2I^-$ | 0.54 | $Li^+ + e^- \rightarrow Li$ | −3.05 |

# Advanced Placement Chemistry Equations and Constants

## ATOMIC STRUCTURE

$E = h\nu$

$\lambda = \frac{h}{m\nu}$

$E_n = \frac{-2.178 \times 10^{-18}}{n^2}$ joule

$c = \lambda\nu$

$p = m\nu$

## EQUILIBRIUM

$K_a = \frac{[H^+][A^-]}{[HA]}$

$E^\Omega =$

$K_w = [OH^-][H^+] = 1.0 \times 10^{-14}$ at 25°C

$\quad = K_a \times K_b$

$pH = -\log[H^+], pOH = -\log[OH^-]$

$14 = pH + pOH$

$pH = pK_a + \log\frac{[A^-]}{[HA]}$

$pOH = pK_b + \log\frac{[HB^+]}{[B]}$

$pK_a = -K_a, pK_b = -\log K_b$

$K_p = K_c(RT)^{\Delta n}$

where $\Delta n =$ moles product gas – moles reactant gas

## THERMOCHEMISTRY/KINETICS

$\Delta S° = \sum S°$ products $- \sum S°$ reactants

$\Delta H° = \sum \Delta H_f°$ products $- \sum \Delta H_f°$ reactants

$\Delta G° = \sum \Delta G_f°$ products $- \sum \Delta G_f°$ reactants

$\Delta G° = \Delta H° - T\Delta S°$

$\quad = -RT \ln K = -2.303 RT \log K$

$\quad = -n \Im E°$

$\Delta G = \Delta G° + RT \ln Q = \Delta G° + 2.303 RT \log Q$

$q = mc\Delta T$

$C_p = \frac{\Delta H}{\Delta T}$

$\ln[A]_t - \ln[A]_0 = -kt$

$\frac{1}{[A]_t} - \frac{1}{[A]_0} = kt$

$\ln k = \frac{-E_a}{R}\left(\frac{1}{T}\right) + \ln A$

$E =$ energy $\qquad \nu =$ frequency

$\lambda =$ wavelength $\qquad p =$ momentum

$\nu =$ velocity $\qquad n =$ principal quantum number

$m =$ mass

Speed of light, $c = 3.0 \times 10^8 \text{ ms}^{-1}$

Planck's constant, $h = 6.63 \times 10\text{-}34 \text{ Js}$

Boltzmann's constant, $k = 1.38 \times 10^{-23} \text{ J K}^{-1}$

Avogadro's number $= 6.022 \times 10^{23} \text{ mol}^{-1}$

Electron charge, $e = -1.602 \times 10^{-19}$ coulomb

1 electron volt per atom $= 96.5 \text{ kJ mol} - 1$

## EQUILIBRIUM CONSTANTS

$K_a$ (weak acid)

$K_b$ (weak base)

$K_w$ (water)

$K_p$ (gas pressure)

$K_c$ (molar concentrations)

$S° =$ standard entropy

$H° =$ standard enthalpy

$G° =$ standard free energy

$E° =$ standard reduction potential

$T =$ temperature

$n =$ moles

$m =$ mass

$q =$ heat

$c =$ specific heat capacity

$C_p =$ molar heat capacity at constant pressure

$E_a =$ activation energy

$k =$ rate constant

$A =$ frequency factor

Faraday's constant, $\Im = 96,500$ per mole of electrons

Gas constant, $R \quad = 8.31 \text{ J mol}^{-1} \text{ K}^{-1}$

$\qquad = 0.0821 \text{ L atm mol}^{-1} \text{ K}^{-1}$

$\qquad = 8.31 \text{ volt coulomb mol}^{-1} \text{ K}^{-1}$

## GASES, LIQUIDS, AND SOLUTIONS

$PV = nRT$

$\left(P + \frac{n^2 a}{V^2}\right)(V - nb) = nRT$

$P_A = P_{total} \times X_A$, where $X_A = \frac{\text{moles A}}{\text{total moles}}$

$P_{total} = P_A + P_B + P_C + ...$

$n = \frac{m}{M}$

$°K = °C + 273$

$\frac{P_1 V_1}{T_1} = \frac{P_2 V_2}{T_2}$

$D = \frac{m}{V}$

$$u_{rms} = \sqrt{\frac{3kT}{m}} = \sqrt{\frac{3RT}{M}}$$

$$KE \text{ per molecule} = \tfrac{1}{2}mv^2$$

$$KE \text{ per mole} = \tfrac{3}{2}RT$$

$$\frac{r_1}{r_2} = \sqrt{\frac{M_2}{M_1}}$$

molarity, $M$ = moles solute per liter solution

molality = moles solute per kilogram solvent

$$\Delta T_f = iK_f \times \text{molality}$$

$$\Delta T_b = iK_b \times \text{molality}$$

$$\pi = MRT$$

$$A = abc$$

## OXIDATION-REDUCTION; ELECTROCHEMISTRY

$$Q = \frac{[C]^c[D]^d}{[A]^a[B]^b}, \text{ where } a\,A + b\,B \rightarrow c\,C + d\,D$$

$$I = \frac{q}{t}$$

$$E_{cell} = E^{\circ}_{cell} - \frac{RT}{n\Im}\ln Q = E^{\circ}_{cell} - \frac{0.0592}{n}\log Q \,@\,25^{\circ}C$$

$$\log K = \frac{nE^{\circ}}{0.0592}$$

$P$ = pressure
$V$ = volume
$T$ = temperature
$n$ = number of moles
$D$ = density
$m$ = mass
$v$ = velocity

$u_{rms}$ = root-mean-square speed

$KE$ = kinetic energy

$r$ = rate of effusion

$M$ = molar mass

$\pi$ = osmotic pressure

$i$ = van't Hoff factor

$K_f$ = molal freezing-point depression constant

$K_b$ = molal boiling-point elevation constant

$A$ = absorbance

$a$ = molar absorptivity

$b$ = path length

$c$ = concentration

$Q$ = reaction quotient

$I$ = current (amperes)

$q$ = charge (coulombs)

$t$ = time (seconds)

$E^{\circ}$ = standard reduction potential

$K$ = equilibrium constant

Gas constant, $R = 8.31 \text{ J mol}^{-1}\text{ K}^{-1}$

$= 0.0821 \text{ L atm mol}^{-1}\text{ K}^{-1}$

$= 8.31 \text{ volt coulomb mol}^{-1}\text{ K}^{-1}$

Boltzmann's constant, $k = 1.38 \times 10^{-23}\text{ J K}^{-1}$

$K_f$ for $H_2O = 1.86 \text{ K kg mol}^{-1}$

$K_b$ for $H_2O = 0.512 \text{ K kg mol}^{-1}$

1 atm = 760 mm Hg

= 760 torr

STP = 0.000°C and 1.000 atm

Faraday's constant, $\Im = 96,500$ coulombs per mole of electrons

## Section II: Free-Response Questions
### Time: 90 minutes
### Number of Questions: 6

Allow yourself no more than 90 minutes to answer these questions. A calculator may be used only on questions 1, 2, and 3. You may refer to the Equations Sheet, the Periodic Table, and to the Standard Reduction Table throughout this section.

## Part A

1. All students must answer this first question. It is worth 20% of Section II. Given:

   (I) $H_2(g) + Cl_2(g) \rightleftharpoons 2HCl(g)$ $K_p = 2.5 \times 10^{33}$

   (II) $NH_3(g) + HCl(g) \rightleftharpoons NH_4Cl(s)$ $K_p = 5.1 \times 10^{15}$

   (III) $N_2(g) + 4H_2(g) + Cl_2(g) \rightleftharpoons 2NH_4Cl(s)$ $K_p = 3.9 \times 10^{70}$

   (a) Show how the equations, (I), (II), and (III) may be arranged to give the chemical equation $N_2(g) + 3H_2(g) \rightleftharpoons 2NH_3(g)$.

   (b) Determine the value for $K_p$ for $N_2(g) + 3H_2(g) \rightleftharpoons 2NH_3(g)$.

   (c) Equation (III) has a much higher value for $K_p$ than the $K_p$ value for equation (I). What is the physical meaning of these differences?

   (d) Assume that reaction (II) above began with solid $NH_4Cl$, what would be the partial pressure of $NH_3(g)$ and of $HCl(g)$ at equilibrium?

   (e) Determine the value $K_c$ for equation (II) at 25°C.

## Part B

Look over Questions 2 and 3 that follow. Then select one of these two and answer it completely. This question (either #2 or #3) is worth 20% of Section II. (A second problem will not be scored.)

2.
   (a) 50.00 mL of 1.000 M HCl is titrated with 1.000 M KOH. Calculate the pH when a total of
      i.    49.99 mL of KOH has been added.
      ii.   50.00 mL of KOH has been added.
      iii.  50.01 mL of KOH has been added.
      iv.   Comment on the pH values you have calculated.

   (b) A chemist wishes to make up 250. mL of a $HNO_3$ solution with a pH of 2.00, using a stock solution of 2.00 M $HNO_3$. What volume of the stock solution is needed?

   (c) A buffer is prepared by adding 20.5 g of acetic acid and 17.8 g of sodium acetate to water to make 500. mL of solution. Determine the pH of this buffer.

   The $K_a$ of $CH_3COOH$ is $1.8 \times 10^{-5}$.

3.
   (a) Use bond energy data to estimate the enthalpy change for this reaction:
      $H_2(g) + F_2(g) \rightarrow 2HF(g)$
      Bond energy data in kJ/mol of bonds: H–H 432    F–F 154    H–F 565

(b) In this case the bond energy approach gives a good agreement with the measured $\Delta H$ for this reaction, but this is not always so. Why is this not always a successful approach to use for estimating reaction enthalpy?

(c) The carbon–carbon bond length in $C_2H_6$ is 154 pm, whereas in $C_2H_2$ it is only 120 pm. Why is there a difference?

(d) Using the above diagram, explain why a H–H bond length of 0.074 nm is most reasonable.

(e) Explain what is meant by the statement "Ionic and covalent bonds are extreme bond types."

NO CALCULATORS ARE TO BE USED FOR THE REMAINDER OF THE TEST.

## Part C

You must answer Question 4. The section II scoring weight for this question is 15%.

4. Write reaction equations that show both reactants and products for five of the following eight selections. All solutions are aqueous. Note that ions are to be shown if a substance is extensively ionized. Omit spectator ions; do not balance.
   (a) Solid lead (II) hydroxide is added to sulfuric acid.
   (b) Acidified potassium dichromate solution is mixed with potassium iodide solution.
   (c) 1 M acetic acid is mixed with an equal volume of 1 M potassium hydroxide.
   (d) Calcium nitrate solution is added to an ammonium sulfate solution.
   (e) 0.5 M lead (II) nitrate is added to an equal volume of 1.0 M hydrochloric acid.
   (f) Solutions of magnesium sulfate and barium hydroxide are mixed.
   (g) Silver nitrate solution is added to ammonia solution.
   (h) Propanol is burned in air.

## PART D

Spend about 40 minutes on this part of the examination. Answering these questions gives you an opportunity to demonstrate your ability to present your material in clear, orderly, and convincing English. Your answers will be graded on the basis of accuracy, the kinds of information you include to support your responses, and the importance of the descriptive material used. Be specific; general, all-encompassing answers will not be graded as well as detailed answers with examples and equations.

The Laboratory Based Question: Answer the following essay question. The Section II score weighting is 15% for each question in this part (total 45 %).

5. A student performs a calorimetric experiment to determine the heat of formation for MgO. The student has the following equipment:
   coffee cup calorimeter
   thermometer
   balance
   chemicals: MgO(s), Mg ribbon, HCl(aq)

   The student takes measurements that will enable the calculation of the value of $\Delta H$ for the following two reactions:

   (1) $Mg(s) + HCl(aq) \longrightarrow MgCl_2(aq) + H_2(g)$    $\Delta H_1 = ?$

   (2) $MgO(s) + HCl(aq) \longrightarrow MgCl_2(aq) + H_2O(l)$    $\Delta H_2 = ?$

   The student is also given the following:

   (3) $H_2(g) + 1/2 O_2(g) \longrightarrow H_2O(l)$    $\Delta H_3^\circ = -285.8$ kJ/mol

(a) Write the equation for the heat of formation of MgO. Show how the three equations above can be used to determine the value of $\Delta H_f$ (heat of formation) for MgO.

(b) What measurements must be taken to find the $\Delta H$ for reaction (1) above?

(c) Without doing any calculations, indicate the equations that must be used to calculate $\Delta H_1$.

(d) The calorimeter constant is the amount of heat that the calorimeter must gain to change temperature by one degree. Why would the use of a calorimeter constant improve the accuracy of the results?

(e) The accepted value for $\Delta H°_f = -602$ kJ/mol. If the experimentally determined value is $-600.$ kJ/mol, show how (without doing any calculations) the percent error would be calculated.

Answer the following essay question. The Section II score weighting value is 15%.

6. When phosphorus trichloride gas is mixed with chlorine gas, a third gas, phosphorus pentachloride, forms. If this reaction takes place in a closed system, equilibrium results.

(a) Write the equation for the reaction described.

(b) What changes in pressure could be expected as this reaction begins (assume constant volume)? Explain your answer.

(c) How could you determine when equilibrium has been established?

(d) What will be the effect on the value of $K_c$ when more $Cl_2(g)$ is added to the system (assume that there is no temperature change and that volume remains constant)?

(e) Assuming that the reaction of $PCl_3$ with chlorine is exothermic, what effect will increasing the temperature have on the value of $K_c$?

Answer either question 7 or 8. Only one of the two will be graded. The Section II score weighting for the question you choose is 15 percent.

7. Explain the following differences (be sure to refer to both substances):

(a) Sodium fluoride has a crystal lattice energy of 923 kJ/mol, whereas magnesium oxide is much higher with a $\Delta H_{CL}$ of 3916 kJ/mo.

(b) The first ionization energy for atoms of Na is 495 kJ/mol, whereas the first ionization energy for Mg is 735 kJ/mol.

(c) Elements at the top of groups 3A–6A in the Periodic Table are nonmetals, but those at the bottom of these same groups are metals.

(d) The shape of $BF_3$ is trigonal planar, whereas the shape of $NH_3$ is trigonal pyramidal.

8. Using principles of atomic structure or bonding, explain the cause of the following differences:

(a) Water is less dense as a solid than it is as a liquid.

(b) The shape of the meniscus is different for water and for mercury when each liquid has been put into separate burets.

(c) Hydrogen peroxide has both a higher viscosity and a higher boiling temperature than water.

(d) Polyethylene, a polymer, has strength and toughness as a solid, yet it can be melted and formed into a new shape (thermoplastic behavior).

(e) The compound, $C_2H_2Cl_2$, has three isomeric forms due in part to *cis–trans* isomerism, whereas $C_2H_3Cl$ has only one structure.

## ANSWERS TO PRACTICE TEST 1

### MULTIPLE-CHOICE ANSWERS

Using the table below, score your test.

Determine how many questions you answered correctly and how many you answered incorrectly. You will find explanations of the answers on the following pages.

| | | | | |
|---|---|---|---|---|
| 1. D | 2. C | 3. A | 4. B | 5. B |
| 6. C | 7. E | 8. E | 9. D | 10. D |
| 11. D | 12. E | 13. A | 14. D | 15. C |
| 16. C | 17. E | 18. C | 19. B | 20. A |
| 21. B | 22. A | 23. C | 24. E | 25. C |
| 26. E | 27. D | 28. B | 29. C | 30. B |
| 31. B | 32. E | 33. B | 34. B | 35. D |
| 36. A | 37. D | 38. E | 39. E | 40. A |
| 41. C | 42. E | 43. A | 44. E | 45. C |
| 46. A | 47. D | 48. E | 49. A | 50. E |
| 51. A | 52. A | 53. A | 54. E | 55. D |
| 56. D | 57. A | 58. C | 59. D | 60. A |
| 61. A | 62. C | 63. D | 64. B | 65. D |
| 66. E | 67. B | 68. A | 69. C | 70. E |
| 71. E | 72. C | 73. A | 74. D | 75. D |

### CALCULATE YOUR SCORE:

Number answered correctly:  _____
Adjust ¼ point for guessing penalty:
   Count the number of questions you
   answered incorrectly, multiply by .25, and subtract:  – _____
Determine your adjusted score:  _____

### WHAT YOUR SCORE MEANS:

Because the test is different each year, the scoring is a little different. But generally, if you scored 25 or more on the multiple-choice questions, you'll most likely get a 3 or better on the test. If you scored 35 or more, you'll probably score a 4 or better. And if you scored a 50 or more, you'll most likely get a 5. Keep in mind that the multiple-choice section is worth 45% of your final grade, and the free-response section is worth 55% of your final grade. To learn more about the scoring for the free-response questions, turn to the last page of this section.

# ANSWERS AND EXPLANATIONS

## MULTIPLE-CHOICE ANSWERS

1. ANSWER: **D**  Acetic acid will produce an acidic solution with a pH below 7 (*Chemistry* 6th ed. pages 139, 657–659 / 7th ed. pages 131–132, 626–628).

2. ANSWER: **C**  The only ion with color in this group is the $Cu^{2+}$ ion which is blue (*Chemistry* 6th ed. page 998, esp. Table 21.11 / 7th ed. pages 955, esp. 21.11).

3. ANSWER: **A**  Freezing temperature lowering is a colligative property; the more particles in solution the greater is the effect. Because $Ca(NO_3)_2$ provides three particles for each unit and the others all provide fewer, this substance in water would lower the freezing temperature the most (*Chemistry* 6th ed. pages 533–535, 542 / 7th ed. pages 506–507, 513).

4. ANSWER: **B**  The solution which is basic (pOH above 7) is the sodium acetate. This is due to a reaction between the acetate ion and water (called hydrolysis) forming $OH^-$ ions: $C_2H_3O_2^- + HOH \rightarrow OH^- + HC_2H_3O_2$ (this is a weak acid, therefore tending to stay mainly in the molecular form) (*Chemistry* 6th ed. pages 688-689 / 7th ed. pages 655–657).

5. ANSWER: **B**  This is an ion with 32 valence electrons to consider, $[(4 \times 7) + (5) - 1]$, which form four pairs around the phosphorus and four pairs around each Cl with four bonds between P and the four Cl. This forms a tetrahedron (*Chemistry* 6th ed. pages 391–395, esp. diagram on lower left p. 394 / 7th ed. pages 369–373, esp. diagram in center left p. 373).

6. ANSWER: **C**  Galvanizing is a process in which steel is dipped into molten zinc, leading a protective coating on the steel. Because zinc is more reactive than steel, this outer coating is sacrificed to corrosion rather than the steel (*Chemistry* 6th ed. page 854 / 7th ed. pages 815–816).

7. ANSWER: **E**  Even though oxygen has six electrons, only two are bonding electrons; the nonbonding pairs force the bonding pairs into a bent configuration rather than a linear structure. This is the same general shape seen in the water molecule (*Chemistry* 6th ed. pages 389–394, esp. Fig. 8.16 / 7th ed. pages 367–373, esp. Fig. 8.17).

8. ANSWER: **E**  Ideal gases have particles of zero volume, no forces between the particles, and are described by the relationship $PV = nRT$ (*Chemistry* 6th ed. pages 194, 198–200 / 7th ed. pages 183, 186–188).

9. **Answer: D** The two lone pairs of electrons on Xe favor a position of 180° from each other, leaving the four bonded pairs pointing toward the four corners of a square (*Chemistry* 6th ed. pages 395–396 / 7th ed. pages 373–374).

10. **Answer: D** The $CO_3^{2-}$ reacts with HOH as: $CO_3^{2-}$ + HOH → $HCO_3^-$ + $OH^-$, which gives a basic solution. You can determine that the salts are acidic, basic, or neutral by determining the parent acid and parent base. For salt $Na_2CO_3$, the parent base NaOH is strong and the parent acid is weak $H_2CO_3$, so the salt is basic. This is a memory aid and does not really explain the chemistry of why this occurs (*Chemistry* 6th ed. pages 687–693 / 7th ed. pages 655–660).

11. **Answer: D** Molar heat capacity refers to the energy required to change the temperature of one mole of the substance, (in this case, steam or water), by 1°C (*Chemistry* 6th ed. page 489 / 7th ed. page 464).

12. **Answer: E** The negative slope of the solid/liquid line is due to the density of the solid (ice in this case) which is less than that of the liquid (water). With almost all substances, the solid state is denser than the liquid state; water is nontypical in this property (*Chemistry* 6th ed. pages 492–497 / 7th ed. pages 467–470).

13. **Answer: A** The molar volume of any ideal gas @ STP is 22.4 L. From 5.6 L × 1 mol/22.4 L = mol we see that response A is correct (*Chemistry* 6th ed. page 203 / 7th ed. pages 190–191).

14. **Answer: D** The $K_{sp}$ is an equilibrium constant with the usual expression, $K_{sp}$ = $[Pb^{2+}][Br^-]^2$. The solid, of course, has a constant concentration and so is not a part of the equilibrium expression (*Chemistry* 6th ed. pages 751–752 / 7th ed. pages 717–718).

15. **Answer: C** Increasing the surface area does increase the amount of solid that will dissolve in a given period of time. However, since ions reform the solid on the surface of the solid, the rate of reforming the solid also increases, and at the same rate. Therefore, solubility is not increased by grinding; only the rate of dissolving is increased (*Chemistry* 6th ed. page 752 / 7th ed. pages 717–718).

16. **Answer: C** If the $Q$ exceeds the $K_{sp}$ a precipitate forms. Think of the ion product, $Q$, as a limit which if exceeded causes the equilibrium to shift toward the solid (*Chemistry* 6th ed. pages 760–761 / 7th ed. pages 724–726).

17. ANSWER: **E** A buffer resists pH change by reaction with either added $H^+$ or added $OH^-$. It may be the weak acid which reacts with $OH^-$; it may be the anion of the salt of the acid which reacts with the $H^+$ (*Chemistry* 6th ed. pages 772–773, 720–721 / 7th ed. pages 736–737, 687–688).

18. ANSWER: **C** Entropy may be thought of as a measure of the disorder of a system. Solids are more ordered than liquids or gases or solutions. Graphite shows less order than diamond, where all the bonds are identical, because graphite has a 2-D covalent network with delocalized electrons between the layers and diamond has highly localized electrons in a 3-D network of covalent bonds. (Note that you must refer to both substances to get full credit on the AP exam on essay questions on such topics) (*Chemistry* 6th ed. pages 784–789, 468–470 / 7th ed. pages 749–754, 443–446).

19. ANSWER: **B** Larger, heavier substances are more subject to becoming polar due to interaction with other molecules (*Chemistry* 6th ed. pages 483–487 / 7th ed. pages 459–462).

20. ANSWER: **A** $H = (+)$ for bond breaking, which is always an endothermic process. $S = (+)$ for an increase in disorder due here to the increase in the number of particles (*Chemistry* 6th ed. pages 794, 797, 802–804 / 7th ed. pages 759, 761, 766–768).

21. ANSWER: **B** Without a salt bridge, one side of the cell would become increasingly positive (on the oxidation side, where electrons leave) and the other side would become increasingly negative. The salt bridge allows for ions to flow between the compartments, to keep the net charge zero as electrons flow through the external circuit (*Chemistry* 6th ed. pages 828–830 / 7th ed. pages 791–793).

22. ANSWER: **A** The reaction $Ag \rightarrow Ag^+$ is more likely at the lower concentration of $Ag^+$; therefore on the left side, the reaction will be $Ag \rightarrow Ag^+ + 1e^-$. Those electrons then flow through the external circuit to the right side (*Chemistry* 6th ed. pages 841–842 / 7th ed. pages 803–804).

23. ANSWER: **C** The reaction is $Ag^+ + 1e^- \rightarrow Ag$ (*Chemistry* 6th ed. pages 828–830 / 7th ed. pages 791–793).

24. ANSWER: **E** All IA compounds, all nitrates, and $CaCl_2$ are soluble in water (*Chemistry* 6th ed. page 152 / 7th ed. pages 143–144).

25. ANSWER: **C** Both 1.0 M HCl and 1.0 M $HC_2H_3O_2$ can deliver the same number of moles of $H^+$ to neutralize the base. Diprotic $H_2SO_4$ will neutralize twice as much NaOH (*Chemistry* 6th ed. pages 153–155 / 7th ed. pages 144–146).

26. ANSWER: **E** $Al(NO_3)_3$ contains four ions per formula unit; each mole of $Al(NO_3)_3$ contributes four moles of ions: $1.2 \text{ mol/L} \times 4 = 4.8$ mol

of ions/L (*Chemistry* 6th ed. pages 142–143 / 7th ed. pages 134–135).

27. **Answer: D** Gases with the lowest masses have the greatest effusion rates (*Chemistry* 6th ed. pages 220-222 / 7th ed. pages 206–208).

28. **Answer: B** As liquid $H_2O$ changes to solid (ice), heat is lost by the system; this is an exothermic process. Both reactions noted in responses (A) and (D) are common demonstration reactions involving the heating of the solid to form oxygen; they are endothermic. Response (C) is a reaction involving only bond breaking which is always endothermic. Response (E) represents the electron affinity of atomic chlorine gas, another endothermic process (*Chemistry* 6th ed. pages 243–245 / 7th ed. pages 231–232).

29. **Answer: C** This law describes the mass ratios for two elements which form two different compounds. These mass ratios can be reduced to small whole numbers (*Chemistry* 6th ed. pages 45–46 / 7th ed. pages 42–43).

30. **Answer: B** Note how the compound must be similar to $CO_2$ and that for purposes of determining shape, a double bond is treated the same as a single bond: $:\ddot{S}=\ddot{S}:$ (*Chemistry* 6th ed. pages 389–397 / 7th ed. pages 367–376).

31. **Answer: B** With a critical temperature above room temperature, the gas can be liquefied with increasing pressure due to intermolecular forces, which become more important as the distance between molecules decreases (*Chemistry* 6th ed. pages 494, 527 / 7th ed. pages 468–469, 500).

32. **Answer: E** The average rate and the instantaneous rate should not be confused. If the concentration is plotted on the $y$-axis vs. time, a curve is generated. For any given point on this curve (i.e., any given time), a tangent may be drawn. The slope of this tangent line is the change in concentration/change in time, which is rate. By convention, this is always expressed as a positive value (*Chemistry* 6th ed. pages 557–562 / 7th ed. pages 527–533).

33. **Answer: B** The addition of a catalyst lowers the activation energy by finding a less energy demanding pathway. It does this by forming different intermediate substances (*Chemistry* 6th ed. pages 588–591 / 7th ed. pages 557–559).

34. **Answer: B** $Cr_2O_7^{2-}$ ions are orange. Other ions and their colors: $Cu^{2+}$ is light blue, $CrO_4^{2-}$ is yellow, and $K^+$ and $Cl^-$ are colorless. Many of the $Cr^{6+}$ complexes have color. Most transition metal compounds are colored because the "d" electrons are easily excited by visible light into other available "d" orbitals (*Chemistry* 6th ed. page 1015 / 7th ed. pages 970–971).

35. **ANSWER: D** 0.20 mol OH$^-$ + 0.10 mol $H_3PO_4$ → 0.40 mol $H_2O$ + 0.10 mol $HPO_4^-$ (*Chemistry* 6th ed. pages 158–161 / 7th ed. pages 149–154).

36. **ANSWER: A** The triple bond between the carbons is made of one sigma bond and two pi bonds. No orbitals are hybridized to form the two pi bonds (two unhybridized "p" orbitals). The sigma bond between the two carbons and the sigma bond between the C and H are formed by two "sp" hybrid orbitals on each carbon (*Chemistry* 6th ed. pages 413–424 / 7th ed. pages 391–401).

37. **ANSWER: D** Because $P_{H_2} + P_{H_2O} = P_{total}$, $P_{H_2} = P_{total} - P_{H_2O} = 757 - 21$. (*Chemistry* 6th ed. pages 210–211 / 7th ed. pages 198–199).

38. **ANSWER: E** $Mg(OH)_2(s) \rightleftharpoons Mg^{2+}(aq) + 2OH^-(aq)$. The added H$^+$ (the pH is 4, therefore acidic) leads to reaction with the OH$^-$ from the $Mg(OH)_2$ to form water. The reaction shifts toward the $Mg^{2+}$ and OH$^-$ ions; some of the $Mg(OH)_2$ dissolves (*Chemistry* 6th ed. pages 759–760 / 7th ed. page 724).

39. **ANSWER: E** The equivalence point will occur when the mol of OH$^-$ from the base and mol of H$^+$ from the acid are equal (*Chemistry* 6th ed. pages 733–741 / 7th ed. pages 700–707).

40. **ANSWER: A** A salt like AgOH ($K_{sp} = 10^{-8}$) and a salt like $Ca(OH)_2$ ($K_{sp} = 10^{-6}$) form a different number of ions in solution. With AgOH, for example, solubility = $\sqrt{K_{sp}}$ ; whereas with $Ca(OH)_2$, the solubility = $\sqrt[3]{K_{sp}/4}$ (*Chemistry* 6th ed. pages 752–756 / 7th ed. pages 718–721).

41. **ANSWER: C** Due to the hydrolysis of the acetate ion, the solution will be basic (therefore H$^+$ and OH$^-$ cannot be equal); because some of the acetic acid formed from this hydrolysis reaction will react with water leaving less acetic acid, the acetic acid and OH$^-$ will not be equal. Because some of the acetate ion reacts with water (the hydrolysis reaction again), the $[C_2H_3O_2^-]$ will be somewhat less than 0.10 M (*Chemistry* 6th ed. pages 688-690 / 7th ed. pages 655–657).

42. **ANSWER: E** It is important to know that temperature is a measure of <u>average</u> molecular kinetic energy. While the exact relationship, $KE_{ave} = 3/2\ RT$, will most likely not be on the test, you should see that temperature is not simply equal to kinetic energy (*Chemistry* 6th ed. page 217 / 7th ed. page 204).

43. **ANSWER: A** The additional Ag$^+$ causes the equilibrium to shift to the left, forming more solid $Ag_2CrO_4$ (common ion effect). Note that the concentration of a solid is constant; therefore response D is incorrect (*Chemistry* 6th ed. pages 756–757 / 7th ed. pages 721–722).

44. **Answer: E** Because half of the acid has reacted, the [acid] = [$A^-$], from HA $\rightleftharpoons$ $H^+$ + $A^-$, $K_a$ =[$H^+$] [$A^-$] / [HA] = [$H^+$]. Because the pH = 5, $K_a$ = 1 × 10$^{-5}$ (*Chemistry* 6th ed. pages 756–757 / 7th ed. pages 721–722).

45. **Answer: C** Entropy or disorder is increasing when a liquid becomes a gas; this is shown with a positive value for entropy (*Chemistry* 6th ed. pages 784-789 / 7th ed. pages 749–754).

46. **Answer: A** Perhaps the quickest way to solve this is to substitute the values given in $\Delta G = \Delta H - T\Delta S$, while being careful with the units. From $\Delta H$ = –10 kJ and $\Delta S$ = –100 J/K, if the temperature is 100K, then (–10,000)–(100)(–100) = (–10,000)–(–10,000) = 0. Any temperature lower than 100K will make $\Delta G$ negative, hence a spontaneous reaction (*Chemistry* 6th ed. pages 802–803, 795–796 / 7th ed. pages 766–767, 759–760).

47. **Answer: D** Equation I plus the reverse of equation II/2 = the desired equation, so (=370)–(–140/2)= –300 kJ (*Chemistry* 6th ed. pages 804–805 / 7th ed. pages 768–769).

48. **Answer: E** The value for $\Delta G$ indicates how far the process is from equilibrium (which direction is favored), not how rapidly it reaches equilibrium (*Chemistry* 6th ed. pages 783-784 / 7th ed. page 749).

49. **Answer: A** From log $K = nE^\circ/0.0591$ = (2)(–0.60)/0.06 = –1.20/0.06 = –20 , you can see (even if you can not calculate the exact value of $K$) that a negative value means a very small value for $K$. Because $K$ in this case is $K_{sp}$, the solubility would be very low (*Chemistry* 6th ed. page 895 / 7th ed. pages 857–858).

50. **Answer: E** The aluminum ion gains 3 electrons and Mg loses 2 electrons. The overall reaction must reflect an equal number of electrons gained and lost (*Chemistry* 6th ed. pages 833–835 / 7th ed. pages 796–798).

51. **Answer: A** If you know the mass of $PbCl_2$, you can determine the moles of $Pb^{2+}$, then the moles of electrons required, and the number of coulombs of charge needed. This must be multiplied by the current (1A = 1C/s) (*Chemistry* 6th ed. pages 856–857 / 7th ed. page 818).

52. **Answer: A** The anode is the site of oxidation. $Zn - 2e^- \rightarrow -Zn^{2+}$ is oxidation (*Chemistry* 6th ed. pages 834–835 / 7th ed. pages 797–798).

53. **Answer: A** In this cell the zinc loses electrons ($Zn - 2e^- \rightarrow Zn^{2+}$) and the $Ag^+$ gains electrons ($Ag^+ + 1e^- \rightarrow Ag$) (*Chemistry* 6th ed. pages 834–835 / 7th ed. pages 797–798).

54. **Answer: E** Molarity = mol solute / L of solution (*Chemistry* 6th ed. pages 140–142 / 7th ed. pages 133–134).

55. **ANSWER: D** The 1.0 mol of $Pb^{2+}$ (1.0 mol/L × 1.0 L) is now occupying 2.0 L of the solution, so the $[Pb^{2+}]$ = 1.0 mol/2.0 L = 0.50 M. The $[K^+]$ can be found in like manner. The $[NO_3^-]$ = (1.0 mol/L × 1.0 L + 2.0 mol/L × 1.0 L) / 2.0L = 1.5 M (*Chemistry* 6th ed. pages 142–144 / 7th ed. pages 134–136).

56. **ANSWER: D** Because 30.9/100 atoms are Cu mass 65, and 69.1/100 atoms are Cu mass 63, the average mass of a copper atom is (0.309 × 65) + (0.691 × 63) (*Chemistry* 6th ed. pages 81–85 / 7th ed. pages 77–81).

57. **ANSWER: A** Units will help explain this calculation: 5.234 mg/1000 mg/g × 1 mol / 40.18 g × (6.02 × 10^{23}) atoms/mol (*Chemistry* 6th ed. page 86 / 7th ed. page 82).

58. **ANSWER: C** Arrhenius restricted his acid-base definitions to water solutions. Acids provide $H^+$ ions (protons) and bases provide $OH^-$ (*Chemistry* 6th ed. page 179 / 7th ed. pages 168–169).

59. **ANSWER: D** Weak electrolytes ionize very slightly, so few ions are formed when they dissolve (*Chemistry* 6th ed. page 138 / 7th ed. page 131).

60. **ANSWER: A** You cannot determine the rate law by just looking at the chemical equation (unlike the equilibrium constant). The rate law and consequently the order are determined by observing how the rate varies with the concentration of reactants (*Chemistry* 6th ed. pages 563–564 / 7th ed. pages 534–535).

61. **ANSWER: A** Because we know that HCl is a strong acid, i.e. it ionizes 100%, there must not be a strong $Cl^- \rightarrow H^+$ affinity (in water). The equilibrium system, $AgCl(s) \rightleftharpoons Ag^+(aq) + Cl^-(aq)$ can be shifted to the left due to the common ion effect (*Chemistry* 6th ed. pages 769–771 / 7th ed. pages 734–736).

62. **ANSWER: C** Note that doubling [A] doubles the rate of this reaction, suggesting that this is a first-order reaction with respect to [A]. Keeping the [A] constant and changing [B] (comparing Exp. 1 to 3), has no effect on the reaction rate, so [B] is not a part of the rate law (*Chemistry* 6th ed. pages 564–566 / 7th ed. pages 535–537).

63. **ANSWER: D** Note from the stoichiometry of the equation, every time a mol of A disappears, two moles of D appear. . . twice the rate of the disappearance of A (*Chemistry* 6th ed. pages 566–568 / 7th ed. pages 537–538).

64. **ANSWER: B** Because this is written in the general form for linear equations ($y = mx + b$), the slope = $-\Delta H^\circ/R$. It is important to note this would allow for the determination of $\Delta H$ (and for $\Delta S^\circ$ for $\Delta S^\circ/R$ is the intercept) for a reaction run at several different temperatures and K noted for each (*Chemistry* 6th ed. pages 813–814 / 7th ed. pages 776–778).

65. **Answer: D** Reactions are spontaneous if $\Delta G^\circ$ (the Gibb's Free Energy) is negative. This must be the case if $\Delta H$ is (–) and $\Delta S$ is (+), (from: $\Delta G = \Delta H - T\Delta S$) (*Chemistry* 6th ed. pages 791–798 / 7th ed. pages 756–762).

66. **Answer: E** The oxidizing agent is itself reduced. In this case, the $PbSO_4$ is both oxidized and reduced (*Chemistry* 6th ed. pages 168–171 / 7th ed. pages 158–162).

67. **Answer: B** The atomic mass for C given in the Periodic Table is the average of the naturally occurring mixture of $^{12}C$, $^{13}C$, and $^{14}C$ (*Chemistry* 6th ed. pages 82–83 / 7th ed. pages 78–79).

68. **Answer: A** Because the reaction for the collar is $Zn - 2e^- \rightarrow Zn^{2+}$, it is undergoing oxidation. The site of oxidation is known as the anode (*Chemistry* 6th ed. pages 851–855 / 7th ed. pages 813–816).

69. **Answer: C** This is the agreed upon standard for potential (*Chemistry* 6th ed. pages 830–835 / 7th ed. pages 794–798).

70. **Answer: E** Using the fact that this is a second-order reaction, rate = $k[X]^2$. Because the concentration of X is tripled (goes from 0.22 M to 0.66 *M*), the rate is $(3)^2$, or 9 times greater (*Chemistry* 6th ed. pages 564–573 / 7th ed. pages 535–542).

71. **Answer: E** Though the phrase generally applies to molecules of like polarity, all of the responses do apply (*Chemistry* 6th ed. pages 135–136 / 7th ed. pages 128–129).

72. **Answer: C** Because each oxygen has an oxidation state of –2 for a total of $3x - 2 = -6$, and the net charge on this ion is 1–, the nitrogen must have an oxidation state of 5+ (*Chemistry* 6th ed. pages 165–167 / 7th ed. pages 155–157).

73. **Answer: A** This weak acid, acetic acid, dissociates (ionizes) only to a slight extent in aqueous solution (*Chemistry* 6th ed. pages 138–139 / 7th ed. page 131).

74. **Answer: D** The use of a 1-liter flask constructed with a hairline thin volume marker (calibration mark) at just the 1-liter volume is usually used (*Chemistry* 6th ed. pages 143–144 / 7th ed. pages 135–136).

75. **Answer: D** Keep in mind that the pH scale is a log scale based on 10; $pH = -\log[H^+]$. $X/Y = 10^{-7}/10^{-9} = 10^2/1 = 100/1$ (*Chemistry* 6th ed. pages 662–665 / 7th ed. pages 631–634).

## SECTION II: FREE-RESPONSE ANSWERS

**Question 1: Answers**
(a) $K_p$
Reverse and double equation II:

$$2\text{NH}_4\text{Cl}(s) \rightleftharpoons 2\text{NH}_3(g) + 2\text{HCl}(g) \qquad 1/(5.1\times10^{15})^2$$

Reverse equation I:

$$2\text{HCl}(g) \rightleftharpoons \text{H}_2(g) + \text{Cl}_2(g) \qquad 1/(2.5\times10^{33})$$

Add equation III:

$$\text{N}_2(g) + 4\text{H}_2(g) + \text{Cl}_2(g) \longrightarrow 2\text{NH}_4\text{Cl}(s) \qquad 3.9\times10^{70}$$

$$\text{N}_2(g) + 3\text{H}_2(g) \rightleftharpoons 2\text{NH}_3(g)$$

(b) $K_p = 1/(5.1\times10^{15})^2 \times 1/(2.5\times10^{33}) \times 3.9\times10^{70} = 6.0\times10^5$.
Note: Chemical equations are added to give the desired equation. Equilibrium constants are multiplied to give the $K_p$ for the overall process.
(c) Generally a large value for an equilibrium constant indicates that, at equilibrium, there is a higher concentration of products than reactants. Chemists say "products are favored" in this situation.

(d) Each time a mole of ammonia is formed, a mole of hydrogen chloride also forms, and because $K_p = (P_{\text{NH}_3})(P_{\text{HCl}}) = 5.1\times10^{15}$,

$$P_{\text{NH}_3} = P_{\text{HCl}} = \sqrt{5.1\times10^{-15}} = 7.1\times10^{-8} \text{ atm.}$$

(e) From $K_c = K_p/(RT)^{\Delta n}$
$(5.1\times10^{-15})/(0.0821\times298)^{-2} = 3.1\times10^{-12}$
(*Chemistry* 6th ed. pages 769–772 / 7th ed. pages 734–736)

**Question 2: Answers**
(a)
   i.   49.99 mL of KOH has been added:
       0.05000 L × 1.000 mol/L = 0.05000 mol H⁺
       0.04999 L × 1.000 mol/L = 0.04999 mol OH⁻
       leaving 0.00001 mol of H⁺ to survive.
       $[\text{H}^+] = 0.00001\text{mol}/0.09999\text{L} = 1\times10^{-4}\text{M}$  pH = 4.0

   ii.  50.00 mL of KOH has been added:
       All the [H⁺] reacts with all of the [OH⁻] to form water.
       $[\text{H}^+] \times [\text{OH}^-] = K_w = 1\times10^{-14}$ and $[\text{H}^+] = [\text{OH}^-] = 1\times10^{-7}\text{M}$    pH = 7.0

   iii. 50.01 mL of KOH has been added:
       0.05000 L × 1.000 mol/L = 0.05000 mol H⁺
       0.05001 Lx 1.000 mol/L = 0.05001 mol OH⁻
       leaving 0.00001mol OH⁻ to survive.
       $[\text{OH}^N] = 0.00001 \text{ mol}/0.10001 \text{ L} = 1\times10^{-4}$  pOH =4.0      pH = 10.0

   iv. There as been a large change in pH (from 4 to 7 to 10) caused by the buffer being overwhelmed by the addition of 0.02 mL of base. This causes the large

change seen on the titration curve graph at what is usually called the inflection point
(*Chemistry* 6th ed. pages 730–733, esp. Fig. 15.1 / 7th ed. pages 697–699, esp. 15.1).

(b) pH = 2.00, so $[H^+] = 10^{-2.00} = 0.010\ M$
0.0100 mol/L × 0.250 L = 0.0025 mol $H^+$ needed.
0.0025 mol / 2.00 mol/L = 0.00125 L = 1.3 mL
(*Chemistry* 6th ed. pages 662–665 / 7th ed. pages 631–634)

(c) Concentrations: $[HC_2H_3O_2] = (20.5\ g/60.0\ g/mol)/0.500\ L = 0.683\ M$
$[NaC_2H_3O_2] = (17.8g / 82.0\ g/mol) / 0.500\ L = 0.434\ M$

$$HC_2H_3O_2 \rightleftharpoons H^+ + C_2H_3O_2^-$$

$[HC_2H_3O_2] = (0.683 - x)$ and $[C_2H_3O_2^-] = (0.434 + x)$ where $x = [H^+]$ and is very very small.
$K_a = [H^+][C_2H_3O_2^-]/[HC_2H_3O_2] = [H^+] (0.434 + x)/(0.683 - x) = 1.8 \times 10^{-5}$
$[H^+] = 2.8 \times 10^{-5}M$     pH = 4.55
(*Chemistry* 6th ed. pages 716–724 / 7th ed. pages 684–691)

## Question 3: Answers

(a) To break one mole of each H–H and F–F bonds requires (+432) + (+154) = +586 kJ.
To form 2 mol of H–F bonds releases (–565)2 = –1130 kJ
The difference (–1130 + 586) gives an estimated $\Delta H$ for this reaction of –544 kJ
(*Chemistry* 6th ed. pages 371–374 / 7th ed. pages 350–353).

(b) Bond energy values are average values for species in the gas phase. For individual bond energies, the nearby bonds sometimes affect the strength of the chemical bonds in question. Sometimes the molecular environment is very important (*Chemistry* 6th ed. page 371 / 7th ed. pages 350–351).

(c) In $C_2H_6$, the C–C bond is a single bond with only one pair of electrons between the carbon nuclei; in $C_2H_2$, there is a triple bond between the carbons with three pair of shared electrons between the nuclei. More electron pairs means more attractive force between the shared electrons and the nuclei, therefore a shorter bond length (*Chemistry* 6th ed. pages 371–372 / 7th ed. pages 350–352).

(d) The graph shows the change in energy of the system (due to repulsion of electrons by electrons, repulsion between nuclei, and attraction between nuclei and electrons) and how it changes with change in internuclear distance. The atoms move to an optimum distance which achieves the lowest overall energy of the system. In the case shown in this diagram, the lowest energy is achieved between two hydrogen atoms at a distance of 0.074 nm (*Chemistry* 6th ed. pages 348–350 / 7th ed. pages 330–331).

(e) Bonds result when electrons are shared by nuclei. When the sharing is exactly equal between two nuclei, the bond is a "pure" covalent bond. Such a bond is found between H–H, or Cl–Cl. As sharing becomes more uneven, the bond is said to be polar covalent. In the extreme case of uneven sharing, where an electron is completely lost by one nucleus and completely gained by the other, an ionic bond has formed; hence ionic and covalent bonds are extreme types of electron sharing (*Chemistry* 6th ed. page 350 / 7th ed. page 331).

**Question 4: Answers**

(a) $Pb(OH)_2 + H^+ + HSO_4^- \rightarrow PbSO_4 + H_2O$

(b) $Cr_2O_7^{2-} + H^+ + I^- \rightarrow Cr^{3+} + I_2 + H_2O$

(c) $HC_2H_3O_2 + OH^- \rightarrow H_2O + C_2H_3O_2^-$

(d) $Ca^{2+} + SO_4^{2-} \rightarrow CaSO_4$

(e) $Pb^{2+} + Cl^- \rightarrow PbCl_2$

(f) $Mg^{2+} + SO_4^{2-} + Ba^{2+} + OH^- \rightarrow Mg(OH)_2 + BaSO_4$

(g) $Ag^+ + NH_3 \rightarrow [Ag(NH_3)_n]^+$

(h) $C_3H_7OH + O_2 \rightarrow CO_2 + H_2O$
(*Chemistry* 6th ed. pages 154–156, 172–179 / 7th ed. pages 145–146, 162–168)

**Question 5: Answers**

(a) $Mg(s) + 1/2\ O_2(g) \rightarrow MgO(s)$

| | | |
|---|---|---|
| Eq. 1 | $Mg(s) + 2HCl(aq) \rightarrow H_2(g) + MgCl_2(aq)$ | $\Delta H_1$ |
| rev. Eq. 2: | $H_2O(l) + MgCl_2(aq) \rightarrow MgO(s) + 2HCl(aq)$ | $-\Delta H_2$ |
| Eq. 3 | $H_2(g) + 1/2O_2(g) \rightarrow H_2O(l)$ | $\Delta H_3$ |

Addition of
above: $\qquad Mg(s) + 1/2\ O_2(g) \rightarrow MgO(s) \qquad\qquad \Delta H_f$

So $\Delta H_1 + (-\Delta H_2) + \Delta H_3 = \Delta H_f$.

(b) Data required: mass of Mg ribbon
　　initial temperature
　　final temperature
　　total mass of the solution

(c) Equations required: $q = S \times M \times \Delta T$
　　　　　　　　　　　　moles Mg = grams Mg / molar mass
　　　　　　　　　　　　$\Delta H = q$ / mol Mg

(d) Including the calorimeter constant in the calculations would account for any heat gained or lost by the apparatus itself (the calorimeter includes the thermometer as well as the coffee cup calorimeter) and therefore provides a more accurate value for $\Delta H$.

(e) Percentage error
% error = (experimental value – accepted value) / accepted value $\times 100\%$
　　　　　=　　　(–600.)　　　–　　(–602)　　/　　(–602)　　$\times 100\%$
(*Chemistry* 6th ed. pages 248–267 / 7th ed. pages 235–252)

**Question 6: Answers**

(a) and (b) From $PCl_3 + Cl_2 \rightleftharpoons PCl_5$ (all gases), it is apparent that there are fewer product molecules (particles of gas) and therefore less pressure . . . pressure drops (*Chemistry* 6th ed. pages 198–199 / 7th ed. pages 186–187).

(c) Look for a constant macroscopic property, such as constant pressure (*Chemistry* 6th ed. pages 610–612 / 7th ed. pages 579–582).

(d) While the equilibrium shifts toward the product, there is no change in the value of $K_c$ (*Chemistry* 6th ed. pages 636–637 / 7th ed. pages 604–605).

(e) Increasing the temperature opposes an exothermic process, therefore driving this reaction toward reactants. The value of $K_c$ ([products] / [reactants]) will decrease. See also the Arrhenius equation (*Chemistry* 6th ed. pages 641–642 / 7th ed. pages 609–610).

## Question 7: Answers

(a) The crystal lattice energy for MgO is more than four times greater than that for NaF, mainly due to the difference in the charges on the ions. The attractive forces holding 2+ and 2– ions together is about 4 times that of the forces of attraction between 1+ and 1– ions (assuming that the distance between ionic centers is equal, which in fact it is not: NaF 231 pm vs. MgO 205 pm) (*Chemistry* 6th ed. pages 360–362, 364–366 / 7th ed. pages 340–342, 344–346).

(b) Consider the electron configurations:  Na  $1s^2$  $2s^2 2p^6$  $3s^1$
Mg  $1s^2$  $2s^2 2p^6$  $3s^2$
Because the valence (outermost) electron is attracted by 11 protons in the nucleus of sodium and 12+ in Mg, greater attraction would be expected within Mg. Also recall the trends in atomic radii and the importance of the distance from this nuclear charge to the outermost electron. Mg is smaller than Na, (actually 160 pm vs. 186 pm).

Electrostatic energy = $E = (kQ_1 Q_2)/r$ (one form of Coulomb's law)(*Chemistry* 6th ed. pages 349, 329–330, 363–364 / 7th ed. page 330, 313–314, 343–344).

(c) Metals are substances that lose electrons easily. This is true for atoms of large radii and small $Z_{eff}$ (*Chemistry* 6th ed. pages 56–61, 329–332 / 7th ed. pages 52–57, 313–316).

(d) Boron has three valence electrons which form a trigonal planar structure as they repel each other. This is the most stable arrangement for just three electrons. They become three electron pairs when bonding with F occurs. With $NH_3$, there are three bonding pairs and one lone pair. As these electrons move as far from each other as they can, they form a tetrahedron; the shape of the nitrogen nucleus and the three bonded hydrogens is a trigonal pyramid (*Chemistry* 6th ed. pages 389–395 / 7th ed. pages 367–373).

## Question 8: Answers

(a) Water forms open crystals as a solid (made of many water molecules) with more space between the $H_2O$ molecules. This arrangement maximizes intermolecular forces in the solid state. As ice melts, the open crystals fall in upon themselves and the liquid state occupies less space, thereby a greater density (*Chemistry* 6th ed. pages 478–479, 459–460 / 7th ed. pages 454–456, 435–436).

(b) Compare the force within the liquid with that between liquid and glass. In the case of water, the attraction for the glass is stronger, making for a concave meniscus. In mercury, the Hg–Hg intermolecular forces are greater than the Hg–glass attraction, making for a convex shape (*Chemistry* 6th ed. pages 450–456 / 7th ed. pages 426–430).

(c) $H_2O_2$ has more hydrogen-oxygen groups, and therefore more hydrogen "bonding" is possible than in $H_2O$. That provides groups of $H_2O_2$ molecules which tangle and resist flow (*Chemistry* 6th ed. pages 454–455 / 7th ed. pages 429–430).

(e) *Cis–trans* isomerism means that when the two Cl groups are on the same side of the double bond, they form a compound different from that when they are on opposite sides of the double bond. There are three structures possible:

whereas the monochloride has only one possible structure:

$$\begin{array}{ccc} H & & H \\ \diagdown & & \diagup \\ & C{=}C & \\ \diagup & & \diagdown \\ H & & Cl \end{array}$$

## SCORING THE FREE-RESPONSE QUESTIONS

It is difficult to come up with an exact score for this section of the test. However, if you compare your answers to the answers in this book, remembering that each part of the test you answer correctly is worth points even if other parts of the answer are incorrect (see the section titled "Scoring for the Free-Response Questions" on page 12 of this book), you can get a general idea of the percentage of the questions for which you would get credit. If you believe that you got at least one-third of the possible credit, you would probably receive a 3 on this part of the test. If you believe that you would receive close to half or more of the available credit, your score would more likely be a 4 or a 5.

**Alkali metals** — 1 / 1A
**Alkaline earth metals** — 2 / 2A
**Transition metals** — groups 3–12
**Halogens** — 17 / 7A
**Noble gases** — 18 / 8A
metals → nonmetals

| 1 1A | 2 2A | 3 | 4 | 5 | 6 | 7 | 8 | 9 | 10 | 11 | 12 | 13 3A | 14 4A | 15 5A | 16 6A | 17 7A | 18 8A |
|---|---|---|---|---|---|---|---|---|---|---|---|---|---|---|---|---|---|
| 1 H 1.008 | | | | | | | | | | | | | | | | | 2 He 4.003 |
| 3 Li 6.941 | 4 Be 9.012 | | | | | | | | | | | 5 B 10.81 | 6 C 12.01 | 7 N 14.01 | 8 O 16.00 | 9 F 19.00 | 10 Ne 20.18 |
| 11 Na 22.99 | 12 Mg 24.31 | | | | | | | | | | | 13 Al 26.98 | 14 Si 28.09 | 15 P 30.97 | 16 S 32.07 | 17 Cl 35.45 | 18 Ar 39.95 |
| 19 K 39.10 | 20 Ca 40.08 | 21 Sc 44.96 | 22 Ti 47.88 | 23 V 50.94 | 24 Cr 52.00 | 25 Mn 54.94 | 26 Fe 55.85 | 27 Co 58.93 | 28 Ni 58.69 | 29 Cu 63.55 | 30 Zn 65.38 | 31 Ga 69.72 | 32 Ge 72.59 | 33 As 74.92 | 34 Se 78.96 | 35 Br 79.90 | 36 Kr 83.80 |
| 37 Rb 85.47 | 38 Sr 87.62 | 39 Y 88.91 | 40 Zr 91.22 | 41 Nb 92.91 | 42 Mo 95.94 | 43 Tc (98) | 44 Ru 101.1 | 45 Rh 102.9 | 46 Pd 106.4 | 47 Ag 107.9 | 48 Cd 112.4 | 49 In 114.8 | 50 Sn 118.7 | 51 Sb 121.8 | 52 Te 127.6 | 53 I 126.9 | 54 Xe 131.3 |
| 55 Cs 132.9 | 56 Ba 137.3 | 57 La* 138.9 | 72 Hf 178.5 | 73 Ta 180.9 | 74 W 183.9 | 75 Re 186.2 | 76 Os 190.2 | 77 Ir 192.2 | 78 Pt 195.1 | 79 Au 197.0 | 80 Hg 200.6 | 81 Tl 204.4 | 82 Pb 207.2 | 83 Bi 209.0 | 84 Po (209) | 85 At (210) | 86 Rn (222) |
| 87 Fr (223) | 88 Ra 226 | 89 Ac† (227) | 104 Rf | 105 Db | 106 Sg | 107 Bh | 108 Hs | 109 Mt | 110 Ds | 111 Uuu | 112 Uub | | 114 Uuq | | | | |

*Lanthanides

| 58 Ce 140.1 | 59 Pr 140.9 | 60 Nd 144.2 | 61 Pm (145) | 62 Sm 150.4 | 63 Eu 152.0 | 64 Gd 157.3 | 65 Tb 158.9 | 66 Dy 162.5 | 67 Ho 164.9 | 68 Er 167.3 | 69 Tm 168.9 | 70 Yb 173.0 | 71 Lu 175.0 |
|---|---|---|---|---|---|---|---|---|---|---|---|---|---|

†Actinides

| 90 Th 232.0 | 91 Pa (231) | 92 U 238.0 | 93 Np (237) | 94 Pu (244) | 95 Am (243) | 96 Cm (247) | 97 Bk (247) | 98 Cf (251) | 99 Es (252) | 100 Fm (257) | 101 Md (258) | 102 No (259) | 103 Lr (260) |
|---|---|---|---|---|---|---|---|---|---|---|---|---|---|

Group numbers 1–18 represent the system recommended by the International Union of Pure and Applied Chemistry.

# Practice Test 2

These questions are representative of the AP Chemistry examination, but keep in mind that it is impossible to predict exactly how well you will do on the actual exam. The first section of this test is 45% of your total test grade. Time yourself so that you finish this part in 90 minutes. Remember that on the actual examination there is a guess factor; scoring is the number of questions answered correctly minus one-fourth the number wrong.

## AP Chemistry Exam
### Section I: Multiple-Choice Questions
### Time: 90 minutes
### Number of Questions: 75

No calculators are to be used in this section; no tables (except Periodic Table) permitted.

## Part A

Directions: The following questions consist of five lettered headings followed by a listing of numbered phrases. For each phrase, you are to select the one heading that is most closely related to it. Headings may be used once, more than once, or not at all.

Questions 1–4 refer to the endings on the names of ions that indicate the formula of the ion.

    (A) -ide
    (B) -ite
    (C) -ium
    (D) -ate
    (E) –ane

Choose the correct ending for each of the following:

1. $CO_3^{2-}$

2. $PO_3^{3-}$

3. $NH_4^+$

4. $ClO_2^-$

Questions 5–7 refer to the following organic functional groups.

    (A) Carboxylic acids
    (B) Alcohols
    (C) Esters
    (D) Amines
    (E) Ketones

To which functional group family does each compound below belong?

5. $CH_3CH_2CH_2OH$

6. $H_3CC(=O)CH_3$

7. $H_3CC(=O)OCH_2CH_3$

**GO ON TO NEXT PAGE**

Questions 8–10 refer to the following phase diagram.

Use the following key to answer the questions.

(A) Sublimation
(B) Condensation
(C) Boiling
(D) Melting
(E) Freezing

8. If the temperature increases from 10 °C to 60 °C at a constant pressure of 0.4 atm, which of the processes occurs?

9. If the temperature decreases from 110 °C to 40 °C at a constant pressure of 1.1 atm, which of the processes occurs?

10. If the pressure increases from 0.5 to 1.5 atm at constant temperature of 50 °C, which of the processes occurs?

Questions 11–14 refer to the following elements.

(A) He
(B) Na
(C) Si
(D) Mg
(E) Ar

11. This element has atoms with the largest atomic radius.

12. This element has the highest first ionization energy.

13. This element is the most metallic in character.

14. This element most easily forms an ion with a 2+ charge.

Questions 15–17 refer to the following elements.

(A) Aluminum
(B) Chlorine
(C) Copper
(D) Oxygen
(E) Tin

15. Forms an amphoteric oxide

16. Found in both brass and bronze

17. Acidifying bleach releases this gas

## Part B

<u>Directions:</u> Each of the questions or incomplete statements below is followed by five suggested answers or completions. Select the one that is best in each case.

18. If two atoms have different atomic numbers but the same mass number, what must be true?
    (A) They must be atoms of the same element.
    (B) They must contain the same number of electrons.
    (C) Each must contain the same total number of neutrons and protons.
    (D) The number of neutrons in both must be the same.
    (E) The number of protons in each atom must be the same.

19. Which one of the following is an example of an intensive physical property?
    (A) Aluminum burns in bromine.
    (B) A balloon of hydrogen and oxygen explodes.
    (C) A bottle of rubbing alcohol has a mass of 900 grams.
    (D) The molar mass of sodium is 22.99 g. mol.
    (E) The volume of a helium balloon is 6.5 liters.

20. What is the arrangement of electron pairs around the central atom in the molecule krypton difluoride, $KrF_2$?
    (A) linear
    (B) trigonal bipyramidal
    (C) trigonal planar
    (D) octahedral
    (E) tetrahedral

21. Which of the following sets of quantum numbers is **not** allowed?

    |     | $n$ | $\ell$ | $m_\ell$ | $m_s$ |
    |-----|-----|--------|----------|-------|
    | (A) | 3   | 0      | 0        | $(-)\frac{1}{2}$ |
    | (B) | 2   | 1      | 0        | $(-)\frac{1}{2}$ |
    | (C) | 4   | 3      | –3       | $(-)\frac{1}{2}$ |
    | (D) | 2   | 0      | –1       | $(-)\frac{1}{2}$ |
    | (E) | 3   | 1      | –1       | $(+)\frac{1}{2}$ |

22. A substance with strong intermolecular forces of attraction would be expected to have
    I.   a low boiling point.
    II.  a low vapor pressure.
    III. a high heat of vaporization.
    IV.  a low melting point.

    (A) I only
    (B) II only
    (C) I and II only
    (D) II and III only
    (E) I, II, III, IV

23. Berkelium-243 is prepared by α-particle bombardment of americium-241. How many neutrons are emitted in the process?
    (A) 0
    (B) 1
    (C) 2
    (D) 3
    (E) 4

24. What is the average bond order in the molecular ion $NO_2^+$?
    (A) 1.0
    (B) 2.0
    (C) 2.5
    (D) 1.5
    (E) 3.0

**GO ON TO NEXT PAGE**

25. Resonance is required to reconcile the Lewis electron representations with the actual or real structure in which of the following?
   I.   Carbon dioxide, $CO_2$
   II.  Ozone, $O_3$
   III. Sulfate ion, $SO_4^{2-}$
   IV.  Nitrate ion, $NO_3^-$
   V.   Azide ion, $N_3^-$
   (A) I only
   (B) II only
   (C) II and V only
   (D) II and IV only
   (E) I, II, III, IV and V

26. An element has two naturally occurring isotopes. One isotope has an abundance of 80% and a mass of 122.0 amu. The other has a mass of 120.0 amu. What is the atomic mass of the element?
   (A) 120.0
   (B) 121.6
   (C) 122.0
   (D) 121.8
   (E) 120.8

27. If 3.0 moles of pentane (vapor pressure = 88.7 torr) are mixed with 2.0 moles of hexane (vapor pressure = 44.5 torr), what is the total vapor pressure, in torr, above the nearly ideal solution?
   (A) 53.2
   (B) 62.2
   (C) 66.6
   (D) 71.0
   (E) 133

28. 0.24 mole of sodium bicarbonate is dissolved in sufficient water to make 300.0 mL of solution. What is the molarity of the sodium bicarbonate solution?
   (A) 0.24M
   (B) 0.72M
   (C) 0.32M
   (D) 0.80M
   (E) 0.66M

29. The concentration of nitrate ions in a $3.6 \times 10^{-4}$ M aqueous solution of lead nitrate, $Pb(NO_3)_2$, is
   (A) $3.6 \times 10^{-4}$ M.
   (B) $1.8 \times 10^{-4}$ M.
   (C) $1.2 \times 10^{-4}$ M.
   (D) $4.8 \times 10^{-4}$ M.
   (E) $7.2 \times 10^{-4}$ M.

30. The Brønsted-Lowry conjugate base of the bisulfate ion $HSO_3^-$ is
   (A) $H_2SO_3$.
   (B) $HSO_4^-$.
   (C) $SO_3^{2-}$.
   (D) $SO_4^{2-}$.
   (E) $OH^-$.

31. Ammonia is converted to nitric oxide at high temperatures by the reaction:

$$4NH_3(g) + 5O_2(g) \longrightarrow 4NO(g) + 6H_2O(g).$$

In one particular experiment, equal molar amounts of the reactants were mixed and concentrations of the reactants and products were plotted against time. The graph below was obtained. Identify the components A, B, C, and D.

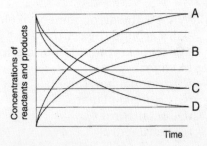

Choose the line where all the components are identified properly.

|     | A      | B     | C      | D      |
|-----|--------|-------|--------|--------|
| (A) | $NH_3$ | $O_2$ | NO     | $H_2O$ |
| (B) | $H_2O$ | NO    | $O_2$  | $NH_3$ |
| (C) | NO     | $H_2O$| $NH_3$ | $O_2$  |
| (D) | $O_2$  | NO    | $H_2O$ | $NH_3$ |
| (E) | $H_2O$ | NO    | $NH_3$ | $O_2$  |

32. Hydrogen sulfide, $H_2S$, in solution (hydrosulfuric acid) has a $K_a$ equal to $1.0 \times 10^{-7}$ at 25 °C. What is the $K_b$ for its conjugate base, $HS^-$?
    (A) $1.0 \times 10^{-7}$
    (B) $1.0 \times 10^{-14}$
    (C) $1.8 \times 10^{-5}$
    (D) $1.0 \times 10^{+7}$
    (E) $2.5 \times 10^{-6}$

For questions 33 and 34, consider the following mechanism for the reaction of nitric oxide and hydrogen:

$2NO(g) \rightleftharpoons N_2O_2(g)$          *Fast*

$N_2O_2(g) + H_2(g) \longrightarrow N_2O(g) + H_2O(g)$   *Slow*

$N_2O(g) + H_2(g) \longrightarrow N_2(g) + H_2O(g)$    *Fast*

33. What is (are) reaction intermediate(s) in this reaction mechanism?
    (A) $NO$ *and* $N_2O$
    (B) $N_2O$ *and* $N_2O_2$
    (C) $N_2O_2$
    (D) $N_2O$ *and* $N_2$
    (E) $N_2O$ *and* $H_2$

34. What is the rate law for the reaction according to this mechanism?
    (A) Rate = $k[NO]$
    (B) Rate = $k[N_2O_2]^2[H_2]$
    (C) Rate = $k[NO]^2$
    (D) Rate = $k[N_2O_2]^2[H_2]^2$
    (E) Rate = $k[NO]^2[H_2]$

35. Which of the following statements concerning the kinetic molecular theory of gases is *false*?
    I. The rate of effusion of a gas is directly proportional to the root mean square speed of its molecules.
    II. The molar volume of a gas depends on the molar mass of a gas.
    III. The average kinetic energy of the molecules in a sample of gas depends only upon the temperature.
    IV. At constant pressure, gases occupy greater volumes at higher temperatures.

    (A) I only
    (B) II only
    (C) III only
    (D) IV only
    (E) I, II, III, and IV

36.

3.0 liters 3.0 atm argon    2.0 liters 5.0 atm helium    3.0 liters 4.0 atm neon

With the stopcocks closed, the three bulbs in the apparatus shown above are filled with the quantities indicated of argon, helium, and neon, respectively. Assume the gases are ideal and that the volumes include the volumes of the connecting tubes. What is the partial pressure of helium inside the apparatus when the taps are opened?
    (A) 1.00 atm
    (B) 1.25 atm
    (C) 1.50 atm
    (D) 2.00 atm
    (E) 2.50 atm

**GO ON TO NEXT PAGE**

37. The *principal* reason(s) why the ideal gas law fails as the pressure increases or as the temperature of a system decreases is
    I. The intermolecular attraction between gas molecules becomes more effective.
    II. The average velocity of the gas molecules increases.
    III. The volume of gas molecules becomes significantly large.
    (A) I only
    (B) II only
    (C) III only
    (D) I and III only
    (E) I, II and III

38. Which one of the following equimolar solutions will act as a buffer solution?
    (A) $H_2SO_4$ and $Na_2SO_4$
    (B) $Na_2S$ and KHS
    (C) $H_2SO_4$ and $H_2SO_3$
    (D) KOH and KCN
    (E) HI and KI

39. Calculate the equilibrium constant $K$ for the following equilibrium

    $$3 F_2(g) + Cl_2(g) \rightleftharpoons 2ClF_3(g)$$

    with the following equilibrium concentrations: $[F_2] = 2.0M$; $[Cl_2] = 2.5M$; $[ClF_3] = 3.0M$.
    (A) 0.40
    (B) 0.90
    (C) 1.2
    (D) 0.25
    (E) 0.45

40. Technetium is a radioactive element of growing importance in medical imaging, prepared from molybdenum by the following equation:

    $$^{98}_{42}Mo + {}^{1}_{0}n \longrightarrow {}^{99}_{43}Tc + ?$$

    What particle is necessary to balance the equation?
    (A) $\alpha$ particle
    (B) helium nucleus
    (C) proton
    (D) $\beta$ particle
    (E) positron

41. In the Lewis dot structure for the molecule $S_2O$, how many lone pairs of electrons are there around the central atom?
    (A) 0
    (B) 1
    (C) 2
    (D) 3
    (E) 4

42. What is the molar solubility of cadmium sulfide CdS? ($K_{sp}$ of CdS = $3.6 \times 10^{-29}$)
    (A) $3.6 \times 10^{-16}$M
    (B) $6.0 \times 10^{-15}$M
    (C) $7.2 \times 10^{-58}$M
    (D) $1.8 \times 10^{-16}$M
    (E) $6.0 \times 10^{-14}$M

43. Below are representations of energy levels (drawn to scale) in the Bohr model of the hydrogen atom. Which electron making a transition, denoted by the arrows, results in the emission of a photon of shortest wavelength?

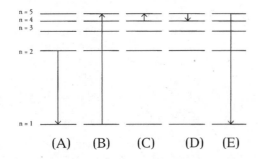

(A)　(B)　(C)　(D)　(E)

44. What is the change in oxidation number of sulfur in the following half-reaction?

$S_4O_6^{2-} + 10H_2O \longrightarrow SO_4^{2-} + 2OH^- + 14e^-$

(A) –2 to –8
(B) +4 to +1
(C) +4 to +6
(D) +2 ½ to +6
(E) +3 to +6

45. A U-tube mercury manometer is open to the atmosphere (1 atm = 760 mm Hg) on the right arm and is connected to a glass vessel containing a gas on the left arm as shown.

The level of the mercury in the tube is 114 cm(1140 mm) higher in the right arm. What is the pressure of the gas in the glass vessel?
(A) 0.50 atm
(B) 1.0 atm
(C) 1.5 atm
(D) 2.0 atm
(E) 2.5 atm

46. The reaction of ammonia with nitric oxide is described by the following equation:

$4 NH_3(g) + 6 NO(g) \longrightarrow 5 N_2(g)$
$+ 6H_2O(g)$

At one stage in the reaction, nitrogen is produced at a rate of 30 mol/L/s. At what rate is nitric oxide used up at this stage?
(A) 5.0 mol $L^{-1}s^{-1}$
(B) 6.0 mol $L^{-1}s^{-1}$
(C) 24 mol $L^{-1}s^{-1}$
(D) 30 mol $L^{-1}s^{-1}$
(E) 36 mol $L^{-1}s^{-1}$

47. For which substance does the standard enthalpy of formation $\Delta H_f^o$ equal zero?
(A) $H_2O(l)$
(B) $Cu(s)$
(C) $N_2(l)$
(D) $O_3(g)$
(E) $Fe(l)$

48. A nonvolatile solute A is dissolved in water. Which of the following affects the vapor pressure of water?
I.　The exposed surface of the water.
II.　The density of A.
III.　The amount of A dissolved in the water.

(A) I only
(B) II only
(C) III only
(D) II and III only.
(E) I, II and III

49. A 2.0-L sample of an ideal gas at 300 K and 3.0 atm constant pressure is heated until the volume increases to 6.0 L. Then the temperature is held constant and the pressure is increased, restoring the volume to 2.0 L. What final pressure is required?
(A) 4.5 atm
(B) 6.0 atm
(C) 9.0 atm
(D) 10. atm
(E) 12 atm

**GO ON TO NEXT PAGE**

50. Adding which of the following solutes will increase the solubility of silver carbonate?
    I.   Silver nitrate, $AgNO_3$
    II.  Nitric acid, $HNO_3$
    III. Sodium carbonate
    (A) I only
    (B) II only
    (C) III only
    (D) II and III
    (E) None, they will all decrease solubility

51. Which statement is correct?
    (A) The pH of pure water is always 7.
    (B) At temperatures greater than $25\,°C$, the pH of a salt solution in water is always less than 7.
    (C) The higher the pH, the more acidic the solution.
    (D) The strongest acid that can exist in aqueous solution is the hydronium ion.
    (E) If the concentration of acid is very high, the acid is said to be strong.

52. When zinc sulfide reacts with 32.0 g of oxygen to yield zinc oxide and sulfur dioxide, how many moles of ZnO are produced?

    $2\ ZnS(s) + 3\ O_2(g) \longrightarrow 2\ ZnO(s) + 2SO_2(g)$

    (A) 0.66
    (B) 1.0
    (C) 1.3
    (D) 2.0
    (E) 2.6

53. Which of the following has the lowest freezing point?
    (A) $0.20\ m\ C_6H_{12}O_6$ (glucose)
    (B) $0.10\ m\ NH_4Br$
    (C) $0.05\ m\ ZnSO_4$
    (D) $0.10\ m\ K_3PO_4$
    (E) $0.20\ m\ Na_2S$

54. Which one of the following choices includes only solutes that are weak electrolytes in aqueous solution.
    (A) $HF$, $NH_3$, $NH_4F$
    (B) $HCl$, $NaOH$, $NaCl$
    (C) $HNO_3$, $H_2SO_4$, $HClO_4$
    (D) $HF$, $HBr$, $HI$
    (E) $HCN$, $NH_3$, $HF$

55. A monoatomic ion contains 15 protons, 16 neutrons, and 18 electrons. What ion of what isotope is this?
    (A) $^{31}P^{3+}$
    (B) $^{34}Se^{2-}$
    (C) $^{18}Ar^{3+}$
    (D) $^{31}P^{3-}$
    (E) $^{34}As^{3-}$

56. The rate law for a chemical reaction between two substances A and B is: Rate of reaction = $k[A]^2[B]$ where $k$ is the rate constant for the reaction. What happens to the rate of the reaction if the concentration of A is doubled and the concentration of B is halved? (The temperature remains unchanged.)
    (A) The rate halves.
    (B) The rate is reduced to one quarter.
    (C) The rate stays the same.
    (D) The rate doubles.
    (E) The rate increases by four.

57. How many structural isomers can be drawn for butane ($C_4H_{10}$)?
    (A) one
    (B) two
    (C) three
    (D) four
    (E) five

58. Calculate the standard cell potential for the cell

Sr(s) |Sr$^{2+}$(aq) Sn$^{2+}$(aq)|Sn(s) given the half-cell standard potentials:

E° Sr$^{2+}$(aq)|Sr(s) = –2.89 V

E° Sn$^{2+}$(aq)|Sn(s) = –0.14 V.
(A) –3.03V
(B) +3.03V
(C) –2.75V
(D) +2.75V
(E) +5.05V

59. Which process is always exothermic?
(A) Evaporation of a liquid
(B) Dissolving a typical salt in water
(C) Breaking a hydrogen molecule into atoms
(D) Sublimation of solid carbon dioxide
(E) Freezing water

60. Given the standard heats of formation: $\Delta H_f^o$ CH$_3$OH (g) = –239 kJ/mol; $\Delta H_f^o$ CO$_2$(g) = –394 kJ/mol; $\Delta H_f^o$ H$_2$O(l) = –286kJ/mol,

what is $\Delta H_{rxn}$ in kJ for the following reaction?

2CH$_3$OH(g) +3 O$_2$(g)→ 2CO$_2$(g)+ 4H$_2$O(l)

(A) –727 kJ
(B) +727 kJ
(C) –441 kJ
(D) +1445 kJ
(E) –1454 kJ

61. In the following reaction 2MnO$_4^-$ (aq) +16H$_3$O$^+$(aq) +5 Zn(s) → 2Mn$^{2+}$(aq) +24H$_2$O(l) + 5Zn$^{2+}$(aq), how many electrons are transferred between the reducing agent and the oxidizing agent in the balanced equation?
(A) 5
(B) 7
(C) 10
(D) 14
(E) 15

62. A reaction takes place within a system. As a result, the entropy of the system decreases, it becomes more ordered. What *must* be true?
I.   The reaction is exothermic.
II.  The entropy of the universe increases.
III. The entropy of the surroundings increases.

(A) I only
(B) II only
(C) I and II only
(D) I and III only
(E) I, II and III

63. Balance the following equation for the reaction of dimethylhydrazine with dinitrogen tetroxide:

___C$_2$H$_8$N$_2$  +  ___N$_2$O$_4$ ⟶

___N$_2$ + ___CO$_2$ + ___H$_2$O

The sum of all the coefficients in the balanced equation is
(A) 9
(B) 11
(C) 12
(D) 14
(E) 15

64. Which of the following is an acid-base reaction?
I.   Ca(OH)$_2$(aq)  + 2HNO$_3$(aq) ⟶ Ca(NO$_3$)$_2$(aq) + 2H$_2$O(l)
II.  2H$_2$(g) + O$_2$(g) ⟶ 2H$_2$O(g)
III. CuBr$_2$(aq)  + 2NaOH(aq) ⟶ Cu(OH)$_2$(s)  + 2NaBr(aq)

(A) I only
(B) II only
(C) III only
(D) I and III
(E) II and III

**GO ON TO NEXT PAGE**

65. A catalyst
(A) increases the amount of product at equilibrium.
(B) changes the route the reaction takes between reactants and products.
(C) increases the activation energy required for the reaction.
(D) increases the kinetic energy of the reactant molecules.
(E) shifts the equilibrium toward the product side.

66. When crystalline ammonium nitrate dissolves in water to make a solution, the solution gets very cold, dropping in temperature about 20°C. What are the signs of ΔH, ΔS, and ΔG for this process?

| | ΔH | ΔS | ΔG |
|---|---|---|---|
| (A) | (−) | (+) | (+) |
| (B) | (−) | (+) | (−) |
| (C) | (+) | (+) | (+) |
| (D) | (+) | (−) | (+) |
| (E) | (+) | (+) | (−) |

67. The graph below shows the speed distributions for three gases, $H_2$, $H_2O$, and $H_2S$ at the same temperature.

Match the gas with the distribution.

| | A | B | C |
|---|---|---|---|
| (A) | $H_2$ | $H_2O$ | $H_2S$ |
| (B) | $H_2O$ | $H_2$ | $H_2S$ |
| (C) | $H_2$ | $H_2S$ | $H_2O$ |
| (D) | $H_2S$ | $H_2O$ | $H_2$ |
| (E) | $H_2O$ | $H_2S$ | $H_2$ |

68. Ammonia solution is usually diluted for household use. What is the concentration in mole/L of an ammonia solution prepared by taking 100. mL of a 6.50 M solution of ammonia and adding it to enough water to make 2.00 L of solution?
(A) 0.0650 M
(B) 0.0875 M
(C) 0.0975 M
(D) 0.130 M
(E) 0.325 M

69. Consider the following disturbances of the equilibrium system shown. In which, if any, direction will the system shift to restore equilibrium in each case? The reaction is endothermic in the forward direction.

$$2\,NOCl_2(g) \rightleftharpoons 2NO(g) + Cl_2(g)$$

| | increase temperature | decrease volume |
|---|---|---|
| (A) | Right | Left |
| (B) | Left | No change |
| (C) | Left | Left |
| (D) | No change | Left |
| (E) | Right | Right |

70. Which of the following statements is true?
(A) Increasing the pressure on an ideal gas at constant temperature slows the molecules down.
(B) Deviation of a gas from ideal behavior is due only to the finite size of real molecules.
(C) Argon gas (Ar) diffuses faster than chlorine gas ($Cl_2$) at the same temperature.
(D) The smaller the gas molecule, the higher its kinetic energy at the same temperature.
(E) All molecules in a sample of a gas travel at the same speed at a particular temperature.

71. Which of the following binary compounds would you expect to form a basic aqueous solution?
    (A) $CO_2$
    (B) CaO
    (C) $NO_2$
    (D) $P_4O_{10}$
    (E) $H_2O_2$

72. What is the abbreviated electron configuration for the phosphide ion, $P^{3-}$?
    (A) $[Ne]3s^23p^3$
    (B) $[Ne]3s^2$
    (C) $[Ar]$
    (D) $[Ar]4s^2$
    (E) $[Ne]3s^23p^5$

73. Which compound contains the metal atom with the highest (most positive) oxidation number?
    (A) $KMnO_4$
    (B) $K_2CrO_4$
    (C) $Cr_2O_3$
    (D) $MnO_2$
    (E) $K_2Cr_2O_7$

74. A system absorbs 2000 J of heat from its surroundings and as a result, does some work on the surroundings. The internal energy $\Delta E$ of the system increases by 500 J. How much work was done?
    (A) zero
    (B) 500 J
    (C) 1000 J
    (D) 1500 J
    (E) 2000 J

75. Of the colors listed below that are part of the visible portion of the electromagnetic spectrum, which one has the highest frequency?
    (A) Red
    (B) Orange
    (C) Green
    (D) Yellow
    (E) Blue

**GO ON TO NEXT PAGE**

## A5.5 Standard Reduction Potentials at 25°C (298 K) for Many Common Half-Reactions

| Half-Reaction | $e°$ (V) | Half-Reaction | $e°$ (V) |
|---|---|---|---|
| $F_2 + 2e^- \rightarrow 2F^-$ | 2.87 | $Cu^+ + e^- \rightarrow Cu$ | 0.52 |
| $Ag^{2+} + e^- \rightarrow Ag^+$ | 1.99 | $O_2 + 2H_2O + 4e^- \rightarrow 4OH^-$ | 0.40 |
| $Co^{3+} + e^- \rightarrow Co^{2+}$ | 1.82 | $Cu^{2+} + 2e^- \rightarrow Cu$ | 0.34 |
| $H_2O_2 + 2H^+ + 2e^- \rightarrow 2H_2O$ | 1.78 | $Hg_2Cl_2 + 2e^- \rightarrow 2Hg + 2Cl^-$ | 0.34 |
| $Ce^{4+} + e^- \rightarrow Ce^{3+}$ | 1.70 | $AgCl + e^- \rightarrow Ag + Cl^-$ | 0.22 |
| $PbO_2 + 4H^+ + SO_4^{2-} + 2e^-$ | | $SO_4^{2-} + 4H^+ + 2e^- \rightarrow H_2SO_3 + H_2O$ | 0.20 |
| $\rightarrow PbSO_4 + 2H_2O$ | 1.69 | $Cu^{2+} + e^- \rightarrow Cu^+$ | 0.16 |
| $MnO_4^- + 4H^+ + 3e^- \rightarrow MnO_2 + 2H_2O$ | 1.68 | $2H^+ + 2e^- \rightarrow H_2$ | 0.00 |
| $2e^- + 2H^+ + IO_4^- \rightarrow IO_3^- + H_2O$ | 1.60 | $Fe^{3+} + 3e^- \rightarrow Fe$ | −0.036 |
| $MnO_4^- + 8H^+ + 5e^- \rightarrow Mn^{2+} + 4H_2O$ | 1.51 | $Pb^{2+} + 2e^- \rightarrow Pb$ | −0.13 |
| $Au^{3+} + 3e^- \rightarrow Au$ | 1.50 | $Sn^{2+} + 2e^- \rightarrow Sn$ | −0.14 |
| $PbO_2 + 4H^+ + 2e^- \rightarrow Pb^{2+} + 2H_2O$ | 1.46 | $Ni^{2+} + 2e^- \rightarrow Ni$ | −0.23 |
| $Cl_2 + 2e^- \rightarrow 2Cl^-$ | 1.36 | $PbSO_4 + 2e^- \rightarrow Pb + SO_4^{2-}$ | −0.35 |
| $Cr_2O_7^{2-} + 14H^+ + 6e^- \rightarrow 2Cr^{3+} + 7H_2O$ | 1.33 | $Cd^{2+} + 2e^- \rightarrow Cd$ | −0.40 |
| $O_2 + 4H^+ + 4e^- \rightarrow 2H_2O$ | 1.23 | $Fe^{2+} + 2e^- \rightarrow Fe$ | −0.44 |
| $MnO_2 + 4H^+ + 2e^- \rightarrow Mn^{2+} + 2H_2O$ | 1.21 | $Cr^{3+} + e^- \rightarrow Cr^{2+}$ | −0.50 |
| $IO_3^- + 6H^+ + 5e^- \rightarrow \frac{1}{2}I_2 + 3H_2O$ | 1.20 | $Cr^{3+} + 3e^- \rightarrow Cr$ | −0.73 |
| $Br_2 + 2e^- \rightarrow 2Br^-$ | 1.09 | $Zn^{2+} + 2e^- \rightarrow Zn$ | −0.76 |
| $VO_2^+ + 2H^+ + e^- \rightarrow VO^{2+} + H_2O$ | 1.00 | $2H_2O + 2e^- \rightarrow H_2 + 2OH^-$ | −0.83 |
| $AuCl_4^- + 3e^- \rightarrow Au + 4Cl^-$ | 0.99 | $Mn^{2+} + 2e^- \rightarrow Mn$ | −1.18 |
| $NO_3^- + 4H^+ + 3e^- \rightarrow NO + 2H_2O$ | 0.96 | $Al^{3+} + 3e^- \rightarrow Al$ | −1.66 |
| $ClO_2 + e^- \rightarrow ClO_2^-$ | 0.954 | $H_2 + 2e^- \rightarrow 2H^-$ | −2.23 |
| $2Hg^{2+} + 2e^- \rightarrow Hg_2^{2+}$ | 0.91 | $Mg^{2+} + 2e^- \rightarrow Mg$ | −2.37 |
| $Ag^+ + e^- \rightarrow Ag$ | 0.80 | $La^{3+} + 3e^- \rightarrow La$ | −2.37 |
| $Hg_2^{2+} + 2e^- \rightarrow 2Hg$ | 0.80 | $Na^+ + e^- \rightarrow Na$ | −2.71 |
| $Fe^{3+} + e^- \rightarrow Fe^{2+}$ | 0.77 | $Ca^{2+} + 2e^- \rightarrow Ca$ | −2.76 |
| $O_2 + 2H^+ + 2e^- \rightarrow H_2O_2$ | 0.68 | $Ba^{2+} + 2e^- \rightarrow Ba$ | −2.90 |
| $MnO_4^- + e^- \rightarrow MnO_4^{2-}$ | 0.56 | $K^+ + e^- \rightarrow K$ | −2.92 |
| $I_2 + 2e^- \rightarrow 2I^-$ | 0.54 | $Li^+ + e^- \rightarrow Li$ | −3.05 |

# Advanced Placement Chemistry Equations and Constants

## ATOMIC STRUCTURE

$E = h\nu$

$\lambda = \frac{h}{m\nu}$

$c = \lambda\nu$

$p = m\nu$

$E_n = \frac{-2.178 \times 10^{-18}}{n^2}$ joule

## EQUILIBRIUM

$K_a = \frac{[H^+][A^-]}{[HA]}$

$E^\Omega =$

$K_w = [OH^-][H^+] = 1.0 \times 10^{-14}$ at 25°C

$\quad = K_a \times K_b$

$pH = -\log[H^+],\ pOH = -\log[OH^-]$

$14 = pH + pOH$

$pH = pK_a + \log\frac{[A^-]}{[HA]}$

$pOH = pK_b + \log\frac{[HB^+]}{[B]}$

$pK_a = -K_a,\ pK_b = -\log K_b$

$K_p = K_c(RT)^{\Delta n}$

where $\Delta n =$ moles product gas − moles reactant gas

## THERMOCHEMISTRY/KINETICS

$\Delta S° = \sum S°$ products $- \sum S°$ reactants

$\Delta H° = \sum \Delta H_f°$ products $- \sum \Delta H_f°$ reactants

$\Delta G° = \sum \Delta G_f°$ products $- \sum \Delta G_f°$ reactants

$\Delta G° = \Delta H° - T\Delta S°$

$\quad = -RT \ln K = -2.303 RT \log K$

$\quad = -n \Im E°$

$\Delta G = \Delta G° + RT \ln Q = \Delta G° + 2.303 RT \log Q$

$q = mc\Delta T$

$C_p = \frac{\Delta H}{\Delta T}$

$\ln[A]_t - \ln[A]_0 = -kt$

$\frac{1}{[A]_t} - \frac{1}{[A]_0} = kt$

$\ln k = \frac{-E_a}{R}\left(\frac{1}{T}\right) + \ln A$

$E$ = energy

$\lambda$ = wavelength

$\nu$ = velocity

$m$ = mass

$\nu$ = frequency

$p$ = momentum

$n$ = principal quantum number

Speed of light, $c = 3.0 \times 10^8$ ms$^{-1}$

Planck's constant, $h = 6.63 \times 10^{-34}$ Js

Boltzmann's constant, $k = 1.38 \times 10^{-23}$ J K$^{-1}$

Avogadro's number $= 6.022 \times 10^{23}$ mol$^{-1}$

Electron charge, $e = -1.602 \times 10^{-19}$ coulomb

1 electron volt per atom $= 96.5$ kJ mol$^{-1}$

## EQUILIBRIUM CONSTANTS

$K_a$ (weak acid)

$K_b$ (weak base)

$K_w$ (water)

$K_p$ (gas pressure)

$K_c$ (molar concentrations)

$\quad S°$ = standard entropy

$\quad H°$ = standard enthalpy

$\quad G°$ = standard free energy

$\quad E°$ = standard reduction potential

$\quad T$ = temperature

$\quad n$ = moles

$\quad m$ = mass

$\quad q$ = heat

$\quad c$ = specific heat capacity

$\quad C_p$ = molar heat capacity at constant pressure

$\quad E_a$ = activation energy

$\quad k$ = rate constant

$\quad A$ = frequency factor

Faraday's constant, $\Im = 96,500$ per mole of electrons

Gas constant, $R \quad = 8.31$ J mol$^{-1}$ K$^{-1}$

$\quad = 0.0821$ L atm mol$^{-1}$ K$^{-1}$

$\quad = 8.31$ volt coulomb mol$^{-1}$ K$^{-1}$

## GASES, LIQUIDS, AND SOLUTIONS

$PV = nRT$

$\left(P + \frac{n^2 a}{V^2}\right)(V - nb) = nRT$

$P_A = P_{total} \times X_A$, where $X_A = \frac{\text{moles A}}{\text{total moles}}$

$P_{total} = P_A + P_B + P_C + \ldots$

$n = \frac{m}{M}$

$°K = °C + 273$

$\frac{P_1 V_1}{T_1} = \frac{P_2 V_2}{T_2}$

$D = \frac{m}{V}$

$$u_{rms} = \sqrt{\frac{3kT}{m}} = \sqrt{\frac{3RT}{M}}$$

KE per molecule $= \frac{1}{2}mv^2$

KE per mole $= \frac{3}{2}RT$

$$\frac{r_1}{r_2} = \sqrt{\frac{M_2}{M_1}}$$

molarity, $M$ = moles solute per liter solution

molality = moles solute per kilogram solvent

$\Delta T_f = iK_f \times$ molality

$\Delta T_b = iK_b \times$ molality

$\pi = MRT$

$A = abc$

## OXIDATION-REDUCTION; ELECTROCHEMISTRY

$Q = \frac{[C]^c[D]^d}{[A]^a[B]^b}$, where $a\,A + b\,B \rightarrow c\,C + d\,D$

$I = \frac{q}{t}$

$E_{cell} = E^\circ_{cell} - \frac{RT}{n\Im}\ln Q = E^\circ_{cell} - \frac{0.0592}{n}\log Q @25°C$

$\log K = \frac{nE^\circ}{0.0592}$

$P =$ pressure
$V =$ volume
$T =$ temperature
$n =$ number of moles
$D =$ density
$m =$ mass
$v =$ velocity

$u_{rms} =$ root-mean-square speed

$KE =$ kinetic energy

$r =$ rate of effusion

$M =$ molar mass

$\pi =$ osmotic pressure

$i =$ van't Hoff factor

$K_f =$ molal freezing-point depression constant

$K_b =$ molal boiling-point elevation constant

$A =$ absorbance

$a =$ molar absorptivity

$b =$ path length

$c =$ concentration

$Q =$ reaction quotient

$I =$ current (amperes)

$q =$ charge (coulombs)

$t =$ time (seconds)

$E^\circ =$ standard reduction potential

$K =$ equilibrium constant

Gas constant, $R = 8.31$ J mol$^{-1}$ K$^{-1}$
$= 0.0821$ L atm mol$^{-1}$ K$^{-1}$
$= 8.31$ volt coulomb mol$^{-1}$ K$^{-1}$

Boltzmann's constant, $k = 1.38 \times 10^{-23}$ J K$^{-1}$

$K_f$ for $H_2O = 1.86$ K kg mol$^{-1}$

$K_b$ for $H_2O = 0.512$ K kg mol$^{-1}$

1 atm = 760 mm Hg
= 760 torr

STP = 0.000°C and 1.000 atm

Faraday's constant, $\Im = 96{,}500$ coulombs per mole of electrons

## Section II: Free-Response Questions
### Time: 90 minutes
### Number of Questions: 6

Allow yourself no more than 90 minutes to answer these questions. A calculator may be used only on questions 1, 2, and 3. You may refer to the Equations Sheet, the Periodic Table, and the Standard Reduction Table throughout this section.

## Part A

All students must answer this first question. It is worth 20% of Section II.

1. For the reaction $PCl_3(g) + Cl_2(g) \rightleftharpoons PCl_5(g)$, $K_p = 0.0870$ at $300°C$. A flask is charged with 0.30 atm $PCl_3$, 0.60 atm $Cl_2$, and 0.10 atm $PCl_5$ at this temperature.
   (a) Determine whether the above conditions (the added gases) are at equilibrium. If not, determine to which direction the reaction must proceed to reach equilibrium.
   (b) Calculate the equilibrium partial pressures of the gases.
   (c) Calculate $K_c$ for the above equilibrium.
   (d) Predict and justify what effect increasing the volume of the system will have on the mole fraction of $PCl_5$ in the mixture.
   (e) The reaction as written is exothermic. Predict and justify what effect that decreasing the temperature of the system will have on the mole fraction of $PCl_5$ in the mixture.

## Part B

Look over Questions 2 and 3 that follow. Then select one of these two and answer it completely. This question (either #2 or #3) is worth 20% of Section II. (A second problem will not be scored.)

2. A 0.495 M solution of nitrous acid, $HNO_2$, has a pH of 1.83.
   (a) Find the $[H^+]$ and the percent ionization of nitrous acid in this solution.
   (b) Write the equilibrium expression and calculate the value of $K_a$ for nitrous acid.
   (c) Calculate the pH of the solution formed by adding 1.00 g of $NaNO_2$ (molar mass = 69.0 g·mol$^{-1}$) to 750. mL of 0.0125 M solution of nitrous acid.
   (d) Determine the pH of a solution formed by adding 1.00 g of $NaNO_2$ to 750 mL of water, $H_2O$.
   (e) Sketch the pH curve that results when 20.0 mL of a 0.0125 M nitrous acid solution are titrated with 0.0125 M NaOH solution. Label the axes, the equivalence point, and the buffer region clearly.

3. Given the following data:
   $$Fe^{2+} + 2e^- \longrightarrow Fe \qquad E°_{red} = -0.44 \text{ V}$$
   $$Ag^+ + e^- \longrightarrow Ag \qquad E°_{red} = 0.80 \text{ V}$$
   Answer the following questions with respect to the reaction
   $$Fe^{2+}(aq) + 2Ag(s) \longrightarrow Fe(s) + 2Ag^+(aq).$$
   (a) What is the cell potential, $E°$, for the reaction?
   (b) Is the reaction spontaneous at standard state conditions?
   (c) What is the value of $E°$ at equilibrium?
   (d) What is the value of the equilibrium constant at $25°C$?
   (e) If $[Fe^{2+}] = 0.100$ M and $[Ag^+] = 0.0100$ M, what is the magnitude of $E$ at $25°C$?
   (f) For the reaction that is spontaneous, what is the maximum amount of work that can be performed?

**NO CALCULATORS ARE TO BE USED FOR THE REMAINDER OF THE TEST.**

## Part C

You must answer Question 4. The section II scoring weight for this question is 15%.

4.  Write formulas that show both reactants and products for five of the following eight selections. All solutions are aqueous. Note that ions are to be shown if a substance is extensively ionized. Omit spectator ions; do not balance.
    (a) Magnesium turnings are added to a solution of iron(III) chloride.
    (b) A solution of copper(II) sulfate is added to a solution of barium hydroxide.
    (c) Water is added to a sample of pure calcium hydride.
    (d) Excess potassium hydroxide is added to a solution of aluminum nitrate.
    (e) Chlorine gas is bubbled through a solution of potassium bromide.
    (f) A small piece of potassium is added to distilled water.
    (g) Solid ammonium carbonate is heated.
    (h) Hydrogen gas is passed over hot copper(II) oxide.

## Part D

Spend about 40 minutes on this part of the examination. Answering these questions gives you an opportunity to demonstrate your ability to present your material in clear, orderly, and convincing English. Your answers will be graded on the basis of accuracy, the kinds of information you include to support your responses, and the importance of the descriptive material used. Be specific; general, all-encompassing answers will not be graded as highly as detailed answers with examples and equations.

5.  The Laboratory-Based Question: Answer the following essay question. The Section II score weighting is 15% for each question in this part (total 45%).

    A student weighs out equal amounts of magnesium hydroxide, calcium carbonate, calcium sulfate, and sodium bicarbonate but carelessly forgets to label the containers in which each sample is placed. If the only chemicals available are a bottle of dilute hydrochloric acid and some distilled water, describe a procedure that could be used to identify each solid. Answer the following essay question. The Section II score weighting value is 15%

6.  Ozone reacts with nitrogen dioxide according to the following equation:

    $$2NO_2(g) + O_3(g) \rightleftharpoons N_2O_5(g) + O_2(g); \Delta H° = -198 \text{ kJ}, \Delta S° = -168 \text{ J•K}^{-1}$$

    a.  Calculate $\Delta H_f°$ for $NO_2(g)$ in kJ•mol$^{-1}$. [$\Delta H_f°$ kJ•mol$^{-1}$: $O_3(g) = 143$, $N_2O_5(g) = 11$]
    b.  Account for the sign of $\Delta S°$.
    c.  State and explain how the spontaneity of this reaction will vary with increasing temperature.

d. Use the rate data below to determine the rate law for the reaction of $NO_2(g)$ and $O_3(g)$.

| $NO_2(g)$, M | $O_3(g)$, M | Rate M•s$^{-1}$ |
|---|---|---|
| 0.0015 | 0.0025 | $4.8 \times 10^{-8}$ |
| 0.0022 | 0.0025 | $7.2 \times 10^{-8}$ |
| 0.0022 | 0.0050 | $1.4 \times 10^{-7}$ |

e. The following mechanisms have been proposed for this reaction. Discuss the suitability of each to account for the rate law obtained.

| Mechanism I | Mechanism II |
|---|---|
| $NO_2 + NO_2 \longrightarrow NO_3 + NO$ slow | $NO_2 + O_3 \rightleftharpoons NO_3 + O_2$ fast |
| $NO_3 + NO_2 \rightleftharpoons N_2O_5$ fast | $NO_3 + NO_2 \longrightarrow N_2O_5$ slow |
| $NO + O_3 \longrightarrow NO_2 + O_2$ fast | |

f. Describe and account for any change expected in the rate of this reaction as the temperature is increased.

Answer either question 7 or 8. Only one of the two will be graded. The Section II score weighted for the question you choose is 15 percent.

7. Account for the following statements or observations in terms of atomic-, ionic- or molecular-level explanations.
a. Magnesium exists as 2+ ions rather than 1+ ions in all of its compounds despite the fact that the second ionization energy of a magnesium atom is more than twice as great as the first ionization energy.
b. Carbon dioxide ($CO_2$) is a gas at room temperature but silicon dioxide ($SiO_2$) is a high melting solid.
c. Nitrogen forms $NF_3$ but not $NF_5$ whereas phosphorus forms $PF_3$ and $PF_5$. The trifluorides are both trigonal pyramidal and the pentafluoride is trigonal bipyramidal.
 d. Calcium oxide has a much higher melting point (2580 °C) than potassium fluoride (858 °C).

8. Use principles of bonding, intermolecular forces and/or kinetic molecular theory to explain the cause of the following:
a. $CH_4$ boils at –164 °C whereas $CH_3Cl$ boils at –24.2 °C.
b. $ICl(l)$ has a higher boiling point than $Br_2(l)$.
c. $CH_3OH(l)$ is more soluble than $CCl_4(l)$ in liquid $NH_3$.
d. Two miscible liquids, methanol, $CH_3OH$, and water, can be separated from each other into layers by the addition of potassium carbonate, $K_2CO_3$.

*END OF EXAMINATION*

# Answers to Practice Test 2

## Multiple-Choice Answers

Using the table below, score your test.

Determine how many questions you answered correctly and how many you answered incorrectly. You will find explanations of the answers on the following pages.

| | | | | |
|---|---|---|---|---|
| 1. D | 2. B | 3. C | 4. B | 5. B |
| 6. E | 7. C | 8. A | 9. B | 10. B |
| 11. B | 12. A | 13. B | 14. D | 15. A |
| 16. C | 17. B | 18. C | 19. D | 20. B |
| 21. D | 22. D | 23. C | 24. B | 25. D |
| 26. B | 27. D | 28. D | 29. E | 30. C |
| 31. E | 32. A | 33. B | 34. E | 35. B |
| 36. B | 37. A | 38. B | 39. E | 40. D |
| 41. B | 42. B | 43. E | 44. D | 45. E |
| 46. E | 47. B | 48. C | 49. C | 50. B |
| 51. D | 52. A | 53. E | 54. E | 55. D |
| 56. D | 57. B | 58. D | 59. E | 60. E |
| 61. C | 62. E | 63. C | 64. A | 65. B |
| 66. E | 67. D | 68. E | 69. A | 70. C |
| 71. B | 72. C | 73. A | 74. D | 75. E |

## Calculate Your Score:

Number answered correctly: _____
Adjust ¼ point for guessing penalty:
   Count the number of questions you
   answered incorrectly, multiply by .25, and subtract: – _____
Determine your adjusted score: _____

## What Your Score Means:

Because the test is different each year, the scoring is a little different. But generally, if you scored 25 or more on the multiple-choice questions, you'll most likely get a 3 or better on the test. If you scored 35 or more, you'll probably score a 4 or better. And if you scored a 50 or more, you'll most likely get a 5. Keep in mind that the multiple-choice section is worth 45% of your final grade, and the free-response section is worth 55% of your final grade. To learn more about the scoring for the free-response questions, turn to the last page of this section.

## ANSWERS AND EXPLANATIONS

### MULTIPLE-CHOICE ANSWERS

1.  **ANSWER:** D  carbon**ate**  (*Chemistry* 6th ed. pages 62–72 / 7th ed. pages 57–67)

2.  **ANSWER:** B  phosph**ite** (*Chemistry* 6th ed. pages 62–72 / 7th ed. pages 57–67)

3.  **ANSWER:** C  ammon**ium** (*Chemistry* 6th ed. pages 62–72 / 7th ed. pages 57–67)

4.  **ANSWER:** B  chlor**ite** (*Chemistry* 6th ed. pages 62–72 / 7th ed. pages 57–67)

5.  **ANSWER:** B  alcohol (*Chemistry* 6th ed. pages 1057–1064 / 7th ed. pages 1010–1016)

6.  **ANSWER:** E  ketone (*Chemistry* 6th ed. pages 1057–1064 / 7th ed. pages 1010–1016)

7.  **ANSWER:** C  ester (*Chemistry* 6th ed. pages 1057–1064 / 7th ed. pages 1010–1016)

8.  **ANSWER:** A (*Chemistry* 6th ed. pages 492–497 / 7th ed. pages 467–470)

9.  **ANSWER:** B  (*Chemistry* 6th ed. pages 492–497 / 7th ed. pages 467–470)

10. **ANSWER:** B  (*Chemistry* 6th ed. pages 492–497 / 7th ed. pages 467–470)

11. **ANSWER:** B  (*Chemistry* 6th ed. pages 324–330, 915 / 7th ed. pages 309–314, 876–877)

12. **ANSWER:** A (*Chemistry* 6th ed. pages 324–330 / 7th ed. pages 309–314)

13. **ANSWER:** B  (*Chemistry* 6th ed. pages 324–330 / 7th ed. pages 309–314)

14. **ANSWER:** D (*Chemistry* 6th ed. pages 324–330 / 7th ed. pages 309–314)

15. ANSWER: A (*Chemistry* 6th ed. pages 657–662 and 928 / 7th ed. pages 626–631 and 889)

16. ANSWER: C (*Chemistry* 6th ed. pages 466 and 931 / 7th ed. pages 442–443 and 891–892)

17. ANSWER: B  (*Chemistry* 6th ed. page 674 / 7th ed. page 643)

18. ANSWER: C Different atomic numbers, therefore different number of protons. Same mass numbers, therefore same total number of protons and neutrons (*Chemistry* 6th ed. pages 53–56 / 7th ed. pages 49–52).

19. ANSWER: D Molar mass is a physical property; it does not involve chemical change. It is also intensive, as it is independent of the amount present (*Chemistry* 6th ed. page 251 / 7th ed. pages 237–238).

20. ANSWER: B

    Kr: 8 valence electrons

    2F: 14 valence electrons

    Total: 22 val. electrons. Around Kr there are 2 bonding pairs and 3 lone pairs = 5 pairs total

    Therefore: trigonal bipyramidal (*Chemistry* 6th ed. pages 389–400 / 7th ed. pages 367–377)

21. ANSWER: D The absolute value for $m_l$ cannot be greater than $l$ (*Chemistry* 6th ed. pages 307–309 / 7th ed. pages 293–294).

22. ANSWER: D Strong intermolecular forces means that additional energy must be supplied in order to increase the physical distance between the molecules when they move from liquid to vapor state. This results in a greater heat of vaporization. It also means that fewer molecules at a given temperature are able to move into the vapor state; thus the vapor pressure is lower than for other liquids with weaker intermolecular forces (*Chemistry* 6th ed. pages 450–453 and 518 / 7th ed. pages 426–429 and 492).

23. ANSWER: C

    $$^{241}_{95}\text{Am} + {}^{4}_{2}\text{He} \longrightarrow {}^{243}_{97}\text{Bk} + 2{}^{1}_{0}\text{n}$$

    To balance the mass numbers, the number of neutrons must be 2 (*Chemistry* 6th ed. pages 878–883 / 7th ed. pages 841–846).

24. ANSWER: B Both N–O bonds are double bonds; the average bond order is 2.0 (*Chemistry* 6th ed. pages 383–388 / 7th ed. pages 362–367).

25. **ANSWER: D** $O_3$ has one single bond and one double bond. The double bond can be drawn in either position. The nitrate ion has two single bonds and one double bond, the double bond is drawn in one of three positions (*Chemistry* 6th ed. pages 383–388 / 7th ed. pages 362–367).

26. **ANSWER: B** Weighted average = $(0.80 \times 122) + (0.20 \times 120) = 121.6$ (*Chemistry* 6th ed. pages 348–352 / 7th ed. pages 330–333)

27. **ANSWER: D**

Mole fraction of pentane in solution = 3/5

Mole fraction of hexane in solution = 2/5

Partial pressure due to pentane = 3/5 × 88.7 torr = 53.2 torr. This is Raoult's law.

Partial pressure due to hexane = 2/5 × 44.5 torr = 17.8 torr and therefore the total vapor pressure = 53.2 +17.8 =71.0 torr (*Chemistry* 6th ed. pages 524–531 / 7th ed. pages 497–504)

28. **ANSWER: D** Molarity = number of moles/volume of solution (in liters) = 0.24 moles/0.300 L = 0.80 M (*Chemistry* 6th ed. pages 140–148 and 512–515 / 7th ed. pages 133–140 and 485–488).

29. **ANSWER: E** There are two nitrate ions per formula unit, so the concentration is twice $3.6 \times 10^{-4}$M (*Chemistry* 6th ed. pages 140–148 / 7th ed. pages 133–140).

30. **ANSWER: C** The only difference in a conjugate acid-base pair is the removal of a $H^+$ going from the acid to the base or the addition of a $H^+$ going from the base to acid (*Chemistry* 6th ed. pages 657–662 / 7th ed. pages 626–631).

31. **ANSWER: E** The reactants diminish in concentration (C and D). D [$O_2$] goes down 5/4 faster than C [$NH_3$]. The products are formed (A and B). A [$H_2O$] is formed 6/4 times faster than B [NO] (*Chemistry* 6th ed. pages 102–112 / 7th ed. pages 96–106).

32. **ANSWER: A**

$K_w/K_a = 1.0 \times 10^{-14}/1.0 \times 10^{-7} = 1.0 \times 10^{-7}$

(*Chemistry* 6th ed. pages 666–682 / 7th ed. pages 635–650)

33. **ANSWER: B**

Reactants are NO and $H_2$.

Products are $N_2$ and $H_2O$.

Intermediates are $N_2O_2$ and $N_2O$; they are formed and used up (*Chemistry* 6th ed. pages 579–582 / 7th ed. pages 549–552).

34. **Answer: E** Slowest step: Rate = $k[N_2O_2][H_2]$.

    Assume first step reaches steady state: $K' = [N_2O_2]/[NO]^2$, so $[N_2O_2] = K[NO]^2$ and Rate = $kK[NO]^2[H_2]$ (*Chemistry* 6th ed. pages 579–582 / 7th ed. pages 549–552).

35. **Answer: B** The molar volume of an ideal gas is independent of its mass. Review the conditions necessary for the change in volume of a gas (*Chemistry* 6th ed. pages 212–222 / 7th ed. pages 199–208).

36. **Answer: B** The partial pressure of the helium = 5.0 atm × (2.0L/8.0L) = 1.25 atm. (*Chemistry* 6th ed. pages 206–219 / 7th ed. pages 194–206)

37. **Answer: A** Review the conditions under which a gas becomes less than ideal and what is responsible for this behavior (*Chemistry* 6th ed. pages 222–224 / 7th ed. pages 208–210).

38. **Answer: B** The pair must be a weak electrolyte and its conjugate partner (*Chemistry* 6th ed. pages 716–726 / 7th ed. pages 684–693).

39. **Answer: E**

$$K = \frac{[ClF_3]^2}{[F_2]^3[Cl_2]} = \frac{(3)^2}{(2)^3(2.5)} = 0.45$$

(*Chemistry* 6th ed. pages 561–563 / 7th ed. pages 532–534)

40. **Answer: D** The mass numbers must balance: 99 on both sides.

    The atomic numbers must balance: 42 on both sides, so the unknown particle has a mass number of zero and an atomic number of –1. This is a β particle (an electron) (*Chemistry* 6th ed. pages 878–883 / 7th ed. pages 841–846).

41. **Answer: B**

Total valence electrons = 2 × 6 for S + 6 for O = 18 total.

Divide 18 by 8, for 2 bonding pairs, leaving 2 electrons or one lone pair (*Chemistry* 6th ed. pages 375–379 / 7th ed. pages 354–358).

42. **Answer: B**

$K_{sp} = [Cd^{2+}][S^{2-}] = 3.6 \times 10^{-29}$; $[Cd^{2+}] = \sqrt{3.6 \times 10^{-29}} = \sqrt{36 \times 10^{-30}} = 6.0 \times 10^{-15}$

(*Chemistry* 6th ed. pages 751–760 / 7th ed. pages 717–724)

43. ANSWER: **E** The arrow must point down to show emission and the largest difference corresponds to the greatest difference in energy which means highest frequency and shortest wavelength (*Chemistry* 6th ed. pages 297–304 / 7th ed. pages 284–290).

44. ANSWER: **D**

In $S_4O_6^{2-}$, $4x + 6(-2) = -2$, so $\times = +2\ \frac{1}{2}$

(*Chemistry* 6th ed. pages 164–171 / 7th ed. pages 154–162)

45. ANSWER: **E** The difference in levels is 1.5 atm. The pressure outside is 1.0 atm; therefore, the total pressure of the gas is 2.5 atm (*Chemistry* 6th ed. pages 190–192 / 7th ed. pages 179–181).

46. ANSWER: **E** Rate at which NO is used = rate of production of $N_2$ × (6/5) = 36 mol/L/s. (*Chemistry* 6th ed. pages 557–561 / 7th ed. pages 527–532).

47. ANSWER: **B** An element in its standard state has a defined standard heat of formation of zero (*Chemistry* 6th ed. pages 260–267 / 7th ed. pages 246–252).

48. ANSWER: **C** Only the mole fraction of the solute and thus the mole fraction of the solvent (Raoult's law) is of consequence (*Chemistry* 6th ed. pages 524–531 / 7th ed. pages 497–504).

49. ANSWER: **C** The volume ultimately remains unchanged. The temperature increases by a factor of 3 and from the given information, one can observe that the volume increases by a factor of three. Thus, to restore the sample to its original volume, the pressure must be increased by a factor of 3 as well (*Chemistry* 6th ed. pages 198–202 / 7th ed. pages 186–190).

50. ANSWER: **B** $Ag_2CO_3 \rightleftharpoons 2Ag^+ + CO_3^{2-}$

Silver nitrate contains $Ag^+$, a common ion; so this will suppress solubility.

Nitric acid will react with carbonate ion, the conjugate base of a very weak acid, so this will increase solubility because the carbonate ion will be removed from solution. Sodium carbonate contains the carbonate ion, a common ion and so this will also suppress solubility (*Chemistry* 6th ed. pages 622–631 and 636–642 / 7th ed. pages 591–599 and 604–610).

51. ANSWER: **D** Review what can happen to $K_w$ as temperature is increased and how this would affect values of pH. Review definitions of strong acids (*Chemistry* 6th ed. pages 662–665 / 7th ed. pages 631–634).

52. ANSWER: **A** 32 grams $O_2$ = 1.0 mol $O_2$

According to the equation, 1.0 mole $O_2$ produces 2/3 mole of ZnO or 0.66 mol (*Chemistry* 6th ed. pages 108–112 / 7th ed. pages 102–106).

53. ANSWER: E  $i = 3$, so the concentration of ions is $3 \times 0.20 = 0.60$. This is the greatest increase in the disorder of the solution (*Chemistry* 6th ed. pages 540–543 / 7th ed. pages 512–513).

54. ANSWER: E  Review what are considered strong acids and strong bases, remembering that any other acid or base is therefore considered weak, by definition. Also remember that all salts are considered strong electrolytes (*Chemistry* 6th ed. pages 136–140 / 7th ed. pages 129–133).

55. ANSWER: D

15 protons means the atomic number is 15, which is P.

16 neutrons means that the mass number is 15+16 = 31.

18 electrons compared to 15 protons means that the charge is 3– (*Chemistry* 6th ed. pages 53–56 / 7th ed. pages 49–52).

56. ANSWER: D

Rate $= k[A]^2[B] = (2)^2(1/2) = 4 \times \frac{1}{2} = 2$ (*Chemistry* 6th ed. pages 563–567 / 7th ed. pages 534–538).

57. ANSWER: B  When drawing structures of organic molecules to determine number of isomers, start by drawing a straight chain hydrocarbon, making certain each carbon atom has four bonds to it. Then remove one methyl ($–CH_3$) from either end and place it on a carbon that is not on the end. When you do this with $C_4H_{10}$, you produce only one additional isomer, so the total number of isomers is two (*Chemistry* 6th ed. pages 1044–1052 / 7th ed. pages 997–1005).

58. ANSWER: D  Change the sign for the oxidation at the anode and add the two half-cell potentials.

+ 2.89 – 0.14V = +2.75 V.

(*Chemistry* 6th ed. pages 830–837 / 7th ed. pages 794–800)

59. ANSWER: E  The freezing of water involves the release of energy since the water molecules are moving very close to one another to maximize the formation of hydrogen bonds. Bond formation is always exothermic (*Chemistry* 6th ed. pages 810–814 / 7th ed. pages 774–778).

60. ANSWER: E

Heat of reaction = heat of formation of products – heat of formation of reactants

$$= [(2 \times -394) + (4 \times -286)] - [(2 \times -239)] = -1454 \text{ kJ}$$

(*Chemistry* 6th ed. pages 256–260 / 7th ed. pages 242–246)

61. ANSWER: **C**  For example: For the oxidation, $5 \text{ Zn}(s) \longrightarrow 5 \text{ Zn}^{2+}(aq)$, 10 electrons are lost. Ten electrons are gained in the reduction, so the equation is balanced (*Chemistry* 6th ed. pages 172–179 / 7th ed. pages 162–168).

62. ANSWER: **E**  II is an expression of the second law of thermodynamics.

    If the entropy of the system decreases, the entropy of the surroundings must increase, which is statement III. This can happen only if the reaction is exothermic, which is statement I (*Chemistry* 6th ed. pages 790–795 / 7th ed. pages 755–759).

63. ANSWER: **C**

    $$1\text{C}_2\text{H}_8\text{N}_2 + 2\text{N}_2\text{O}_4 \longrightarrow 3\text{N}_2 + 2\text{CO}_2 + 4\text{H}_2\text{O}$$

    (*Chemistry* 6th ed. pages 104–108 / 7th ed. pages 98–102)

64. ANSWER: **A** (*Chemistry* 6th ed. page 148 / 7th ed. page 140)

65. ANSWER: **B**  (*Chemistry* 6th ed. pages 588–595 / 7th ed. pages 557–563)

66. ANSWER: **E**  The process is endothermic, so $\Delta H = +$.

    There in an increase in disorder so $\Delta S = +$.

    The process happens, so $\Delta G = -$ (*Chemistry* 6th ed. pages 795–806 / 7th ed. pages 759–770).

67. ANSWER: **D**  Average speeds increase as the molecular masses decreases (*Chemistry* 6th ed. pages 212–222 / 7th ed. pages 199–208).

68. ANSWER: **E**

    $$100 \text{ mL} \times 6.50 \text{ M} = 2000 \text{ mL} \times ??\text{M}.$$

    new molarity = 0.325M

    (*Chemistry* 6th ed. pages 140–148 and 512–515 / 7th ed. pages 133–140 and 485–488)

69. ANSWER: **A**  The reaction is endothermic in the forward direction, so increasing temperature by adding heat shifts the reaction to the right.  Decreasing volume will force the reaction to the side with

fewer moles of gas, the left (*Chemistry* 6th ed. pages 636–642 / 7th ed. pages 604–610).

70. ANSWER: C  Review the kinetic molecular theory of gases. Remember that temperature is a measure of average kinetic energy. Also remember that kinetic energy depends on both the mass of the molecules as well as their average root mean square speed. Hence, two gases at the same temperature will be traveling at speeds that are in inverse relationship to their masses (*Chemistry* 6th ed. pages 212–219 / 7th ed. pages 199–206).

71. ANSWER: B (*Chemistry* 6th ed. pages 695–696 / 7th ed. pages 662–663)

72. ANSWER: C (*Chemistry* 6th ed. pages 317–324 / 7th ed. pages 302–309)

73. ANSWER: A  In $KMnO_4$, Mn has an oxidation number of +7 whereas the Cr in both $K_2CrO_4$ and $K_2Cr_2O_7$ has an oxidation number of +6. The Cr on $Cr_2O_3$ has an oxidation number of +3 (*Chemistry* 6th ed. pages 164–171 / 7th ed. pages 154–162).

74. ANSWER: D

$\Delta E = q + w$

500 J = 2000 J + $w$; $w$ = –1500 J or 1500 J of work was done *by* the system (*Chemistry* 6th ed. pages 242–248 / 7th ed. pages 229–235).

75. ANSWER: E (*Chemistry* 6th ed. pages 290–292 / 7th ed. pages 275–277)

## SECTION II: FREE-RESPONSE ANSWERS

**Question 1: Answers**
(a)

$$Q = \frac{P_{PCl_5}}{P_{PCl_3} \times P_{Cl_2}} = \frac{0.10}{(0.30)(0.60)} = 0.56$$

Since 0.56 (Q) > 0.0870 (K), the reaction proceeds to the left.
(b)

|          | $PCl_3(g)$ + | $Cl_2(g)$ ⇌ | $PCl_5(g)$ |
|----------|--------------|-------------|------------|
| Initial  | 0.30 atm     | 0.60 atm    | 0.10 atm   |
| Change   | +$x$         | +$x$        | –$x$       |
| Equil    | (0.30 + $x$)atm | (0.60 + x) atm | (0.10 – x) atm |

Since the reaction proceeds to the left, $PCl_5$ must decrease and the reactants increase.

$$K_p = \quad 0.0870 = \quad \frac{(0.10 - x)}{(0.30 + x)(0.60 + x)} \quad = \quad \frac{(0.10 - x)}{0.18 + 0.90x + x^2}$$

Using the quadratic equation and solving for $x$, $x = 0.078$.

Therefore, at equilibrium, $P_{PCl_3} = (0.10 - 0.078) = 0.022$ atm $= 0.02$ atm

$$P_{PCl_5} = (0.30 + 0.078) = 0.378 \text{ atm} = 0.38 \text{ atm}$$

$$P_{PCl_2} = (0.60 + 0.078) = 0.678 \text{ atm} = 0.68 \text{ atm}$$

(c) Since $K_p = K_c(RT)^{\Delta n}$ where $\Delta n = 1 - 2 = -1$, so $K_c = K_p(RT) = 0.0870(0.08206)(573)$
$$= 4.09$$

(d) Increasing the volume of the container favors the process where more moles of gas are produced, so the reverse reaction is favored; the equilibrium shifts to the left; the mole fraction of $PCl_5$ decreases.

(e) For an exothermic reaction, decreasing the temperature increases the value of $K$, favoring the products. The partial pressure of $PCl_5$ increases (*Chemistry* 6th ed. pages 617–620 and 622–642 / 7th ed. pages 586–588 and 591–610).

## Question 2: Answers

(a) pH = 1.83  $[H^+] = 1.5 \times 10^{-2}M$

So percent ionized $= \dfrac{0.015 \times 100}{0.495} = 3.0\%$

(b)

$$K_a = \frac{[H^+][NO_2^-]}{[HNO_2]} = \frac{(1.5 \times 10^{-2})^2}{0.495} = 4.5 \times 10^{-4}$$

The answer will vary slightly if the ionized acid is subtracted from the initial amount of acid present.

$$K_a = \frac{(1.5 \times 10^{-2})^2}{0.480} = 4.7 \times 10^{-4}$$

(c)

$$1.00 \text{ g NaNO}_2 \times \frac{1 \text{ mole}}{69.0 \text{ grams}} = \frac{0.0145 \text{ mol NO}_2^-}{0.750 \, L} = 0.0193 \text{ M NO}_2^-$$

Then using the equilibrium expression, solve for $[H^+]$:

$$4.5 \times 10^{-4} = \frac{[H+][0.0193]}{0.0125} \quad [H^+] = 2.9 \times 10^{-4} \quad pH = 3.54$$

If you use the other value for $K_a$ ($4.7 \times 10^{-4}$), then pH = 3.52

(d) This is a hydrolysis problem: $NO_2^- + H_2O \rightleftharpoons HNO_2 + OH^-$.

$$K_b = K_w/K_a = (1.0 \times 10^{-14})/4.5 \times 10^{-4} = 2.2 \times 10^{-11}$$

$$2.2 \times 10^{-11} = \frac{[HNO_2][OH^-]}{0.0193}$$

$[OH^-]^2 = 4.2 \times 10^{-13}$; $[OH^-] = 6.5 \times 10^{-7}$; so pOH = 6.19 or pH = 7.81

(e)

(*Chemistry* 6th ed. pages 729–751 / 7th ed. pages 696–716)

## Question 3: Answers

(a)
$$Fe^{2+} + 2e^- \longrightarrow Fe \qquad\qquad E^\circ_{red} = -0.44 \text{ V}$$
$$2(Ag \longrightarrow Ag^+ + e^-) \qquad\qquad E^\circ_{oxid} = -0.80 \text{ V}$$

$$Fe^{2+} + 2Ag \longrightarrow 2Ag^+ + Fe \qquad E^\circ = -1.24 V$$

(b) The reaction is not spontaneous because $E^\circ < 0$.

(c) $E = 0$ for any reaction at equilibrium.

(d) You can calculate log $K$ by using the equation

$$\log K = \left(\frac{nE^\circ}{0.0592 \text{ V}}\right) V$$

$$= \frac{2(-1.24 \text{ V})}{(0.0592 \text{ V})} = -42.0$$

Taking the antilog of both sides yields $K = 1.0 \times 10^{-42}$

(e) $E$ for the reaction is calculated using the Nernst equation:

$$E = E^\circ - \left(\frac{0.0592 \text{ V}}{2}\right)\log\frac{\left[Ag^+\right]^2}{\left[Fe^{2+}\right]}$$

$$-1.24 \text{ V} - (0.0296 \text{ V}) \, \log\left[\frac{(0.0100)^2}{(0.100)}\right] = -1.15 \text{ V}$$

(f) The reaction that is spontaneous is the reverse of the one above, with $E = 1.24$ V.
$W_{max} = -nFE = -(2 \text{ mol } e^-)(96,500 C \bullet mol^{-1})(1.24 \text{ V})(1J/C \bullet V) = 2.39 \times 10^5$ J.
(*Chemistry* 6th ed. pages 837–846 / 7th ed. pages 800–808)

## Question 4: Answers

(a) $Mg + Fe^{3+}(aq) \longrightarrow Mg^{2+}(aq) + Fe^{2+}(aq)$

(b) $Cu^{2+} + SO_4{}^{2-} + Ba^{2+} + OH^- \longrightarrow Cu(OH)_2 + BaSO_4$

(c) $CaH_2 + H_2O \longrightarrow Ca^{2+} + OH^- + H_2$

(d) $Al^{3+} + OH^- \longrightarrow Al(OH)_4^- \text{ [or Al(OH)}_3]$

(e) $Cl_2 + Br^- \longrightarrow + Cl^- + Br_2$

(f) $K + H_2O \longrightarrow K^+ + OH^- + H_2$

(g) $(NH_4)_2CO_3 \longrightarrow NH_3 + CO_2 + H_2O$

(h) $H_2 + CuO \longrightarrow Cu + H_2O$

(Chemistry 6th ed. pages 154–156, 172–179 / 7th ed. pages 145–146 and 162–168)

## Question 5: Answers

Add water to each solid. The calcium sulfate and the calcium carbonate will be insoluble; however, the magnesium hydroxide and sodium bicarbonate will each be soluble to some degree. Then add the acid to a sample of each insoluble solid; the one which produces bubbles of gas (carbon dioxide) is the calcium carbonate. Then take a sample of each solid that dissolved in water and add acid to each one. The sodium bicarbonate will also produce bubbles of gas (carbon dioxide). The magnesium hydroxide will react with the acid and increase in solubility, but will not produce any gaseous products (Chemistry 6th ed. pages 158–164, 226–227 and 749 / 7th ed. pages 149–154, 212–213 and 715).

## Question 6: Answers

(a) $\Delta H^\circ_{rxn} = \Delta H_f^\circ(N_2O_5) + \Delta H_f^\circ(O_2) - [2\Delta H_f^\circ(NO_2) + \Delta H_f^\circ(O_3)]$

$-198 \text{ kJ} = 11 \text{ kJ} + 0 - [2\Delta H_f^\circ(NO_2) + 143 \text{ kJ}]$.

So $H_f^\circ(NO_2) = 33 \text{ kJ}$

(b) $\Delta S^\circ < 0$ because 3 moles of gas are converted into 2 moles of gas. There are less degrees of freedom in the products compared to the reactants.

(c) Remember that $\Delta G^\circ = \Delta H^\circ - T\Delta S^\circ$. Spontaneity (when $\Delta G^\circ < 0$) will decrease as the temperature increases because the entropy change in the process is negative. This makes the contribution of the $T\Delta S^\circ$ term more positive. As temperature increases, this term contributes more to the free energy, and because it is positive, spontaneity must decrease.

(d) Looking at the table we see from comparing experiments 1 and 2 that when $[NO_2]$ increases by a factor of 1.5, the rate of reaction also increases by 1.5, so the rate is first order with respect to $[NO_2]$. Then, comparing experiments 2 and 3, we see that an increase in $[O_3]$ by a factor of 2 increases the rate or reaction by 2.0 also, so the rate is also first order with respect to $[O_3]$. Thus the rate law is rate = $k[NO_2][O_3]$.

(e) Mechanism #1 would suggest a rate law of rate $=k[NO_2]^2$ and mechanism #2 would suggest a rate law of rate $= k[NO_2]^2[O_3][O_2]^{-1}$. Neither of these mechanisms is consistent with the observed rate law and so they are not valid possible mechanisms.

(f) The rate of reaction would be expected to increase with an increase in temperature. The collision rate increases as temperature increases and a larger fraction of collisions have the energy necessary for reaction (the $E_a$) (Chemistry 6th ed. pages 261–267, 563–567, 579–588 and 809–814 / 7th ed. pages 246–252, 534–538, 549–557 and 773–778).

## Question 7: Answers

(a) The $Mg^{2+}$ ion is smaller and has a higher charge than the $Mg^+$ ion, so the lattice energy that arises when $Mg^{2+}$ ions form compounds is much greater than what would be observed if $Mg^+$ ions formed compounds. The increase in lattice energy more than offsets the larger ionization energy of the $Mg^{2+}$ ion (Chemistry 6th ed. pages 364–366 / 7th ed. pages 344–346).

(b) Carbon dioxide (O=C=O) molecules are nonpolar and interact with each other only through weak dispersion forces. These weak forces are easily overcome so $CO_2$ is a gas at room temperature. $SiO_2$ doesn't have the same molecular structure, because Si does not form double bonds as readily as carbon does. Si–O form single bonds that lead to a network solid held together with strong, covalent bonds, so it is a solid that has a high melting point (*Chemistry* 6th ed. pages 470–472 / 7th ed. pages 446–448).

(c) Nitrogen can form three bonds ($NF_3$) but not five ($NF_5$) because it lacks *d* orbitals that are energetically available for the formation of hybrid orbitals (or alternatively, because it is too small to accommodate five atoms). Both $NF_3$ and $PF_3$ are trigonal pyramidal because the central atom has three bonding pairs and one lone pair of electrons (leading to $sp^3$ hybridization). $PF_5$ is trigonal bipyramidal because it has five bonding pairs (leading to $dsp^3$ hybridization) (*Chemistry* 6th ed. pages 389–400, 414–416, 421 and 959 / 7th ed. pages 367–377, 391–393, 397–398 and 917–918).

(d) $Ca^{2+}$ and $O^{2-}$ ions are attracted about four times more strongly than $K^+$ and $F^-$ ions. Ions with a 2+ charge are attracted more strongly than ions with a 1+ charge. In addition, the calcium-to-oxygen distance is less than the potassium-to-fluoride distance, leading to an increased force of attraction for the shorter bond (*Chemistry* 6th ed. pages 364–366 / 7th ed. pages 344–346).

**Question 8: Answers**

(a) The difference in boiling point exists because $CH_4$ molecules are nonpolar, whereas $CH_3Cl$ molecules are polar. The additional dipole-dipole forces among the $CH_3Cl$ molecules compared to only London dispersion forces among the methane molecules results in this observed increase in boiling point (*Chemistry* 6th ed. pages 450–453 / 7th ed. pages 426–429).

(b) $ICl(l)$ has a higher boiling point than $Br_2(l)$ because ICl is polar and bromine is nonpolar. The polarity in the ICl covalent bond is due to the electronegativity difference between the two atoms. This polarity contributes to a greater intermolecular attraction than exists with the bromine molecules, which are held together only by London dispersion forces (*Chemistry* 6th ed. pages 450–453 / 7th ed. pages 426–429).

(c) $CH_3OH(l)$ is more soluble than $CCl_4(l)$ in liquid $NH_3$ because the solvent is polar and methanol contains polar bonds; in fact, it can form hydrogen bonds with the solvent ammonia molecules. Because of its structure (trigonal pyramidal), ammonia is a polar molecule; in fact, because of the H-atoms bonded to it, it can form hydrogen bonds with other substances that have –OH, –NH or –FH groups. Even though $CCl_4$ has very polar bonds within the molecule, its symmetrical structure contributes to the molecule's overall lack of polarity (*Chemistry* 6th ed. pages 515–519 / 7th ed. pages 488–492).

(d) The strong hydration of the electrolyte (the potassium carbonate) ties up the water and makes it unavailable for the relatively weak (in comparison to the hydration process) hydrogen bonding with the nonelectrolyte (the methanol). Because it is the hydrogen bonding between water and the nonelectrolyte that keeps it in solution, the solubility of the nonelectrolyte decreases when the hydrogen bonding is disturbed. (This process is commonly referred to as *salting out*) (*Chemistry* 6th ed. pages 389–400 and 515–519 / 7th ed. pages 367–377 and 488–492).

## SCORING THE DIAGNOSTIC TEST FREE-RESPONSE QUESTIONS

It is difficult to come up with an exact score for this section of test. However, if you compare your answers to the answers in this book, remembering that each part of the test you answer correctly is worth points even if the other parts of the answer are incorrect (see the section titled "Scoring for the Free-Response Questions" on page 12 of this book), you can get a general idea of the percentage of the questions for which you would get credit. If you believe that you got at least one-third of the possible credit, you would probably receive a 3 on this part of the test. If you believe that you would receive close to half or more of the available credit, your score would more likely be a 4 or a 5.